NATO ASI Series

Advanced Science Institutes Series

A series presenting the results of activities sponsored by the NATO Science Committee, which aims at the dissemination of advanced scientific and technological knowledge, with a view to strengthening links between scientific communities.

The Series is published by an international board of publishers in conjunction with the NATO Scientific Affairs Division

A Life Sciences **B Physics**	Plenum Publishing Corporation London and New York
C Mathematical and Physical Sciences **D Behavioural and Social Sciences** **E Applied Sciences**	Kluwer Academic Publishers Dordrecht, Boston and London
F Computer and Systems Sciences **G Ecological Sciences** **H Cell Biology** **I Global Environmental Change**	Springer-Verlag Berlin Heidelberg New York London Paris Tokyo Hong Kong Barcelona Budapest

NATO-PCO DATABASE

The electronic index to the NATO ASI Series provides full bibliographical references (with keywords and/or abstracts) to more than 30000 contributions from international scientists published in all sections of the NATO ASI Series. Access to the NATO-PCO DATABASE compiled by the NATO Publication Coordination Office is possible in two ways:

- via online FILE 128 (NATO-PCO DATABASE) hosted by ESRIN, Via Galileo Galilei, I-00044 Frascati, Italy.
- via CD-ROM "NATO-PCO DATABASE" with user-friendly retrieval software in English, French and German (© WTV GmbH and DATAWARE Technologies Inc. 1989).

The CD-ROM can be ordered through any member of the Board of Publishers or through NATO-PCO, Overijse, Belgium.

Series F: Computer and Systems Sciences Vol. 84

The ASI Series Books Published as a Result of Activities of the Special Programme on ADVANCED EDUCATIONAL TECHNOLOGY

This book contains the proceedings of a NATO Advanced Research Workshop held within the activities of the NATO Special Programme on Advanced Educational Technology, running from 1988 to 1993 under the auspices of the NATO Science Committee.

The books published so far as a result of the activities of the Special Programme are:

Vol. F 67: Designing Hypermedia for Learning. Edited by D. H. Jonassen and H. Mandl. 1990.

Vol. F 76: Multimedia Interface Design in Education. Edited by A. D. N. Edwards and S. Holland. 1992.

Vol. F 78: Integrating Advanced Technology into Technology Education. Edited by M. Hacker, A. Gordon, and M. de Vries. 1991.

Vol. F 80: Intelligent Tutoring Systems for Foreign Language Learning. The Bridge to International Communication. Edited by M. L Swartz and M. Yazdani. 1992.

Vol. F 81: Cognitive Tools for Learning. Edited by P.A.M. Kommers, D.H. Jonassen, and J.T. Mayes. 1992.

Vol. F 84: Computer-Based Learning Environments and Problem Solving. Edited by E. De Corte, M. C. Linn, H. Mandl, and L. Verschaffel. 1992.

Vol. F 85: Adaptive Learning Environments. Foundations and Frontiers. Edited by M. Jones and P. H. Winne. 1992.

Vol. F 86: Intelligent Learning Environments and Knowledge Acquisition in Physics. Edited by A. Tiberghien and H. Mandl. 1992.

Computer-Based Learning Environments and Problem Solving

Edited by

Erik De Corte

University of Leuven
Center for Instructional Psychology and Technology (CIP&T)
Vesaliusstraat 2, B-3000 Leuven, Belgium

Marcia C. Linn

University of California at Berkeley, Graduate School of Education
Berkeley, CA 94720, USA

Heinz Mandl

Universität München
Institut für Empirische Pädagogik und Pädagogische Psychologie
Leopoldstrasse 13, W-8000 München 40, FRG

Lieven Verschaffel

University of Leuven
Center for Instructional Psychology and Technology (CIP&T)
Vesaliusstraat 2, B-3000 Leuven, Belgium

Springer-Verlag
Berlin Heidelberg New York London Paris Tokyo
Hong Kong Barcelona Budapest
Published in cooperation with NATO Scientific Affairs Division

Proceedings of the NATO Advanced Research Workshop on Computer-Based Learning Environments and Problem Solving, held in Leuven, Belgium, September 26–29, 1990

CR Subject Classification (1991): K.3.1, J.4, I.2

Additional material to this book can be downloaded from http://extra.spring.com

ISBN-13:978-3-642-77230-6 e-ISBN-13:978-3-642-77228-3
DOI: 10.1007/978-3-642-77228-3

Softcover reprint of the hardcover 1st edition 1992

Typesetting: Camera ready by authors
45/3140 - 5 4 3 2 1 0 - Printed on acid-free paper

Table of Contents

Part II. Stimulating Higher-Order Thinking and Problem Solving

Part III. Creating Learning Environments

Editors' Preface

Most would agree that the acquisition of problem-solving ability is a primary goal of general education. Yet, recent international assessments of student achievement reveal that, despite the growing interest in this ability, students' problem-solving performance often remains disturbingly poor. This volume documents that a large amount of research carried out in different parts of the world and in a variety of content domains, has resulted in a series of significant findings and principles, that provide a fairly sound basis for improving the learning and instruction of problem solving.

An important force in this improvement in teaching and learning problem-solving skills was the emergence of computer learning environments in the early 1980s. Due to the unprecedented possibilities for data presentation and handling, for high-level interactivity, and for quick and individually adapted feedback, the computer was expected to become a unique instrument in the hands of the teacher for enhancing students' cognitive skills.

A substantial number of studies has been conducted relating to the hypothesis that computer-based learning environments can significantly facilitate the acquisition and transfer of higher-order thinking and learning skills. These investigations have been done from different theoretical perspectives (e.g., discovery learning versus guided instruction), using different kinds of software (programming languages, educational games, and subject-matter related software), and with learners from different ages and cultural backgrounds. This research has produced divergent, sometimes even conflicting results relating to the cognitive-effects hypothesis: While some researchers have reported highly significant positive effects of computerized learning environments on subjects' ability to apply valuable cognitive skills, others have found no significant gains. A substantial body of theoretical, methodological, and developmental knowledge has accumulated and is summarized in this volume.

The present volume emerges from a NATO Advanced Research Workshop that aimed at assembling, discussing and reviewing this knowledge in a multidisciplinary confrontation of experts in cognitive science, computer science, educational technology, and instructional psychology.

The volume includes three related parts: I. Encouraging knowledge construction; II. Stimulating higher-order thinking and problem solving; III. Creating learning environments.

In the first contribution of Part I, Elshout describes and critically discusses the growing interest of educational psychologists and philosophers in everyday life as the ideal learning environment. He argues that the recent enthusiasm for informal educational settings is not wholly rational and shows that there is a heavy price attached to adopting this approach. He reminds readers that the criticized formal educational settings have important positive sides. Elshout expects that this current direction for research on learning and instruction has reached a point of diminishing returns and anticipates that researchers will soon seek a balance between formal and informal learning.

DiSessa discusses the current images of learning offered by research groups and argues that the activities of learners have not received sufficient attention. He describes a potential theory of activities by examining how learners generate new ideas and insights. To illustrate the argument, diSessa analyzes the activities of a group of learners who, working in a science class, grapple with alternative ways to represent motion. He argues that these students are acting as designers and illustrates how they eventually agree that graphing speed versus time is the best representation for motion.

Scardamalia and Bereiter outline the architecture and the major characteristics of a hypermedia system built around a student-generated data-base, called CSILE (Computer-Supported Intentional Learning Environments). In CSILE students work cooperatively to elaborate and upgrade information on-line with several support systems within knowledge-building environments, including data exploration, explanatory coherence, analogy, and publication environments. The authors sketch the educational philosophy underlying this kind of computer-supported learning environment, and discuss the practical implications of using it in schools.

Linn, Katz, Clancy, and Recker explain why more and more instructors teach Lisp in introductory courses and explore ways to facilitate knowledge construction. They seek to identify the "templates" constructed by Lisp programmers, as well as the skills these programmers use to solve complex problems. Templates are generalized, reusable programming building blocks. They describe why a programming environment called the Perspective Library supports students as they construct programming knowledge and explain how "case studies" help students learn to solve complex problems.

Ploetzner and Spada describe a computer simulation model called KAGE (Knowledge Acquisition Governed by Experimentation), that models how students learn the physics of elastic impacts as a part of classical mechanics. KAGE reconstructs the acquisition of qualitative and quantitative knowledge about functional relationships between physical variables. Thereby it predicts the knowledge states that result when particular learning mechanisms are applied to certain instructional information. Using KAGE as a computer-supported learning environment takes advantage of research findings on knowledge acquisition and enables knowledge-based adaptation to the student's needs.

Starting from a discussion of the use of examples in human problem solving and learning, and of the difficulties involved in learning from examples, Reimann concentrates on the question of how to foster the development of effective learning-from-examples skills in students. The strategy that he proposes can be characterized as an active, hypothesis-driven, explanation-oriented approach to studying examples. Based on this analysis, he presents a conceptual framework that serves as the foundation for the design of an intelligent tutoring system to help students improve their example-analysis skills.

Vosniadou draws on a program of research on knowledge acquisition in astronomy to make recommendations about designing instruction in general and designing computer-based learning environments in particular. In her view, knowledge acquisition in the domain of science results from actively restructuring one's prior understanding of the physical world. This understanding stems from a set of fundamental ontological beliefs, synthesized into mental models, that students use in a relatively consistent fashion during problem solving. For instruction to be successful, it must make students realize the inadequacy of their beliefs and provide a different explanatory framework to replace the one they constructed on the basis of their everyday experience. Computer-based learning environments offer opportunities for the exploration of alternative, counter-intuitive hypotheses, and the modelling of expert performance which are difficult to create in traditional learning environments.

Campione, Brown, and Jay report on investigations of computers as tools for sustained learning in the science classroom. Students aged 10 to 14 compose illustrated books about science topics and share them with their peers. Students, working in groups of 5 to 7 at one computer, gather, synthesize, and communicate information. They learn to select relevant information, to summarize, and revise their ideas and to report using desktop publishing. The teachers working with these students engage in some direct instruction but primarily support and guide students in their own explorations.

In the opening chapter of Part II, Mayer discusses the issues of teaching problem solving and transfer within the domain of computer programming. He starts from the following question: how can we create environments in which novice programmers can learn to apply what they have learned to the solution of new programming problems? Three effective methods for promoting such (near) transfer are discussed and illustrated with examples from recent research: (1) to provide conceptual models of the computer during instruction, (2) to pretrain the users in relevant prerequisite skills, such as comprehending a list of directions, and (3) to encourage users to develop the problem-solving strategies of experts, such as breaking a problem into separate parts.

De Corte, Verschaffel, and Schrooten report an empirical study that aimed at the development of a powerful Logo environment for the acquisition and transfer of four higher-order thinking skills in sixth-grade children. The results showed that the thinking skills were mastered very well within the Logo environment and that positive transfer effects were obtained for three out of the four thinking skills. Starting from the results of their own study and of other recent successful studies, some crucial characteristics of powerful Logo learning environments are identified and suggestions for further research are formulated.

Clements and Nastasi review three separate studies examining the role of social interaction processes as mediators of the effects of Logo on children's higher-order cognitive skills. The results of these studies suggest that the enhancement of these cognitive skills may indeed be mediated by engagement in specific conflict-resolution strategies that are particularly engendered by the Logo environment.

Salomon contrasts several different theoretical and methodological approaches to the study of computer-based learning environments and problem solving. First, he argues that one should make a clear distinction between two different ways in which intelligent technologies like the computer may have an effect on human cognitive capacities: they may affect problem-solving during interaction with computer programs (effects *with* technology) and they may leave a more lasting cognitive residue as a consequence (effects *of* technology). Second, two different research approaches of the cognitive effects of intelligent technologies are contrasted, illustrated and discussed; namely, the analytic approach, leading to experiments in which one single instructional variable is manipulated and the other components are controlled, and the systemic approach studying a whole instructional environment the components of which are systemically interrelated, reciprocally influencing each other, therefore making it impossible to single out any one component so as to leave everything else unchanged.

Clark's chapter reviews the research on transfer of problem-solving skills between knowledge domains. He makes the point that there is but scarce empirical evidence in favor of the role of computers and instruction in computer programming in either specific or general transfer. He recommends that future studies focus on cognitive processes that are required for realizing transfer and on how these processes can be supported by transfer-oriented instruction. In this respect, he distinguishes two types of cognitive processes that are engaged when domain-general transfer occurs: the selecting of structural features that are shared by the source and target schema, and the connecting of the features in two (or more) domains during transfer. Based on this review of the research, a number of instructional prescriptions are offered for the design of instruction intended to facilitate domain-general problem solving.

According to Vergnaud, a major theoretical question of cognitive and instructional psychology is the relationship between conceptual knowledge on the one hand and problem-solving capacity on the other. Computer-based environments make available new possibilities for establishing a better relationship between problem solving and the development of a specific knowledge base. But one must analyse carefully the kind of situations that can be provided fruitfully (and the limitations), the kind of interaction that can be managed (and the limitations), the kind of symbolic representations and manipulations that can be used (and the limitations). Vergnaud presents his "theory of conceptual fields" as an appropriate framework for doing such an analysis.

Pea opens Part III with a sketch of the key concepts of a new perspective on learning and problem solving, namely the situated learning and cognition paradigm, in which learning and problem solving is considered as entering into a web of social relations and actions that are constituted by various practices, accountabilities and duties that make up the discourse of scientific knowing. He articulates some of the specific implications for designers of this new perspective and of the charge that computer tools should serve to augment students' sense-making capacities and their learning conversations. Illustrations are taken from his own research project called the "Optics Dynagrams Project", in which small groups of students work with a software simulation of phenomena of geometrical optics.

The report by Glaser, Schauble, Raghavan, and Zeitz describes studies of students engaging in self-directed exploration with computer-based laboratories that simulate phenomena in microeconomics, basic circuit laws and the reflection of light. Each has an intelligent coach that monitors and guides students' experimentation activity and includes discovery tools that support activities like recordering, sorting, and graphing of data; the generation of hypotheses;

and the creation and evaluation of expressions stating the laws among three or more variables. The implications for understanding scientific discovery and science instruction are discussed.

Mandl, Bollwahn, Hron, Oestermeier, and Tergan report on a project in which an automatic diagnosis system on knowledge acquisition and on misconceptions was developed in the context of a computer-based physics-learning environment on superposition of motion. The learning environment was designed according to the principle of inductive learning. It consists of a sequence of similar tasks of increasing complexity. The diagnosis system is able to compare correct and incorrect solutions with the learner's actual solution and thus to infer his or her conceptual knowledge base.

Goodyear's chapter considers a number of key issues concerned with supporting simulation-based learning through the provision of appropriate tutorial interventions. It focuses on sources of pedagogical knowledge that have the capacity to inform real-time decision-making in such contexts. It looks both to empirical studies of learning and to research on the action and thinking of teachers as potential sources of practical pedagogical knowledge.

De Jong and Njoo report on a part of a research project the goal of which is to develop an authoring tool that will enable the creation of an Intelligent Simulation Learning Environment (ISLE). An ISLE can be described as a computer simulation embedded in an environment that includes a diversity of types of instructional support. According to De Jong and Njoo, four themes are essential for instructional use of simulations: simulation models, instructional goals, learning processes, and learner activity. The significance of these themes for designing an ISLE is assessed by combining them with the classical components of intelligent tutoring systems, namely the domain, the learner, the instruction, and the learner- interface components.

Moar, Spensley, O'Shea, Singer, Hennessey, and Scanlon describe the unique range of functions that computers may provide for the learning of mathematics and science. Of particular importance they consider: their interactivity (exemplified by supporting direct manipulation), memory augmentation, qualitative reasoning, conflict resolution, and presentation of counterfactual examples. Two ways in which these themes have been explored are discussed. First, in the design of an alternate-realities simulation of exploring horizontal motion, and second, in the design of an educational animation/modelling system.

The first part of Teodoro's chapter examines the unique role of the computer in science and mathematics education and outlines an approach to computer use in these subject matters and its implications for software development. He introduces the concept of metabook, a teaching/learning tool for exploring formal subject-matter domains like math and science which intimately relates a book with one or more pieces of software. Afterwards he describes an example of a computer exploratory environment for exploring Newtonian dynamics as an example of the implementation of this approach, followed by some preliminary findings on how students solve problems within this environment.

According to Allen, the social context motivating and sustaining the use of computer-based learning environments is typically weak. To address this problem, she examines and develops the concept of communities-in-practice. She further exemplifies this concept by presenting a current research project on small-group research, composing, and presentation with multimedia computing technologies, and discusses the implications of communities-of-practice theory and findings for the design of successful learning environments.

We should like to express our thanks to all those who have contributed in some way to the NATO Advanced Research Workshop on computer-based learning environments and problem solving, and to the production of the present volume.

We are especially indebted to the NATO Scientific Affairs Division for its substantial financial support which made the organization of the workshop and the publication of the present volume possible. We also acknowledge the additional support of the other sponsors: the Belgian National Fund for Scientific Research, Apple Computer Europe, Inc., and the University of Leuven. We thank Apple Computer Belgium, Inc., and IBM Belgium for making available computer equipment throughout the workshop; the demonstrations of software enhanced the amount and the quality of the interactions between the participants.

Our special thanks go to Hilde Schrooten for her assistance in organizing the workshop, and especially for her valuable and painstaking help in preparing the camera-ready manuscript of this volume.

Leuven
January 1992

Erik De Corte
Marcia Linn
Heinz Mandl
Lieven Verschaffel

Part I. Encouraging Knowledge Construction

Introduction to Part I

Marcia Linn
Graduate School of Education, University of California at Berkeley, U.S.A.

How can technology foster knowledge construction?

The chapters in this section raise a variety of important issues concerning knowledge construction and technology. Knowledge construction has a relatively long history in educational thinking [e.g., 1]. These chapters address the goals of the constructive process, the mechanisms that foster knowledge construction, as well as the role of technology in constructing understanding.

The advent of powerful technological tools and the availability of low-cost classroom computer systems has led educators to rethink a broad range of solutions to educational problems [e.g., 2]. While some view computers as a panacea for education, others see these same technologies as a catalyst for change, and still others fear that technological tools will divert resources from important educational problems and deter educators from grappling with serious issues.

In addition, Elshout (this volume) warns of the cyclic nature of educational reform movements pointing out that frequently seemingly "innovative" ideas are no more than the last decade's unsuccessful reform movement in new rhetoric. He warns, for example, that enthusiasm for group learning and technology should be tempered with caution, since this approach has an uneven history.

To this controversy, the chapters in this section offer a refreshing array of existence proofs and success stories concerning technology and knowledge construction. They also provide considerable candor concerning both the strengths and limitations of technological tools as resources for knowledge construction.

Goals of the constructive process

The chapters in this section take both broad and narrow stances toward the goals of the constructive process. diSessa; Campione, Brown, and Jay; Linn, Katz, Clancy, and Recker as well as Scardamalia and Bereiter look broadly at the constructive process and analyze the sorts of goals that one might choose for educational programs. These authors address the tradeoffs between depth versus breadth of understanding and the need for integrated rather than isolated knowledge. They suggest ways to impart a deep and integrative understanding. Consistent with a focus on broad educational goals, these chapters address such issues as formal versus informal learning, domain-specific versus domain-general understanding, and the need for autonomous as well as interactive learning skills.

In contrast, Vosniadou, Ploetzner and Spada, and Reimann address specific educational problems having to do with such activities as learning from examples, modeling specific scientific phenomena, or imparting generative views of scientific events.

When education is viewed broadly, an important issue concerns homogeneity versus heterogeneity of learning goals. Several of the technological environments described in this section offer the opportunity for students to select learning goals and tailor their learning experiences to their own needs. In contrast, when learning is construed more narrowly, it is common for investigators to focus on imparting a specific set of concepts and dispositions.

Mechanisms of knowledge construction

The chapters in this section raise a variety of issues about mechanisms that foster construction, and in addition they offer a number of different perspectives on the constructive process. All the contributors favor a constructive process for knowledge acquisition as opposed to a didactic one. They vary considerably concerning their views of the guidance that students need in order to engage in productive knowledge construction. diSessa, Campione et al., and Linn et al. specifically address the resources that students bring into the learning situation and argue that educators often assume that students have fewer intellectual resources that in fact they possess. diSessa offers convincing evidence that under ideal circumstances, students can bring considerable insight and creativity to bear on ambigious, complex, and ill-defined problems. Linn et al. demonstrate that students construct diverse paths to similar goals and that this diversity is beneficial.

Several of the chapters in this section illustrate ways to take advantage of the social nature of learning. These projects encourage groups of students or individuals to pursue specific topics and then to share their expertise with other students. Campione et al. use the "jigsaw" approach, while Bereiter and Scardamalia take advantage of a network environment to encourage individuals to access the expertise of others.

Several of the projects reported in these chapters take a direct approach to diagnosing difficulties and providing feedback in order to foster knowledge construction. Vosniadou, Reimann, and Ploetzner and Spada all focus on diagnosis and feedback as components of the constructive process.

Thus, these chapters illustrate different perspectives on the constructive process. Some focus on direct feedback while others postulate advantages for quite unguided discovery. Some emphasize self-assessment while others stress external assessment. Some focus on autonomous learning while others foster and emphasize group learning. Some focus on the conceptions that students develop and change while others focus on the activities that students engage in. These chapters illustrate a healthy diversity of approaches to fostering knowledge construction and investigating appropriate mechanisms for knowledge integration.

Using computers to foster knowledge construction

The chapters in this section employ computers to foster knowledge construction in two quite different ways. One approach is to create a computer learning environment that facilitates autonomous and group knowledge construction. Another is to create an intelligent tutor that provides guidance, feedback, and direction to the learner. Those developing computer learning environments target these environments to broad educational goals, while those focussing on intelligent tutors target their developments to acquisition of specific knowledge. Similarly, the mechanisms represented in the computer learning environments are supportive and facilitative while the mechanisms characterizing intelligent tutors focus on feedback and diagnosis.

diSessa, Campione et al., Scardamalia and Bereiter, and Linn et al. describe computer learning environments to foster integrated knowledge construction. Campione et al. advocate the computer as a tool of sustained learning, pointing out that experts use a vast array of computer tools to support their learning processes and that students would benefit from the same resources. diSessa sees the computer learning environment as enabling students to use a variety of representations and computational methods to explore problems. Scardamalia and Bereiter explore the use of a networked environment to support responsible communication among individuals and groups. Linn et al. illustrate how the tools of experts can be made useful for students. In addition, they describe adaptations and augmentations to the tools of experts that make those tools more suited to students. Linn et al. also advocate curriculum materials in the form of case studies that help model how and why problem solving decisions are made.

Reimann takes a different approach to helping students understand how problem-solving decisions are made. He approaches this goal by creating an intelligent tutor that helps students generalize from worked-out examples. He concludes, however, that the information in case studies would be difficult to provide on-line and instead focuses on stimulating students to create their own explanations of the steps required to solve complex problems. The difficulties of diagnosing student problems and providing feedback also constrain to some extent the activities of intelligent tutors described in this section.

Overall, a broad range of uses for technology in fostering knowledge construction are described in these chapters. It appears that computer learning environments or intelligent tutors offer promise in augmenting the implementations of a variety of constructive mechanisms.

Summary

In summary, the chapters in this section offer varied perspectives on the goals of the constructive process, the mechanisms governing knowledge construction, and the roles technology can play in fostering knowledge development. The chapters provide empirical evidence for different approaches to fostering knowledge construction. This evidence combined with the articulation of key issues in knowledge development will help readers construct their own perspective on the role of technology in the development of scientific understanding.

References

1. Inhelder, B., & Piaget, J. (1969). *The early growth of logic in the child.* New York: Norton.
2. Pea, R.D., & Soloway, E. (1988). *The state of the art in educational technology R & D: Issues and opportunities.* Washington, DC: The Office of Technology Assessment.

Formal Education Versus Everyday Learning

Jan J. Elshout

Laboratory of Psychology, University of Amsterdam, Roeterstraat 15, 1018 WB Amsterdam, The Netherlands

Abstract: Educational philosophy seems to be locked into a pendular motion, in some periods favoring rational formal schooling as its ideal, then swinging to the position that the best of learning is to be found in everyday life. When the latter position is in the ascendance, stress is laid on what formal education does *not* accomplish and, on the other hand, on the wonderful feats of learning, transfer and performance we may see of persons of who we would not have expected them; not expected, that is, if we take the view that excellence can only result from a transfusion of well designed knowledge from the outside. When, again, formal schooling is in favor, the stress is laid on the beauty of the efficient mass production of knowledgeable people, freed of the ignorance and the irrationality persons left to their own devices are forced to succumb to. One may ask of what energies this swinging motion feeds (e.g. why do we periodically have outbursts of research on the question, whether two learn better than one?). My suggestion is, that it is the combination of the attractiveness with the inconsistency of *both* positions that makes us feel the tug of the pendulum whenever we get too much committed to one of them. The attractiveness beckons from afar, while to see the inconsistency one has to come close up. Our overall conclusion will be, especially as research on intelligent tutoring systems is concerned, that a pendulum is not a good compass.

Keywords: Decontextualization, Transfer.

An Analogy With Different Forms of the Wason Task

It fits the occasion, I think, to introduce the two protagonists of this paper (formal education vs. everyday learning) by analogy, rather than by formal definition. The analogy than is the following. Formal education is to everyday learning as the well known Wason task in its original formal form is to the logically identical task to check whether a well understood rule of permission is violated [4, 24]. In its original form four cards were placed in front of the subjects:

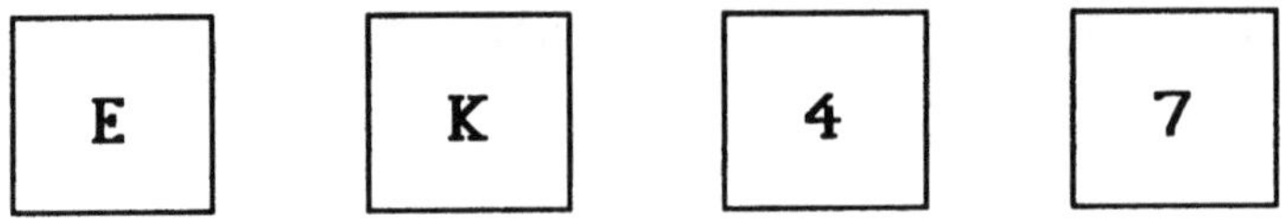

Figure 1. Wason task in original form.

Subjects were told that a letter was placed on one side of each card and a number on the other. The task was to verify whether the following rule held for these four cards: 'If a card has a vowel on one side, then it has an even number on the other side'. The subjects' task was to turn over only those cards that had to be turned over for the correctness of the rule to be judged. A typical result in this bare-bones condition is that only about 4 percent of University level students make the correct choices, turning over the E and the 7.

College level training in logic has very little beneficial effect, adding a meager 3 percent [5]. Explaining a more concrete version of the task in detail to the subjects (involving men and women, and skirts and trousers, situated in Scotland) has a somewhat better effect. This near-transfer intervention raises the percentage of subjects who correctly solve the original problem to 18 percent. Those who did, proved to be somewhat more intelligent [19].

Now for the other side of our continuum. It turns out that students can do much better on the Wason task if the rule to be tested takes the form of a reasonable rule of permission, such as the following: 'If a person is drinking beer then the person must be over 19'. Presented with four cards like these, representing people sitting around a table, 74 percent selected the right cards (beer and 16 years):

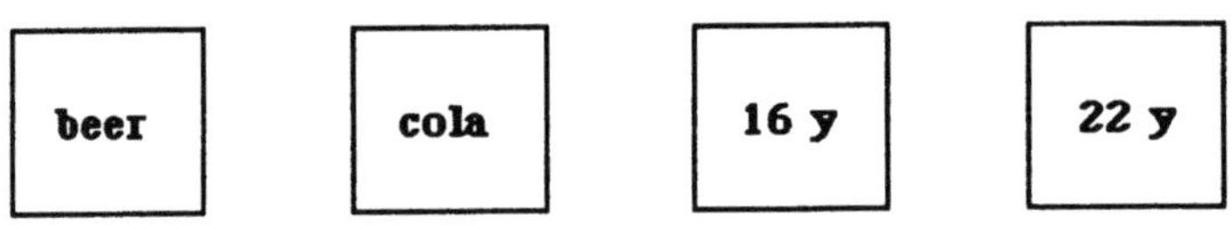

Figure 2. Wason task in rule-of-permission form.

This result of Griggs and Cox [10] has been replicated and it has been shown by Cheng and Holyoak [4] that it is not the familiarity with the rule that is the main operative factor (though familiarity helps, adding about 10 percent [19]) but whether the rule makes good sense or is just senseless. So while students do not know how to generalize their classroom training to the original Wason task, they do know to reason about rules of permission and they do so spontaneously.

Having introduced formal education and everyday learning this way as the contrasting poles of a continuum of sensibleness, I must immediately enter a caveat. I am not implying that pure forms of either pole are to be found anywhere in the world. In fact I doubt they can be found. What I do hope to accomplish introducing this polarity is to characterize the two postures in a debate. It seems to me that educational philosophy is locked into a pendular motion, in some periods favoring rational formal schooling then swinging the other way to the position that the best of learning is to be found in everyday life.

At the moment everyday learning seems to be in the upswing. Stress is laid on what formal education does *not* accomplish (e.g. transfer) and, on the other hand, on the wonderful but surprising feats of learning, transfer and performance we witness outside the formal educational context. These feats are, however, surprising ones only if we take the view that excellence always results from an infusion of well designed knowledge from the outside. I suspect that today there are many who, unlike myself, are perfectly willing to indeed call much of present day formal education senseless, as senseless as the original Wason task.

Learning Alone Versus Learning Together

One aspect the critics of formal education stress is that in formal instructional settings, people are expected to learn and to perform individually, while, in fact, man is essentially a social animal, geared to perform his mental activities in the context of some task that is shared with others; shared implying that the total mental activity of necessity is distributed over several interdependent individuals [20]. Stressing individual learning, as does formal education, would then just be unnatural, going against the grain of our nature. To bolster this argument they point to various instances of successful learning in real life, in thoroughly social and therefore natural contexts. Forms of apprenticeship that still exist in so-called traditional societies, untouched by formal education, are discussed as instances of what natural learning situations can look like.

On a more theoretical level the work of Vygotsky is often referenced. He held that thinking arises in the verbal exchanges with others, especially adults and older children. Further support the critics of formal education find in philosophers of a constructionist bent that hold that knowledge is of essence a social construction. Fully explicated knowledge, knowledge that you can lay on the table as a given or gift for others just cannot be: implicitness and inexplicable subtlety reign, making the possession and sharing of knowledge a thoroughly social undertaking that involves everything meant by 'reaching agreement'. References can be found to the older Wittgenstein and even to the unreadable Heidegger.

Before turning to other points of criticism of formal education let me try to come to grips first with the line of criticism concerning the point of individual versus shared cognition. Let us start with philosophy. In my opinion psychologists should beware of what philosophers have to say about the nature of our knowing and learning. Philosophy has been described, by philosophers, as an unending conversation, of infinite subtlety. We as outsiders, just sampling the run of the arguments from time to time, have a great chance to misunderstand. I am convinced that, for instance, if we would listen better and longer to this conversation, we will find arguments, often from the same voice, not only for the socially and culturally constructed nature of knowledge but also for our essential aloneness as knowers: I can help you make your decisions, help you learn and understand, but in the end of the analysis I can not decide, learn, or understand for you. We may try to arrange things around each other in certain persuasive ways, but in the end what is perceived, what insights we have, what convictions are reached is the responsibility of each individual alone. That lesson too can be drawn from the same unending conversation.

Of course, both types of arguments are perfectly in keeping with what psychology as an empirical science has to say. Perception, one root of all knowing, is both an individual and a social matter. The innate perceptual system constantly works to arrive at the most economical representation of a stable three dimensional world of objects in certain spatial arrangements out there, on the basis of the two dimensional and instable flow of stimulation to the retina. The resulting perception is best called an individual construction. Its relation to the physical world is less one of faithful recording than one of validity. As an example let us take our perception of a rolling wheel in the dark, for the experiment fitted out with one light on the rim, and one light in the center. Then let it roll. The great majority of subjects report seeing two kinds of movement at the same time. One is a rotating movement of the light on the rim around the light in the center. The second movement is a horizontal displacement of this rotating structure as a whole in the direction the wheel goes. One could call this perception an illusion, because physically the actual trajectory of a point on the rim is not one of rotation around the centre of the wheel but a *cycloid*, while the center moves in a straight line [8].

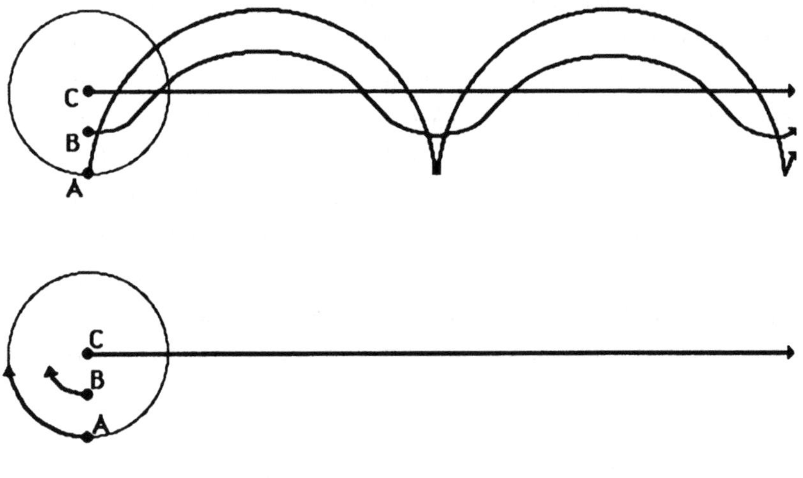

Figure 3. Real and perceived motion of points on a rotating wheel (adapted from Cutting & Profitt [8]).

When, however, we do a mathematical analysis of the double movement, as perceived, it turns out to be totally valid in the sense that we can accurately calculate, predict where each point of the wheel is in its real movement, given its coordinates on the two movements we perceive. Our perceptual system does not compute an illusion but performs some sort of component analysis that does justice to the out-there (to a degree: when subjects are asked to *draw* the trajectory of the valve of a tyre of a rolling wheel they fail) and is much more economical than a faithful recording. Perception construes only two types of trajectory: rotational and straight line. While in fact all points on the spoke of a wheel have different trajectories, from the straight line of the axis, via progressively less flattened cycloids to the perfect cycloid of a point on the rim.

One other important point can be made from this experiment. One needs naive subjects for the illusion of the two separate movements to be compelling. Explanation or suggestive instruction may spoil it. Perception is not immune to other forms of psychological construction, like understanding what others have to say. One can, for instance, make subjects see the two lights on the rolling wheel as the two ends of a 'tumbling stick'.

The lesson we draw from perception (that the individual system constructs, but that the result is neither solipsistic, immune from the outside world, nor immune to the rest of the individuals activity, including interacting with others) can easily be drawn from other parts of psychology too. Even from those theorists that are often called upon as authorities on the inherent social and cultural nature of human knowing, learning, and understanding. For example, according to Vygotsky the basic mechanism in the development of higher order mental activities like thinking, is the internalization (the making mental) of activity patterns that are initially external and material. The pivotal role in the process of internalization is taken by speech. Speech can mediate between the material and the mental because of its hybrid nature. On the one hand it is thoroughly material involving the muscles of the throat and sound; also, without others the individual will not acquire speech. On the other hand speech partakes in the mental. Its utterances may serve as symbols of things not present or as signals for actions yet to perform [23]. If we accept this account (we know now the role it assigns to speech is certainly overrated), we also have to accept that it is the student-child and not the teacher-adult that interiorizes. The child has the responsibility to take the right learning stance. The helping adult has to initiate and maintain useful verbal interactions, but the helper is not a sufficient cause of the development of the child. Thus, seeing the acculturation of children by adults as a process of developing, where the adult is the *actor* (a metaphor used by Vygotskian educators), would seem to overrate the role of the environment. 'Helping to develop' would, if I understand Vygotsky correctly, be a much better characterization, because, however essentially social the process, the child still carries the burden - and in fact as we may learn from eastern educational research, may even refuse to carry it.

Piaget would seem to be an even less appropriate source of arguments for the point of view that individual learning and performance is an unnatural context for instruction. Outside the area of the development of social and moral behaviour (e.g. learning to be a chairman), it would seem that Piaget relegates peer interaction to the role of *possibly* important help! He speaks of the importance of 'cooperations' [18]. The term obviously refers to more than simple social interaction. Considering the possible causes of the frequently observed 'décalages horizontaux' (lack of transfer?) and of stagnant development, Piaget notes that it is obvious that certain environments are less helpful than others [18]. Contradiction between the things we hold as true and the failure of activities that are rationalized by those convictions in the real world are what drives development. The environment should play along. A 'cooperative' environment, both physical and social, provides the stuff where the child's constructions can operate on, and produces the necessary feedback. For instance, you need water to think about water, when you are still in a concrete stage. When the child has reached the formal stage and can verbalize about things not there but just possible, the cooperative environment may have verbal elements also. It is in that stage that others - pointing out contradiction, prodding the child to activity, helping to accumulate facts, etc. - may become important helpers. But at no stage, as I read Piaget, can anybody take over the task of inventing the new construction that is necessary to bridge the contradiction and that leads to a better rationalized new activity.

The line of my argument thus far could bring to mind the infamous nature - nurture debate. In this analogy the individual side of cognition takes the role of nature, its social side that of nurture. The lesson we, hopefully, all learned from the nature-nurture debate is that to explain the phenotype you *always* need to know both the genotype and its history of interactions with the environment. That remains true regardless of the level of correlation, high or low, between differences in environment and differences in the phenotype. The lesson from this analogy in essence is that, if human learning is from a philosophical and anthropological view utterly social and utterly individual at the same time, this distinction as such does not help to illuminate the matter of formal education versus everyday cognition *at all*. We have to leave the lofty plane, step down, and start looking for differences in the social aspects of educational forms that are correlated with differences in achievement. This is an empirical matter.

Is There Evidence in Favor of Shared Learning?

Because I myself am interested in how students learn from computer-simulated environments I turned to the literature to search for studies that compared students working alone to students working together in pairs, with individual achievement as the criterion. I found that around 1970 there was considerable interest in this matter. This interest was partly economical and partly theoretical in nature. At the time, programmed instruction which sometimes used teaching machines but more often programmed texts, was in the process of being replaced with computer assisted instruction. The acceptance of CAI would obviously be much enhanced if it could be shown that pairwise studying leads to equivalent or even superior results than individual studying. From the point of view of the conception of teaching from which programmed instruction originated, however, one would expect individual studying to be far superior. According to the Skinnerian position students should always be required to be active, to make responses. Furthermore the instruction given should begin where the student is and then proceed at a pace that is consistent with the students ability to learn by providing hints and questions that help the student to find answers (responses) that are very likely to be correct and thus to be reinforced [21, 22]. It would seem that in pairwise studying at least some of these requirements are in danger of not being fulfilled. In particular, the requirements of continuous learner activity and of student pacing. With the help of Hoogstraten [13] (who did the review and contributed four of the experiments) I located 22 experiments (in 15 publications from both sides of the Atlantic) that address our question; all in the context of late programmed instruction or early CAI. The students levels range from the seventh grade to first year university. The subject-matter was mostly mathematical or drawn from the natural sciences; in some cases the subject-matter was what has aptly been called vaguely scientific, as in the case where students had to learn to handle a railway time-table. A typical experiment in this collection has between 70 and 80 subjects. About half of them work in pairs, the others individually. Assignment to the conditions was random or quasi random. In some studies the way the pairs were formed was systematically varied to test some hypothesis. No diabolic pairing systems (worst enemy) were found.

Of the 22 experiments only two yielded a significant difference between the conditions of interest. One experiment gave a result favoring working alone, the other one favoring working with a partner. Inspection of the signs of the non-significant differences does nothing to change the opinion that we are in Null-hypothesis country here. No proper meta analysis will lead us out.

At the time Hoogstraten [13], quite correctly, just concluded from these results that pairwise instruction was as good as individual instruction and therefore could be given a go ahead as far as CAI was concerned. A different way to construct these null results is as yet another nail in the coffin of the theory of Behavioral Analysis, at least as applied to education. In 1974 McKeachie summed up the research done comparing different variations within programmed instruction (some quite perverse, such as scrambling the frames) in an article with the ominous and selfexplanatory title 'The decline and fall of the laws of learning' [16]. Bower and Hilgard [3] by way of criticism of this generation of studies offer the hypothesis that they lacked in statistical power. But that type of argument cuts two ways. If more than thousands of subjects are needed to show an effect one had better not speak of a law. That brings us back to our own concern. Was collaborative instruction, pairwise, in the context of programmed instruction or CAI *better* than individual instruction? Obviously not. The review proves it not to be worse but also not to be better. It was a draw in 1976. And I find nothing in the literature of the 14 years since, that indicates that things have changed; or that they are different in other educational contexts. In 1989 two independent meta analyses of meta analyses of educational effectiveness research have appeared. They were published in two separate special issues of the International Journal of Educational Research [9, 14]. Because IJER has guest editors for its special issues, it happens that they cover the same ground: all the meta analyses on what makes a difference in education, and how much of a difference that is. Happily, for all of us, the results of both these meta-meta analyses are in very good agreement. There are two conclusions to be drawn that are of special interest to our own discussion. The first is that after student related factors - such as prior achievement, intelligence, specific interests, etc. - have been taken into account, only a relatively small part of the total variance in achievement can be credited to all educational differences taken together, say 20 to 30 percent. That makes it understandable that educational factors taken one at a time, if they make a difference, typically have an effect, expressed as a point bi-serial correlation, not larger than .20.

Such an effect, though by no means useless, should, if we find it, not lead to bandwagon fanfare, but rather to the attitude that made Shakespeare's King Henry proudly speak: 'Tell the constable we are but warriors for the working day.' Another conclusion we can draw is that collaborative learning (working together as in pairwise learning) is not on the list of factors studied. Peer tutoring is on the list (r=.25), so is rewarding the group vs individual reward (r=.35). However, the field just seems to have lost its appetite for research that tests the hypothesis that working together helps. Peer-based instructional methods have sprung up in classrooms around the world in the last few years. Research is following quickly as a recent issue of IJER shows [25]. Yet, all this research seems to be directed to the important questions of how peer interaction functions in cognition, but not to the basic question of whether it is true, as Resnick [20] and Pea (this volume) and a growing number of other authors contend, that formal education is somehow unnatural in stressing individual cognition. Given all this, you will accept that I consider the position taken by those authors as not to be motivated by the shared facts.

Other Points of Criticism

Let us now turn to the other points of criticism of formal education. These points concern the stress on decontextualization, on the abstract, on verbal or symbolic representation, on explicitness, and on independence of outside help.

Let us return to the Wason task which opened this discussion. The literature makes it clear that formal logical training does not transfer to the task in its symbolic, decontextualized form and also that people *can* solve the identical problem when it is worded in such a way that rule-of-permission logic can be applied. To conclude from these data to the superiority of everyday learning, however, is not warranted. From the same data we may conclude that rule-of-permission logic does not transfer to the problem in its original form either! We do not know how the few students that solve the problem, do it (they could be applying a rule of permission) but the point is that they are few, some 3 to 4 percent. And, on the other hand, we cannot be sure that those 75 percent that do solve the problem in its real-life clothing, never use anything that can be traced to formal training. Perhaps they even use formal logic. Lehman, Lempert, and Nisbett [15] have shown that students from some academic disciplines (e.g. psychology) are better at rule-of-permission problems than students from other disciplines (e.g. chemistry), who presumably are not less intelligent. So something is transferred from academic learning after all! At the moment we can only guess at what is transferred. As far as the study of psychology is concerned we can see several possibilities. Students take courses in scientific reasoning, in experimental design, in correlational analysis, and perhaps even in formal logic. Each course has something to offer that, at least in principle, could be useful for solving the Wason task in whatever clothing. All courses mentioned either directly or indirectly show students how to construct the set of all possible outcomes and to decide which of those outcomes are important for the conclusion we want to draw. And even if the students manage to forget all of this because it is too abstract, they will not be able to escape being familiar with the type of reasoning used in more practical areas, such as clinical diagnosis and test validation. They know for instance that there is something wrong with the statement that the Rorschach test is a valid test for diagnosing paranoia because 90 percent of the paranoiacs see a flying bat in the fifth card. They know they should ask what normal people make of that ink-blot.

I grant that even with all this training, psychology students do not perform better than anybody else when it comes to the formal Wason task. The point I wanted to make, however, was that formal education helps in solving everyday rule-of-permission problems, and that this form of transfer not only does occur as Lehman et al. showed, but that it also is not at all mysterious.

The relationship of helping and of transfer is not symmetrical, as is shown by research by Hayes and Simon [11] on Tower of Hanoi isomorphs. Mastery of the more difficult isomorph first produces more transfer than mastering the easier one first. Let me try some generalizations: *The first generalization* is that *when we rank different forms of logically similar tasks in their order of difficulty, we will find that they scale from the 'does not make everyday sense' to the 'makes perfect everyday sense'*, like with the Wason task in its different formats. We could then for each level of this scale find out what knowledge is used by those who are successful at that level. The *second generalization* I propose is that *the higher up the difficulty ladder knowledge is successfully applied the further down the scale we will find it applied too.* (Not by everybody, but that is another matter). In other words: knowledge that is sufficient for solving more difficult problems has a larger range of transfer than knowledge that only serves to solve easier ones. The *third generalization* I propose is that *the higher up the difficulty scale we get, the more the knowledge that is sufficient to function well will prove to have exactly those features that are in dispute: more decontextualised, abstract, more verbal, symbolic, more explicit, more being itself a tool than needing props in the outside world.* As a concrete example, think of the domain of Weight Watcher problems. The task is to measure off a fraction of a fraction of some amount of foodstuff, one measuring cup being available. One instance of this task family is often referred to: measure off three fourths of yesterday's portion of cottage cheese, which was two thirds of a measuring cup. Put this version in the middle of the difficulty scale. Below it (less difficult) would be to measure off one half of two thirds. While above it would be the task of measuring off two thirds of three fourths of a measuring cup. Still more difficult would be a version when it is not cottage cheese that has to be measured but some sticky fluid. Note that the formal, abstract approach to this type of problem (i.e. to multiply the fractions before using the cup) serves for the whole difficulty range, while the situated improvisation (making a pancake of yesterday's amount and cutting it into four parts, then removing one part) breaks down above a certain level. For each different problem domain we may call experts those who solve all problems, and that tells us something about what it means to be an expert. Experts may know more of the specifics of their domain, they know and use much more of the abstract and the general *in their domain* also. That too has been established [1, 6, 12, 17]. Am I not forgetting that specific, context bound knowledge has been shown to be very much more important than general reasoning schemes in problem solving? Not at all. I am just speaking of transfer within one specific scalable problem domain like that of the Wason task in different formats. There may be N such specific scales existing independently, though I think they cluster.

The large shift in difficulty level of the Wason task (depending on its clothing) is known in the Piagetian literature as a 'décalage horizontal'. Such shifts pose great difficulty for Piagetian theory because that theory holds that it is the underlying logical structure that is developed; the surface therefore should not matter much. Children that conserve quantity for gasoline but not for colored water, for instance, pose a problem. Those that stress the context specificity of thinking have proposed that where such shifts occur, we in fact have not one problem in different formats, but different problems, each solved in its own specific way. The problem that best fits in your cultural niche is the one solved most confidently. That analysis to me seems to be correct. Logical isomorphy indeed is first in the eye of the logicians. Still we also find children who don't show the décalage, who seem to be less bothered by what to the logicians are surface features. For them the logical isomorphy probably is a reality. I propose that *the knowledge applied by those who successfully solve all problems differs in kind from the knowledge applied by those for whom the problems are really different, the one natural, the other strange*. I venture, as the *fourth generalization*, that *the differences again will point to more decontextualisation, etc., etc., in the knowledge of those with a broader range of success*. A study by Cole, Gay, Glick, and Sharp [7] is of interest here. In their research in Liberia with the K'pelle they presented children with a puzzle box: you had to figure out that to get a cookie out here, you had to do something arbitrary there. This formal version led to failure by most, the thing made no sense to the K'pelle children. Cole's group then made another contraption that operated on logically the same rule but in its choice of materials and operations was made compatible with K'pelle culture. There was a dramatic rise in success. Back in California however, American children proved to be as successful on either contraption. Anthropologists attribute this lack of a difficulty shift to *schooling*. Schooling is a catch-all term that anthropologists use and that I do not need to unpack for you. Schooling versus traditional education makes a far greater difference than all other educational differences put together, regardless of whether you take average income or academic performance as the criterion. Anthropological studies certainly give us no grounds to berate schooling per se.

Nor do psychological studies justify looking down upon the decontextualized scientific knowledge that schools try to dispense. It seems that those who criticize formal education favor comparisons far different from those I just proposed. My comparisons were each time strictly within the same task family. Concerning the burning matter of how scientific knowledge is best acquired, it is not illuminating at all to compare the frequently failing efforts of formal education to introduce youngsters to a literate and scientific culture with the traditional apprenticeships or other examples of successful out of school and in the middle of life learning. The aims of these two approaches are different and what is learned differs, so the comparison is wrong. Both more fair and more illuminating would be a comparison along the full scale of the types of knowledge we may wish students to acquire. Let us compare vocational training in the West with traditional apprenticeships in Liberia, and so on along the full scale from the most implicit know how to the most abstract know why.

How Illuminating is Lack of Transfer?

In my opinion we are unduly fascinated with that classic ill of formal education called lack of transfer. Not only does formal education not have exclusive rights on the phenomenon but also failure to transfer often is the result of a failure to think through information received and to elaborate on it, to contextualize it in our memory system. The cause of this failure may not lie with the educational format at all. Sometimes this lack of elaboration (and therefore lack of transfer) results from our own free choice. We do not want to understand everything. More often, however, lack of elaboration is forced upon us because the structure we have to integrate overwhelms our working memory. Dividing a fraction by a fraction dazzles and so do recursion and gravity. Momentarily our working memory is overwhelmed. If the condition is serious enough and we do not choose to give up, we may use one of two backup strategies. One is memorizing verbally, the other is memorizing conceptually, even with partial understanding of what it means, but without elaboration since the person has his hands full as it is. Both backup strategies result in compartimentalization. Verbal memorizing leads to successful recitation and failure to solve the problems in the book. Conceptual memorizing leads to successful solving of the problems in the book and failure to solve so-called transfer problems in a different context. Both strategies have been described by Boschowitsch, a student of Vygotsky [2]. She calls students that use these strategies 'formalists'. I think, however, it is as mistaken to think that 'formalistic thinking' is a trait of a type of a student (though some will more often be forced to think this way than others) as it is to think it typical of formal education as an instructional genre. At the root of failure to transfer lies failure to comprehend complex information. And as long as our goal is to help students take part in our scientific and technological culture there will be students dazzled by its complexities, as we ourselves are dazzled everyday. By way of illustration of this last point let me briefly describe an informal study of mine. I found that none of the psychologists I asked, was able to explain for example why we have high water two times a day while the moon passes over only once; or to explain the phases of the moon, or the passing of the seasons, or what makes the end of a bicycle pump burning hot when we inflate a tyre. Topics like these are typically not discussed in school though the relevant physical principles are discussed. When it was explained which principles to use and how, they proved not to have been forgotten, but often to never have been really understood. It seems that formal education lets you get away with a great amount of half digested knowledge. Most important, however, for explaining these instances of lack of transfer, is that both the principles involved and the phenomena themselves (however everyday they may be) are complex. It took a Newton, for instance, to unravel the question about the two tides. It would be unrealistic to expect easy transfer. Most of my subjects, by the way, were a bit ashamed. Not so much because they could not answer these questions, but because they recognized them as questions they should have put to themselves long ago. Though it does not help in everyday life to know the answers, my subjects felt they should have noted the problems. This feeling shows that they do have the epistemological attitude cherished by formal education: disinterested interest, wanting to know because you like knowing.

Could We Stop the Pendulum?

It will be clear from what I had to say this far that I consider the recent enthusiasm for 'informal education' as not wholly rational. I do sympathize up to a point with those who see formal education as a form of wasteful force-feeding and are disillusioned with the slow pace of progress of the sciences on which education should be based. The promises of the cognitive revolution, of artificial intelligence, and of ITS were certainly overrated. If only we could learn from all those instances outside the force-feeding educational industry where the education comes natural, without force, without student apathy or rebellion and with transfer! As all this is considered, however, it will become clear that there is a heavy price attached to adopting the various ways of natural, situated cognition. Apprenticeship systems, for instance, tend to be highly selective, non democratic, and rigid. They tend not to adapt to changes in the outside world, such as changes in supply and demand. On the other hand, industrial production on a scientific basis, both of goods and of knowledgeable people, has its positive sides. It is to be expected, therefore, that the pendulum will start to swing the other way again. Hopefully, that will not lead to new unrealistic claims of high-tech breakthroughs in education, but to a more balanced view, a more humble one too. The King saying 'we are but warriors for the working day' meant they were but foot soldiers, not knights in armour. Remember that in the battle of Azincourt it was the foot soldiers that won the day.

References

1. Adelson, B. (1984). When novices surpass experts: The difficulty of a task may increase with expertise. *Journal of Experimental Psychology: Learning, Memory, and Cognition, 10*, 483-495.
2. Boschowitsch, L. I. (1970). *Die Persönlichkeit und ihre Entwicklung im Schulalter*. Berlin: Volk und Wissen Verlag.
3. Bower, G. H., & Hilgard, E. R. (1981). *Theories of learning*. Englewood Cliffs, NJ: Prentice Hall.
4. Cheng, P. W., & Holyoak, K. J. (1985). Pragmatic reasoning schemas. *Cognitive Psychology, 17*, 391-416.
5. Cheng, P. W., Holyoak, K. J., Nisbett, R. E., & Oliver, L. M. (1986). Pragmatic versus syntactic approaches to training deductive reasoning. *Cognitive Psychology, 18*, 293-328.
6. Chi, M. T. H., Feltovich, P. J., & Glaser, R. (1981). Categorization and representation of physics problems by experts and novices. *Cognitive Science, 5*, 121-152.
7. Cole, M., Gay, J., Glick, J., & Sharp, D. (1971). *The cultural context of learning and thinking*. New York: Basic Books.
8. Cutting, J. E., & Profitt, D. R. (1982). The minimum principle and the perception of absolute, common, and relative motions. *Cognitive Psychology, 14*, 211-246.

9. Fraser, B. J. (1989). Research syntheses on school and instructional effectiveness. *International Journal of Educational Research, 13*, 707-718.
10. Griggs, R. A., & Cox, J. R. (1982). The elusive thematic-materials effect in Wason's selection task. *British Journal of Psychology, 73*, 407-420.
11. Hayes, J. R., & Simon, H. A. (1977). Psychological differences among problem isomorphs. In N. Castellon, K. Pisoni, & G. Potts (Eds.), *Cognitive theory* (Vol. 2, pp. 21-41). Hillsdale, NJ: Lawrence Erlbaum Associates.
12. Holyoak, K. J., & Koh, K. (1987). Surface and structural similarity in analogical transfer. *Memory and Cognition, 15*, 332-340.
13. Hoogstraten, K. (1976). *Alleen of met zijn tweeën*. Doctoral dissertation. University of Amsterdam, The Netherlands. ["Alone or Pairwise. Five fieldexperiments with programmed material". Dutch text with English summary]
14. Kulik, J. A., & Kulik, C. C. (1989). Meta-analysis in education. *International Journal of Educational Research, 13*, 221-340.
15. Lehman, D. R., Lempert, R. O., & Nisbett, R. E. (1988). The effects of graduate training on reasoning: Formal discipline and thinking about everyday-life events. *American Psychologist, 43*, 431-442.
16. McKeachie, W. J. (1974). The decline and fall of the laws of learning. *Educational Researcher, 3*, 7-11.
17. Novick, L. R. (1988). Analogical transfer, problem similarity, and expertise. *Journal of Experimental Psychology: Learning, Memory, and Cognition, 14*, 510-520.
18. Piaget, J. (1972). Intellectual evolution from adolescence to adulthood. *Human Development, 15*, 1-12.
19. Raaheim, K. (1988). Intelligence and task novelty. In R. J. Sternberg (Ed.), *Advances in the psychology of human intelligence*. Hillsdale, NJ: Lawrence Erlbaum Associates.
20. Resnick, L. B. (1989). Introduction. In L. B. Resnick (Ed.), *Knowing, learning, and instruction*. Hillsdale, NJ: Lawrence Erlbaum Associates.
21. Skinner, B. F. (1954). The science of learning and the art of teaching. *Harvard Educational Review, 24*, 86-97.
22. Skinner, B. F. (1958). Teaching machines. *Science, 128*, 967-977.
23. Vygotsky, L. S. (1987). Thinking and speech. In R. W. Riebes & A.S. Carton (Eds.), *The collected works of L. S. Vygotsky: Vol. 1. Problems of general psychology*. New York: Plenum Press.
24. Wason, P. C., & Johnson-Laird, P. N. (1972). *Psychology of reasoning: Structure and content*. Cambridge, MA: Harvard University Press.
25. Webb, N. M. (Ed.) (1989). Peer interaction, problem solving, and cognition: Multidisciplinary perspectives (special issue). *International Journal of Educational Research, 13*, 1-119.

Images of Learning

Andrea A. diSessa

Graduate School of Education, University of California at Berkeley, Berkeley, CA 94720, USA*

Abstract: Images can be powerful devices for communicating and for organizing activities. In this chapter, I examine the status of images of good learning practice, formulate some ways to judge them, and advocate images associated with our Boxer Project's goals. Along the way, I propose some first steps toward understanding the structure of activities and present a brief case study of a remarkable classroom activity in which a group of sixth grade children essentially invented graphing as a way of representing motion.

Keywords: Computational media, Structure of activities

Introduction

In this chapter, I examine images of learning and competent performance that many education researchers and developers have been advocating. I present and advocate some images that have guided my group in its work over the last few years. But the more general aim is to understand what kind of thing these images are and how we may evaluate them.

The following list exemplifies what I am referring to as images of learning, and identifies some of their advocates. Some of these are more directly images intended to portray "good learning"; others serve as images of our goals as educators, for example, students' becoming "good problem solvers."

1. Problem Solving — Newell and Simon and many mathematics educators (see, for example, the chapter by De Corte, Verschaffel, and Schrooten, this volume)
2. Cognitive Apprenticeship — the flagship image of "situated cognition"
3. A Community of Learners — Brown (this volume)

---

*This work was supported, in part, by the National Science Foundation under grant number MDR 88-50363. The opinions expressed here are those of the author and do not necessarily reflect those of the Foundation.

4. Intentional Learning — Scardamalia and Bereiter, instantiated in the CSILE system (see Scardamalia's chapter, this volume)
5. Communities of Practice — Lave (see also Allen's chapter, this volume)
6. Mindful Learning — Salamon
7. Exploratory Learning — Many in the Logo community; many science educators, especially at the elementary school level (see chapters by de Jong and Glaser, this volume)
8. The Child as Scientist — Associated with Piaget; some constructivist subcommunities of science educators
9. The Child as Designer, Builder — The Boxer group at the University of California at Berkeley; Papert and the Learning and Epistemology Group at MIT

I intend here to take these images as a class (though not one-by-one) very seriously. I believe that they are most times intended to be taken seriously, as methods or educational targets, and also I believe that we can learn a lot by treating them that way. The last item in the list is one I will describe and advocate to represent my own work, especially in contrast to some of the other images.

An Analysis of Images

In this section: In preparation for later analysis, I identify two functions for an image of learning. The one I will pursue is as an emblem for a scientific program that, in the end, must justify itself by scientific criteria. I maintain also that images are best understood as activities, involving both cultural and more specific levels of analysis. Images thus entail implicit claims about activities, so knowing more about the structure of activities is a prerequisite to evaluating them.

To begin thinking about images,[1] we must ask about their ontological status. What are they? What work are they intended to do? To the latter question, I think there are two evident and very different answers.

In the first instance, images may be intended to organize practical action. We all understand intuitively the power of a compelling image in human activities. Political movements often center on slogans that carry with them images, like "the New Deal" or "a kinder, gentler nation." Though many of us consider ourselves scientists and do not come quickly to think of

[1] I do not mean to be restrictive in using the term "image." "Metaphor" might do as well. The focus is any compact and graphic expression of value and orientation, regardless of how we might schematize how such expressions do their work.

our writings in such terms, I do not take this function at all lightly. Most of us are also educators and would dearly love to have our work have extensive practical influence. My impression is that frequently such influence has been achieved only through effective images of this sort, rather than through the popular (but probably radically oversimplified) "results and applications" view of science's influence. It behooves us to try to understand what it is that makes for an effective image for the communities we intend to influence, and how and how well those images actually organize action. Unfortunately, I know of very few who have taken understanding this process very seriously. In contrast to regarding images as a promising means of productive practical influence, Sheldon White has offered an extreme, and perhaps somewhat cynical view of the psychological research community trying to have practical educational influence. He contends that researchers do nothing more than provide a technical language that is used to describe what people advocate as "good practice." Nothing of the research and deep meaning of those terms ever manages to cross the gap.

A second function for these images is as an emblematic announcement of a scientific orientation. For example, the "child as scientist" may emerge from observations of children behaving in surprisingly scientific ways. Or, perhaps more deeply, that image may represent the contention that it is profitable to think of conceptual change generally as theory change, so that there is a fundamental similarity between children's advances and historical developments in science. Naturally, to serve this function, we have a great obligation to unfold our images into what is evidently scientific. Situated cognition is the self-declared scientific perspective of those advocating cognitive apprenticeship, and that community intends to deliver on the promise of scientific program. Later in the paper, I will try to elucidate the scientific claims and program that lie behind the images associated with my own research group.

It is altogether appropriate, sometimes, for images to play both of these roles. It may be we intuitively evaluate both functions in choosing images for ourselves. Problem solving names a cluster of research programs that have a scientific core (at least to the standards of contemporary cognitive and psychological research). Think of *Human Problem Solving*, the General Problem Solver, and the set of research lines connected to the basic claim that problem solving is a natural kind of human activity (if it is not synonymous with intelligent action!), that it has a definable structure, and that good problem solving is teachable. Indeed, problem solving, even by name, inspired the conference that brought this volume together. On the side of organizing practical action, problem solving has been taken up by mathematics educators with such a fervor that it is unusual these days not to find it listed among the first few curricular targets. Ideally, one might hope that the scientific and practical sides of images are well-aligned, that practice in some way implements our scientific understanding. All too often, and I think this is true in the case of problem solving, the connection may be tenuous.

More generally, even if we intend both, it is important in principle to separate the scientific side of our images from the practical one. We need to be clear on the basis for evaluating an image. If we intend practical influence, it is nearly irrelevant that the image might be intriguing to members of our own scientific community. Indeed, I feel sometimes we, the research community, are too inbred and appeal to the "practical instincts" of our fellow scientists, which may be very impractical.[2] Conversely, I don't believe that sweeping the country's schools is, by itself, a particularly good measure of scientific advance.

Images may, then, function scientifically or "politically." But still, what *are* images? What sort of thing can serve as an image?

I believe it is fundamental that images seem always to be *activities*, or very closely related to activities. Problem solving is certainly an activity. Apprenticeship is an activity. Exploratory, mindful and intentional learning are activities. A community of learners is defined by what they do and how they do it.

On the one hand, this is not at all surprising. Learning activities are the heart and soul of instructional design. For my own part, I took it as a profound realization when, a few years ago, it occurred to me that most of what I did as a designer of microworlds was to think about what children would do with them. What child goals would the microworld engage? What specific activities would follow from those goals, and what competences would be relied on and engaged? Would there be blocks to extended engagement with the microworld, and how could those blocks be overcome? Centrally, how would what I wanted them to learn emerge from their activities? Microworld design is the design of mediated activities with learning intentions.

What is much more surprising is that the structure of activities, at the scale of a successful or unsuccessful learning activity, has, as far as I can tell, essentially no status as a scientific focus of attention.[3] I advocate thinking of activities — extended, coherent human action — as a natural kind. We want to understand the structure and development of activities in the same way that we want to understand the structure and development of knowledge. Since our concern is with learning, we want to understand how knowledge and competence emerge in, are dependent on and reflected in activity.

---

[2] The astute reader will note I am suppressing interesting complexities. Images certainly play an important role in the internal politics of scientific communities, as well as serving as easy handles for scientific orientations.

[3] When I began to think of myself as a designer of activities, I sought to find relevant literature on the structure of activities that would be useful for instructional design. I have yet to find such a literature. The things that come closest I have still found inadequate to the task of design in one way or another. For example, what is called activity theory by the mostly Soviet and Scandinavian developers of it seems much too global to afford effective instructional design. And, to be honest, some ostensible theories of activity I find simply incomprehensible. To my eyes, more promising and recent lines are represented, for example, by Geoffrey Saxe [9].

I believe teachers and practitioners are generally much more sensitive to the structure and dynamics of activities than researchers. That is because teachers' bread and butter *is* their repertoire of robust and reliable activities: activities that regularly engage a breadth of different children, that are manageable, that contribute to the long-term development of productive orientations, that build on established competences and interests but extend those in profitable ways. Teachers cultivate their skills at instigating and nurturing particular good activities, and they "trade" activities as a common coin of the realm. Narrow measures of learning effectiveness, which so often dominate the researcher's view, simply will not capture the richness of these practices.

Culture may be an alternative to activities as the category of images. Communities of practice sound more like cultures than activities per se. A community of learners is a learning-oriented culture probably more than it is any particular set of activities.

Culture is certainly helpful in raising our sights beyond individual activities and restrictive definitions of them. The style of interaction, values and concerns of participants probably tells more of educational relevance about an activity than any specification of what the activity is overtly about. Let me elaborate an example.

There is a grain of truth in saying that the point of discovery learning is to let children uncover for themselves, say, a scientific law. Yet so much that is called discovery learning falls within a paradigm where students are sat down in front of a narrowly focused system, which is constrained to particular operations aimed at having the student discover some particular law or fact. In fact, the focus may be the "discovery" of a particular representational form for the law, say, an equation. This does violence to my sense of authentic-feeling activities for children, and to the epistemological problems I believe some version of discovery learning is well-adapted to solve.[4] In contrast, discovery learning for me is as much instilling an orientation toward uncovering regularities. I am much more concerned that the child buys into the enterprise than that he or she finds what I might have intended. In my view, it is useless to discover a law if that discovery does not satisfy a child's felt need for explanation. These "ancillary" conditions and values are so much the point that I would say the activity of discovery itself is subordinate to them. Discovery, per se, is useful to the extent that it advances my aims to change a student's orientation toward explaining the world in small or large ways. Indeed, I don't find it hard to imagine ways of telling students about laws that are more conso-

[4] For example, my work on the role of intuitive notions in physics implies we must let children evolve a new *sense of mechanism* [6] about the world in a way that makes "discovering this new stance" seem a category error. In [7] I give another example of a class of constructions that learners must make, discovering new modes of coordinating "views of the world," that is similarly only very awkwardly described as "discovery."

nant with my intentions than many "discovery" contexts. In short, it makes no sense to take the activity out of the context of its cultural intent.

I will have more to say about the need to focus on culture-level issues shortly. But there are a number of reasons I want to keep the focus predominantly on activities or classes of activities rather than to skip quickly to cultures. First, the term culture comes with a load of theoretical or pre-theoretical commitments, depending on who uses the word, of which I believe we should be wary. In particular, cultures are too often seen as monolithic and global. But most classrooms clearly contain multiple cultures; it is not evident which ones or which interactions are relevant to determining whether a learning activity will be effective or not. An activity may succeed with one class, but fail in a neighboring one or with the same teacher and a different set of "culturally similar" children. To explain these happenings, I believe a finer knife than culture is necessary. Most fundamentally, learning occurs within particular activities relevant to particular subject matter, activities that may be fostered by cultural characteristics, but cannot be taken to be identical to them. The learning design task, I maintain (though cannot argue extensively here), cannot reside only, or even mainly, at the cultural level.

Thus, I argue that to understand and evaluate images we must understand activities as well as the higher level, more cultural aspects of learning groups that foster or undermine productive activities.

Steps Toward a Theory of Activities

In this section I provide a sketch of important features of activities that will be used to analyze particular images later in the paper. We must consider the resources *participants have to engage in particular activities as well as the* criteria *that we have for evaluating success. Of prime importance is to separate aspects of those resources and criteria that relate to conceptual matters from those that relate to pragmatic, interactional or activity-sustaining matters.*

Here, I can take only the most modest steps toward an analytic framework for understanding activities. I begin with a short list of constraints on such a future frame that are part of my orientation toward activities.

Generativity. I maintain that there is a fundamental richness in human activity. However long and detailed a list we prepare of particular activities any group or individual can engage in, that same group or individual can also engage in substantially different activities as well. This is a central, valuable contribution that the concept of culture adds to the concept of activity. The cultural aspect of activities may be precisely this generativity, the relatively invariant characteristics of individuals and groups that serve as generators of possible activities and constraints on their properties.[5]

A role for diversity. To some extent because of the empirical work done in my group, I expect children can contribute to productive group activities in quite diverse ways. I am suspicious, therefore, of descriptions of cultures or activities that do not take into account this kind of diversity. Needless to say, activities that are largely individual may (perhaps must) also be quite diverse, and we must respect this fact.

No necessarily shared central goal. In contrast to implicit suggestions in some of the listed images, I believe profitable activities rarely flow from a single goal or orientation, much less a pervasive, shared communal goal such as "learning."

No global criterion of communal evaluation of individuals. In contrast to, for example, the presumption that communities of practice maintain an implicit but shared ranking of expertise, I believe that frequently, if such rankings exist, they are multi-valent and furthermore play no single, invariant role in the success or failure of learning activities.

I believe generativity is the most important of these vectors. It is particularly important if we take ourselves to be designers and nurturers of classes of activities, rather than discoverers and advocates of particular ones. Generativity is, in particular, the only constraint I will build into the sketch of an analysis of the structure of activities presented here.

A simple and productive way to build generativity into a view of activities is simply to leave the description of particular activities out! Following this line, let us first think to examine (1) the *resources* out of which activities may flow, and (2) the *criteria* they may or must satisfy. Because of their (usually) very different roles, it is appropriate to look at teachers and students separately. We have found it convenient to separate student resources into *conceptual* and *pragmatic* categories. Conceptual resources are something like "what students already know," although a quick admonition is that we cannot identify this dimension with instructional sequencing. In much of our work we have found important student conceptual resources that are not obviously ever directly instructed and are, in any case, most certainly not part of standard curricular goals. In recent work, for example, we have identified important uninstructed skills in dynamic geometric reasoning, and another such uninstructed pool of resources will be discussed later.

---

[5] Again there is a parallel to my work in intuitive physics. I maintain that the way people can think about the physical world is not particularly rigid, resistant or theory-like. In some ways their intuitive physics is notably flexible and generative of particular other ways of thinking. So, similar to my advocacy of "knowledge in pieces," I advocate a kind of "activities in pieces" predisposition toward the structure of activities.

The category of pragmatic resources includes, for example, interactional skills and dispositions that can make possible, or prevent, particular kinds of activities. The disposition and capability to sustain mutually respectful listening may be prerequisite to, for example, collaborative design. Pragmatic resources also include things like existing goals and interests, and the capability and disposition to effectively adapt those goals and interests to instructional constraints. Children like to play, and as a teacher I hope to nurture students who can enjoy this creativity without undermining the learning possibilities inherent in an activity. Similarly, children of a certain age spend a lot of time defining themselves socially, and it would seem adaptive if they can find ways of doing that along with the things that are more salient goals for us as teachers.

Although it makes good sense to look at teachers' resources in a similar light, I will in the principal example presented here consider only teachers' *stance and strategies*, which is a particular version of pragmatic resources, omitting conceptual resources.

In terms of criteria, a basic and important distinction is between *criteria for sustaining* an activity and *criteria related to learning*. This can be motivated by the commonsense observation that children frequently become fanatically attached to activities that have no relation whatsoever to the learning goals we have for them.[6] For the most part, I take learning goals to reside in us, educators and designers of activities, though, of course, learning goals (as pragmatic resources) in our students may make the design of productive activities much easier.

The *articulation of sustaining and learning goals* may be the central-most determiner of whether a successful activity by sustaining criteria works also to satisfy us as educators with commitments to having students learn particular things. In particular, the conceptually most difficult part of instruction may be exactly where it is most difficult to sustain activity. Strategies simply to maintain contact with these tough targets may be the first and most difficult task in getting them learned.

I believe an appropriate method for elaborating these preliminary ideas is by working out case studies of successful and unsuccessful activities. In particular, we may build a kind of perturbation theory by understanding the range of circumstances under which a certain activity can maintain its essential characteristics and possibly its "homeostatic" mechanisms. Conversely, we may seek to understand some of the limits beyond which the activity loses its attractive qualities.

The plan for the rest of this chapter is as follows. First, I will return to images with which we are working in our group and motivate two central ones. I will take one of those, the child

[6] I think the best move in this case may sometimes be to adjust our learning goals to some closer to what the children have implicitly in their activities rather than to abandon an activity that is evidently successful by sustaining criteria.

as designer or builder, and unfold it into a form compatible with this preliminary sketch of a frame for understanding activities. Then I will briefly elaborate a case study of a very successful activity to make more plausible with empirical input some of the things I say about the child as designer. Finally, I will comment on this activity from the point of view of some other images to help define its characteristics by contrast.

Images for Boxer

The child as scientist, in my view, over-estimates the extent to which classroom cultures and practices can and should be like those of scientists. It can fall prey to a cultural imperialism that may misunderstand both scientists and children. In contrast, I argue that children viewed as designers can be a more productive image.

Boxer [5] is designed to constitute the first effective example of a computational medium that can support an extended and more powerful form of literacy, *computational literacy*. Consistent with this literacy image, we intend Boxer to have a very broad range of functionality, from everyday and "menial" uses to deep and extensive productions that, like text books or full curricula, occupy well-endowed teams for substantial amounts of time. Text production is the present-technology version of such a medium. Text is at the core of the literacy that constitutes not only the foundation for many educational processes, but also pervades our society in hundreds of subtle and not so subtle ways. A computational medium like Boxer starts with a computationally enhanced ability to process text, but adds much more extensive capabilities of organization, typical of hypertext systems. In addition, graphical productions in the medium can be dynamic and interactive. Not only can pictures move, but they can be manipulated by the user so that not only presentation, but also interaction becomes part of the communicative channel. Simulations and interactive tools, from very simple to very complex, are familiar examples of this extension. Fundamentally, I believe that programming, in some form, is necessary. Without the capability to define action and interaction, this would be only a one-way medium, the equivalent of literacy constituted of reading without writing. Instead we want a medium that may be truly owned by any of its users in creating their own tools and toys, and also in modifying, extending and combining any creations supplied to them in this medium.

In [5] I describe the relation of a computational medium to activity structures. I maintain that to know a medium is to know the genres that it supports, and the social niches (community supported and long-term patterns of activity) that make those genres meaningful to the people that generate and use them. I try to imagine Boxer's contributions to new genres in various social niches that have particularly good properties related to learning activities. We have some empirical support for these conjectures about the form and function of future literacies, but I wish to pursue a more manageably-scaled image here.

The second image for Boxer, the one I will pursue in more depth, is the child as designer. I can motivate this idea by considering what I take to be problematic about a more familiar image, the child as scientist.

To begin, my own disposition is not to believe the hints at a scientific rationale for the child as scientist I provided earlier. For example, I do not believe that the process of conceptual change that children undergo when they learn science is much like theory change in adult scientists. Science is richly situated. It is articulate; it has a long cultural history both in its general methods and epistemological stance; every scientific change of theory has a substantial particular setting involving the prior set of ideas and the difficulties that need to be overcome to move to the new theory. In short (for lack of space rather than for lack of respect for the substantial individuals and arguments tending to the opposite conclusion), I find little reason a priori to expect adult science will be much like conceptual change in children, and I do not find compelling the particular analyses that have been offered to "cash out" the claimed commonalities (e.g., [3]).[7]

A second approach to the child as scientist, in crude form, is that if we want to teach children science, it is best we do this just by making them scientists. There is a very profitable move here, which I take to be a recognition that activities are epistemologically fundamental. Rather than putting scientific knowledge in children's heads, we should attempt to construct scientific activities as part of their ways of being. A large part of science is a way of acting.

But we must put a finer point on this. As in my discussion of "discovery," it is absurd to teach "the scientific method" as an isolated skill outside of a broader context that makes the notion at all meaningful, even if there is such core method or set of methods. This particular insight, I believe, is fairly broadly accepted and supported in the contemporary literature. However, there is a deeper and more important point that is very little acknowledged: We cannot expect to impose scientific culture or practice on children in order to remedy, at a higher level, the unsituated practice of scientific skills and methods. For brevity, I will here cite one reference to stand in for a broader neglect of this fact. Brown, Collins and Duguid [2] thoroughly embrace the notion that it is essential to situate scientific practice and tool use. But they participate unabashedly in a "cultural imperialism" of assuming the answer is to import scientific cultural surround into the learning context.

---

[7] Some of the arguments against viewing children as scientists in this way appear in [4] and in more detail in [6]. The extensive literature on the "unscientific" nature of everyday logic and reasoning is also relevant.

"Too often the practices of contemporary schooling deny students the chance to engage in *the* relevant domain culture."[8] "The ways schools use dictionaries are very different from the ways practitioners use them." "[Students] need to be exposed to the use of a domain's conceptual tools in authentic activity — to teachers acting as practitioners and using these tools in wrestling with problems of the world."

Authentic activity, the antidote to unsituated teaching of science, is defined by Brown et al. as "the ordinary practices of the culture," presumably the ordinary practices of the working scientist "wrestling with problems of the world." They go on to say, "Many of the activities students undertake are simply not the activities of practitioners." "... school activity is very different from ... authentic activity, because it is very different from what authentic practitioners do."

Brown, et al, later describe and laud a program of mathematical instruction in problem solving created by Schoenfeld as teaching students "how to see the world through mathematician's eyes," giving students "the opportunity to enter the culture of mathematical practice." However laudable Schoenfeld's program is, it is not an uncritical importation of mathematicians' culture to the classroom. I can make the point, perhaps somewhat polemically, by pointing out that the very focus of the program is problem solving. But, problem solving is to a mathematician something like finger exercises are to a concert pianist: mainly something imposed on students, or at best a marginal part of the domain practice. Mathematicians do not solve posed problems to earn their living so much as they pose fruitful directions to pursue, uncover mathematical structure, make appropriate definitions, and centrally, prove theorems and publish results that, hopefully, will make a crowd of highly aesthetically tuned colleagues envious.

My point is not at all to criticize Schoenfeld's program. It is to try to point out where the real intelligence may be, and to undermine the unproductive notion that cultures and even activities are portable enough unproblematically to bring into the classroom. Our job in designing classroom activities and cultures is both easier and harder. It is harder because we cannot simply import, or even select from mathematicians' practice. Whatever we try to cultivate in the classroom must work on the basis of constraints and resources in *that* situation. We must be much more responsive to the people we find there, to the web of practice *they* can build. If a scientific activity works in the classroom, it almost certainly works for different reasons than those that determine it as part of a scientist's practice. Or, if there are large-scale invariants of culture that transcend, these are substantial results research has not yet established.

---

[8] Emphasis is mine.

At the same time, our job is easier because we find resources, goals and establishable activity patterns in children and classrooms that don't make sense to professional scientists or mathematicians. We have a more open hand to invent activities and cultures that are not piece-by-piece scientific, but do lead toward scientific practices,[9] or at least serve to form more scientific knowledge, skills and stances.

It is exactly here that the image of the child as designer and builder fits. Briefly, the claim is that this image does unpack productively into classroom practices that are substantially contiguous both with children's propensities for meaningful action as we find them, and with scientific practice and the understandings we need to foster. It is, I claim, a good representative of the kind of negotiation we need to engage in to promote science without denying children.[10] With respect to the scientific basis for such claims, we look to an analysis of the structure, dynamics and development of activities, with emphasis on continuity.

Let me unfold the image in a bit more detail. The claim is that viewing and cultivating children as designers and builders will prove profitable for reasons such as the following:

- *Continuity.* Within the ordinary range of skills, interests and goals of children we will find important resources for activities like design and construction. These resources must be both pragmatic and conceptual.
- *Externalization.* Design and construction that involves a major externalization, a physical artifact, presents many opportunities to support productive activities. For example, it presents opportunities to reflect, debug and improve by focusing on the use of the partially constructed object. It is not trivial that there is "a place to look and tinker" in sustaining an activity.
- *Investment.* The development of product can generate important sustaining effects through personal or group pride in ownership.
- *Goal Clarity.* Activities may frequently degenerate for lack of a clear target. Design and building can avoid this.
- *Use.* Feedback from better or worse functioning has an essential generating and stabilizing effect with respect to knowledge and skills. Producing a functioning product avoids a plague of typical school-based activities: Only the teacher knows and can tell "what's right." Responsibility for judging can be distributed in more reasonable ways if the principles to be learned determine how well the designed object works by criteria that are understood.

---

[9] It is not evident entering adult scientific practice should be our goal at all. Certainly the vast majority of children will not do this. Instead, they will enter different roles with respect to scientific practice, such as science-literate citizen, or skeptical consumer, for which mimicking scientific practice per se is not evidently the point.

[10] For some relevant data on "scientific" versus "engineering" instincts in children, see [10].

- *Cooperation and Sharing*. Products provide multiple opportunities for students to cooperate and share. Obviously, some design and construction projects are too big for any individual to accomplish alone. Such big projects allow many slots for individuals with different skills and interests to participate effectively. More than this, there are importantly different modes of cooperation around design. Some students may take primary responsibility for design (or only some aspect of it), others for construction, and still others may share in use and evaluation.[11]

A Case Study: Inventing Graphing

In the following, I will briefly recount a case study of what I take to have been an exceptional learning activity. The analysis will follow the discussion above, both with respect to the general frame I introduced to understand activities and with respect to the particular issues enumerated concerning the child as designer. I argue that this activity is unambiguously an example of children working as designers, that it appears to work well for reasons such as those listed directly above, and that there is little overlap between this activity and other images, such as cognitive apprenticeship or intentional learning.

As part of a year-long course on motion for sixth grade students that we have designed and taught, we instigated a group design of representational forms for capturing motion. A much more complete exposition of what happened in this activity will appear in [8]. This occurred early in the course, but after children had already been programming in order to simulate various motions. Specifically, we challenged the children to come up with a way to represent motion statically, and to do so as simply and completely as possible. That is, we asked them to produce a system for representing motion on a sheet of paper, which therefore could not involve genuine motion as their simulations had done. Over a week of intense group design, the class had collected a strikingly inventive set of representations. More impressive, they gradually improved their representations and narrowed their choice until, by consensus, the group agreed their constructed version of graphing speed versus time was the best representation they could come up with.

---

[11] In our work, we have seen what we judged to be important exemplars of each of these modes of sharing around design and construction.

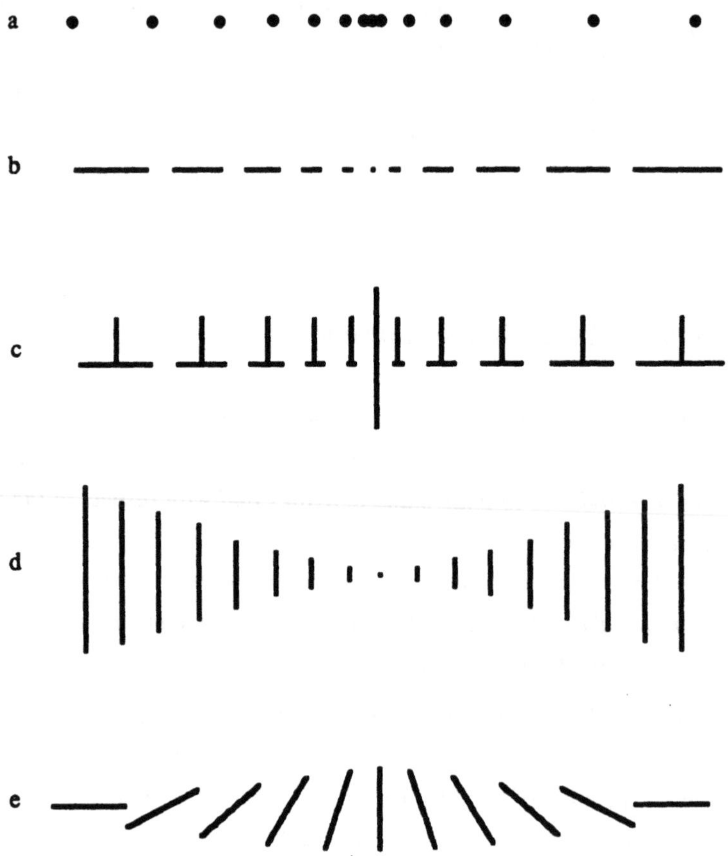

Figure 1. Five Representations of motion. a. Dot spacing represents speed or distance travelled during a constant interval. b. Dash lengths represents speed. c. Vertical line represents time duration; horizontal is speed. d. Vertical line represents speed; horizontal represents distance. e. Slope represents speed; when segments were hooked together, this became graphing.

Figure 1 shows a selection of the representations the children invented. They all portray a motion that starts at a given speed, slows to a period of rest, then gradually regains its speed. In describing these representations, I hope (1) to provide evidence that this was a high-quality and profitable activity, (2) to illustrate the framework for understanding activities, and (3) to provide some evidence that what I have said about the child as designer may have empirical validity.

Figure 1a shows dots spaced farther apart as the speed of the motion increases, and vice versa. This may seem primitive and "concrete." But it had an interesting history. The earliest version of this representation in the motion course occurred when a pair of children decided to show the "speeding up" of a falling ball in their computer simulation of it. Surprisingly, these two were the only children initially to think speeding up was an important aspect of a fall. Their first representation of speeding up showed more dots, closer together, to indicate more speed! A researcher pointed out that this caused the simulation actually to slow down (to draw more dots) toward the bottom of the fall. The children replied they had not wanted to make it actually go faster, but only intended to depict "more speed."[12] This was a striking early example of how abstractly these children could come to the problem of representing motion.

Figure 1b shows a representation where longer lines indicate greater speed. This representation was genuinely a communal construction. Its early version, presented by one student, was gradually improved by the introduction by another student of a concrete metaphor: This representation is as if produced by dropping chalk dust regularly but intermittently from a moving vehicle. Among other things, the move to interpreting lines and spaces as representing constant intervals helped establish a uniform, clear meaning for the representation.

Figure 1c shows a representation that uses a vertical line to represent duration and a horizontal line to represent speed. While we have no space for details, this representation emerged in trying to debug and improve another representation. Early representations, generally speaking, were problematic with respect to depicting the duration of stop. Consider both 1a and 1b. Separately representing speed and duration as in 1c solved that problem. Note that this representation also provides a good, discrete model for the mathematical process of integration — summing the products of speed times time interval to give distance travelled. These children came to understand that integration could easily be accomplished with this representation.

Figure 1d was the first representation to show speed cleanly on the vertical dimension. Note that the T representation (1c) had time doubly represented in the sequence of Ts and also in the vertical line, representing duration. Figure 1d was also one of the first representations that was presented without remnants of concrete features, such as a "road" drawn underneath early representations.[13]

Figure 1e was a stunning achievement. The originator explained he could choose to represent speed by the slant of the line, similar to what occurs in a distance versus time graph

---

[12] In other contexts, such representations have been interpreted as "misconceptions." Many children and naive adults will show a greater density of dots to indicate more speed in an imagined situation where a dot is made at regular intervals.

[13] The standard test motion for representations was one abstracted from a story about a car driver.

(although, of course, none of the students knew this latter fact). Typical of many of the representations, this one had a long developmental trajectory. One of the critical points occurred when its originator decided he could hook his "slants" end to end to produce a continuous representation. Indeed, connected slants formed one of the first truly continuous representations. Connected slants was quickly improved ("appropriated and reinterpreted" might be a better description) by another student, who proposed adding a grid to it to make it easier to read quantitative information out of it. The addition of a grid led directly to graphing in a conventional sense, although there were many conceptual and practical problems that needed settling before the representation converged to graphing velocity versus time, and before the group would adopt it as the consensus best representation.

These discussions were generally highly animated and showed considerable investment on the part of the children. There were few lags or lulls, and, in fact, the teacher had more often to subdue than stimulate. To take a single indicator, when it came time to close the whole discussion, students continued to re-open it, and complained that they were having to move on to another activity.

Let me begin fitting this recounting into the framework for understanding activities presented earlier. Start with conceptual resources. It should be evident simply from the partial list of representations in Figure 1 that these children had a substantial set of skills in inventing representations. Just as impressive, they employed a long list of criteria to critique existing representations, and, in reaction, to stimulate improvement. The following list of criteria is representative of those these children spontaneously applied.

- *Transparency*. A representation should need as little explanation as possible.
- *Homogeneity*. Use as few symbol types as possible.
- *Compactness*. Representations that are spatially compact are, generally, better.
- *Conceptual Clarity*. A representation should cleanly represent the different aspects of interest.
- *Appropriate Abstraction*. Omit "concrete" aspects irrelevant to the motion itself.
- *Completeness*. The representation should allow the determination of all relevant aspects.
- *Economy*. The representation need only show enough to deduce related aspects.
- *Quantitative Precision*. A representation is better if quantities can be precisely read out.
- *Consistency*. Don't adjust conventions for particular motions.
- *Objectivity*. In general, it is better to omit personal and idiosyncratic additions to this kind of representation.

The ideas of each child constituted *conceptual resources* for others. In addition, the tremendous richness of their ideas in itself contributed important resources for *sustaining* the activity.

Pragmatically, it was vital that these children could engage each other productively, take ideas from each other and, together, create a product with rationale that, quite evidently, no one child could have produced. This involved a global stance on the part of the children concerning cooperating and learning from each other. It also involved more micro-interactional skills, which were constantly in evidence in the discussions. Consider the following exchange, where the inventor of slants (here called Mitchel), now adding the notion that the length of the slanted line could represent distance, discovers in interaction with another student an important limitation of his representation: It has a problem with the stopped state and cannot show its duration. Mitchel had drawn a long line to represent a long stop, contrary to his own declaration that length means distance.

Steve: Mitchel, won't your stop be nothing because there's no distance?

Mitchel: There's got to be a little line, because otherwise you wouldn't be able to represent it [slope]. But it would still ... It would just be a dot!

Similarly, in another exchange in the same context, Mitchel proposes to add a separate mark to show time in his representation, even though speed and distance were already represented. Then he vaguely suggests you would not really need the time marker. Charlie picks up the point, drives it home, but Mitchel takes over again to explain the rationale for the move to the teacher.

Charlie: Right, but so you wouldn't need time.

Mitchel: The time would still be there.

Charlie: Why would you need it?

Mitchel: I mean, you wouldn't really.

Charlie: I know, so you wouldn't need it, so why are you saying you need the time?

Teacher: You wouldn't even need time?

Mitchel: You wouldn't even, because it's the speed and the distance which makes up the time.

Typical of many engagements, Mitchel could stand his ground reasonably when criticized, but could understand and gracefully absorb productive critiques as well.

There is a lot to be said with respect to the *teacher's stance and strategies*. In the first instance, she was responsible for initiating the activity itself, for nurturing its development with appropriate sub-activities, and most particularly, for arranging a productive *articulation of learning and sustaining goals*. As an example of the last, she kept ownership prominent by keeping in active consideration a number of representations championed by different children. As well, she returned to difficult conceptual points, even when the natural dynamic led quickly away from them. For example, she returned several times to the particularly sticky issue of whether the horizontal dimension should represent time or distance. She rejected some suggestions from students for how they should proceed as a group, presumably on the basis of a judgment about what would be more profitable in learning terms. On the other hand, she sometimes pursued emergent goals rather than her own preset ones, again on the basis that such pursuit would sustain or energize the discussion while encountering unplanned, but interesting and important learning possibilities. For example, on the last day of discussion, she completely abandoned her graphing goals in order to pursue a furious student-initiated debate about whether an object that slows a forward motion, and then reverses it, must stop between forward and reverse segments.

Consider the list produced earlier of productive features of design activities with respect to this particular activity.

- *Continuity*. Without doubt, the activity engaged very substantial interests and skills of the students. Both of these may be surprising. We know of no other published suggestion or documentation that children possess substantial meta-representational skills with respect to scientific phenomena, the ability to invent and critique representations, as displayed in this activity.[14] Our take on the history of the pragmatic skills evidenced is more complex than I can present here. But, we believe such skills are not automatic, even (perhaps, especially!) in academically-oriented classrooms where the teacher and right answers dominate genuine child give-and-take. At the least, we can say this teacher worked very hard to instill those skills as a background that allowed inventing graphing to happen.

  Continuity in terms of interest may be even more difficult to pinpoint. Where do children get an interest in inventing representations? But see the discussion of products and ownership, below.
- *Externalization*. Most of the activity of the design took place in generating and reacting to the visual display of the representations. I maintain, but won't argue, that this was critical. However, one indicator is particularly interesting. Frequently interpretations waxed and waned and shifted whereas the physical form under discussion remained to provide a continuity. Crudely, "we're talking about this (physical) thing now," organized the discussion much more at some points than representational criteria or even strategy of representation.

[14] Of course, these children might be very special and the case, therefore, a fluke. We don't believe this. For example, Jeanne Bamberger [1] has many lovely examples of children's invented representations for music. Many of these are from much younger and "at risk" children. In addition, Ricardo Nemirovsky (personal communication) reports surprising replication of many of the features of the sequence of inventions (including, for example, the gradual move from discrete to continuous representation) with even younger children. We have also rerun this activity with a rather differently constituted group of high school students.

- *Investment*. Ownership was evidently a powerful contributor to the success of the activity. Nearly every student contributed substantially with representations, critiques, or other ideas. Especially individual productions were strongly valued. For example, one child consistently put a copyright sign and her name on her productions, and commented frequently that she thought they were special. On the other hand, ownership was negotiated reasonably in terms of the overall goals. No student stuck doggedly to his or her representations. They all contributed to others' representations, and improved their own by taking ideas from others.
- *Goal Clarity*. The overarching goal of designing the best static representation constantly oriented and provided continuity to all of the activities during the five day discussion.
- *Use*. While it may be surprising, the goal of generating a representation that could effectively be used by others was consistently maintained. Essentially all of the criteria used to critique and improve representations emerged out of these considerations. It is perhaps a little mysterious that these children could effectively do this without having direct empirical feedback from design-naive others. Then again, most design (architectural, educational, etc.) is carried out only by "simulating" the users' point of view. And students could use their own responses to others' representations as a fair gauge of success or failure.

  During the discussion, the teacher almost never explicitly approved or disapproved of particular representations. Instead, she proposed challenging motions to test the representations; she asked about reasons for individual preference; and she pushed toward consensus. Responsibility for judging resided almost entirely with the students.
- *Cooperation and Sharing*. As we suggested, sharing occurred in multiple and complex ways. Children incorporated ideas from others' productions into their own. They appropriated others' criteria for judgment. At the teacher's instigation, they "tried out" each other's products. Many times, they shifted allegiance to representations that did not originate with them. In additions, some students took recognizably different roles in the design from time to time. For example, some took primarily to acting as critics and advocates for "outsiders" who could not understand in virtue of being present at the design. Most obviously, we believe it was apparent to all that their accomplishments could not have happened without the group's cooperation.

In contrast, we ask if it is possible to assimilate this activity to other images. Was this, for example, intentional learning? It was not in any evident way. Children took their goal to be designing a representation and measured their success in those terms. They almost certainly did not realize they were learning important meta-representational concepts, so could have no clear content learning goals in this regard. Nor was the point (for them) to learn about motion, though this happened incidentally to design. There were no evident and explicit learning goals, or at least no more (and probably less) that in an ordinary classroom situation.

Was this an instance of cognitive apprenticeship? The teacher did not model or fade. She scaffolded not content or skills issues, but mostly on pragmatic issues, like properly articulating learning and pragmatic goals by returning deliberately to knotty issues. She was not engaged in studying "problems of the world," nor did she really exercise her own design and representational expertise at all, and certainly not for her apprentices to see.

In a similar way, the activity is peculiarly situated. There were never any real consumers of the design. Instead, children had apparently internalized enough of the use context that they could substitute for a genuinely practical test. Children did not think of themselves, as far as we could tell, as little scientists or apprentice designers. Instead, the activity was well situated in the way we identified as most important, as a continuation of their existing skills and interests. And it was well situated, *in our eyes*, with respect to developing important scientific knowledge and capabilities.

Were these children learning to be good problem solvers? Perhaps they were, but not in any obvious way. They were not coached in any general heuristics, strategies of problem solving, or in means-ends analysis. They did no evident planning. Neither were they prompted to manage their time appropriately or to monitor their progress. Surely some of the things they were prompted to do could be schematized as good problem solving skills, but they were not schematized in such a way explicitly, and what they did could be schematized (by the children, as well as by us) in many other ways as well. Perhaps most problematic, *nothing* we highlighted as special about the activity is particularly notable if this is *just* an example of good problem solving. Self-sustaining activities, continuity in goals and interests, articulation of pragmatic and conceptual issues, and even the issue of prerequisite pragmatic skills in any respect are all subordinated if not completely neglected in characterizing this as a problem solving experience.

There are certainly community issues here. But this activity was far more interesting both to the children and to us — "better," we would judge — than what happened immediately before and immediately after in the course. As in our general discussion of culture, it was definitively not just the community that made this activity special.

In sum, we believe this was a highly successful learning activity. Children learned about motion, about representing motion, about scientific representations in particular, and about inventing representations more generally. (See [8] for a more thorough discussion of what they learned.) We claim to have provided at least suggestive evidence that understanding how the activity worked was in some degree illuminated by our preliminary sketch of considerations for the structure of activities. And, we maintain that this activity shows, in action, some of the reasons we believe the child as designer is a productive image of good learning activities.

Looking at what children know and are interested in doing as the platform on which to develop scientific thinking, we may even have come full circle to understand science and scientists in a different way. The accomplishments of these children encourage us to realize a side of science that is frequently neglected. We should perhaps think of "the scientist as designer"[15] more than "the child as scientist." In retrospect, the design of representations seems quite near the core of important scientific practice. But it would probably not be among our first choices for scientific practices to emulate in the classroom if we did not know that children could do it so well and with such enthusiasm.

Conclusion

The overt goal of this chapter has been to understand the images of good educational practice we create for ourselves and for others, and to explain and advocate images we have chosen for our project. But the deeper enterprise, which we approached in a somewhat indirect way, is understanding the structure of activities in their own terms and, as well, in terms of their relation to knowledge, skills and the development thereof. That is a profound enterprise that this chapter barely serves to introduce.

References

1. Bamberger, J. (in press). *The mind behind the musical ear.* Cambridge, MA: Harvard University Press.
2. Brown, J. S., Collins, A., & Duguid, P. (1989). Situated cognition and the culture of learning. *Educational Researcher, 18* (1), 32 - 42.
3. Carey, S. (1985). *Conceptual change in childhood.* Cambridge, MA: MIT Press.
4. diSessa, A. A. (1988). Knowledge in pieces. In G. Forman & P. Pufall (Eds.), *Constructivism in the computer age*, 49-70. Hillsdale, NJ: Lawrence Erlbaum Associates.
5. diSessa, A. A. (1990). Social niches for future software. In M. Gardner, J. Greeno, F. Reif, A. Schoenfeld, A. diSessa, & E. Stage (Eds.), *Toward a scientific practice of science education*, 301-322. Hillsdale, NJ: Lawrence Erlbaum Associates.
6. diSessa, A. A. (in press). Toward an epistemology of physics. *Cognition and Instruction.*
7. diSessa, A. A. (in press). Epistemological micro-models. *Genetic psychology and cognitive science: Proceedings of the eleventh advanced course of the Archives Jean Piaget Foundation.*
8. diSessa, A. A., Hammer, D., Sherin, B. & Kolpakowski, T. (in press). Inventing graphing: Children's meta-representational expertise. *Journal of Mathematical Behavior, 10* (2).
9. Lave, J. & Wenger, E. (in press). Situated learning: Legitimate peripheral participation. Cambridge, UK: Cambridge University Press.

---

[15] I thank Lieven Verschaffel for this provocative emblem of a central tenet of our approach, reconceptualizing scientific practice and knowledge by looking at their genetic roots in the ideas and activities of children.

10. Newell, A., & Simon, H. A. (1972). *Human problem solving*. Englewood Cliffs, NJ: Prentice Hall.
11. Saxe, G. B. (in press). *Culture and cognitive development: Studies in mathematical understanding*. Hillsdale, NJ: Lawrence Erlbaum Associates.
12. Schauble, L., Klopfer, L. E., & Raghavan, K. (1991). Students' transition from an engineering model to a scientific model of experimentation. Unpublished manuscript, Learning Research and Development Center, University of Pittsburgh.

An Architecture for Collaborative Knowledge Building

Marlene Scardamalia and Carl Bereiter

Centre for Applied Cognitive Science, The Ontario Institute for Studies in Education, 252 Bloor Street West, Toronto, Ontario, Canada M5S 1V6

Abstract: Conscious, cooperative development of shared knowledge is the focus of the CSILE (Computer Supported Intentional Learning Environments) project. Results to date, both from our own work and from similarly-motivated work, convince us that elementary school students can profitably make knowledge construction the focus of their efforts, although it is a novelty to them (and to their teachers) and requires a great deal of support. On the basis of what has been learned from four years of experimentation with an initial version, CSILE 1.0, we are designing a much more powerfully supportive system, which will be embodied in a second generation of CSILE. A major change in CSILE 2.0 will be the inclusion of distinct environments for different kinds of knowledge-building operations. In this paper we set out the principal features of the new knowledge-building architecture and the design principles that are guiding its development--principles that we believe are applicable to any technology aimed at restructuring classrooms as places for sustained, collaborative inquiry.

Keywords: Knowledge building, collaboration, computer-supported learning, interactive learning environments, hypermedia

Introduction

CSILE (Computer-Supported Intentional Learning Environments) is a family of developing systems intended to support the collaborative construction of knowledge in and beyond the classroom. The focus of this chapter is on the transition from a relatively simple system, providing little structuring of user activity, to a more elaborately interactive system, and on the principles guiding this transition.

The guiding idea in current designs and applications of CSILE is that of the knowledge-building community [12]. Knowledge-building communities are familiar to modern societies in the form of thriving scientific disciplines, such as cellular microbiology. Such disciplines embody a number of concepts that are now gaining currency in educational thought: constructionism--the idea that knowledge is a human construction, not something that exists to be revealed and transmitted; sociocultural activity, as the medium through which knowledge construction takes place [2]; and apprenticeship, in that much of the skill of the young scientist is acquired by working along with a more mature scientist [3]. So successful have scientific research groups been that they are now being used as models for sales and management groups in industry [7]. In principle, they could serve as models for education as well, but virtually the whole structure of conventional schooling is incompatible with such a model. Our aim in the development of CSILE is to support a new kind of environment that will make it possible for schools to function as knowledge-building communities.

CSILE, accordingly, is not designed for teaching any particular subject or skill. Instead, it is a general-purpose environment intended to affect the curriculum in its entirety. Technically, it is a networked hypermedia system in which the database is wholly created by the students. (For descriptive details, see [14] and [11].) For the last several years, versions of CSILE have been in use throughout the school day in grades 1 through 6 in a local public school. Teachers have used it to teach units spanning the entire range of academic subjects. The system has recently been installed in a high-school serving students from grades 7 through 13. Research results with elementary students indicate gains in a variety of areas, including question asking, writing, setting and pursuing cognitive goals, language abilities (based on a composite reading, vocabulary and spelling score from the language sub-scales of a Canadian standardized achievement test), and graphical communication [13].

Currently we are at a significant juncture in CSILE development, working on both a more economical version, suitable for dissemination, and a more advanced version (CSILE 2.0). This has been an occasion for reconsidering all aspects of CSILE 1.0 and also for establishing design principles congruent with the idea of a knowledge-building community. We will deal with both of these considerations in this chapter, because they are intimately connected. Everything we would hope to support in CSILE 2.0 happens occasionally or partially in CSILE 1.0 environments [11, 12]. Thus the goal for design of CSILE 2.0 could be formulated as making the good things that happen in CSILE 1.0 happen more fully and regularly. This does not rule out the possibility, however, that radical design changes may be needed in order to bring this about. Accordingly, there needs to be a dialectical interaction between critical examination of CSILE 1.0 and formulation of design principles for CSILE 2.0.

This chapter is not a narrative account of this dialectical interaction, but rather an account of its results so far. The chapter is divided into three parts:

(1) New design principles. In this section we identify design principles that we have elaborated after four years of classroom implementation and briefly sketch their motivation. These principles are superordinate to those we began with [14], which is to say that the original principles are a partial instantiation of the newly formulated ones. While the principles we elaborate here point to areas where our own designs have fallen short, they also point to a new architecture for collaborative knowledge building. We believe, however, that these principles will apply not just to CSILE architecture but to any knowledge medium aimed at restructuring classrooms as places for sustained, collaborative inquiry.

(2) Evaluation of CSILE 1.0 in light of the new design principles. In this part of the chapter we briefly describe the major features of the current system, and comment on the extent to which each feature implements or falls short with respect to the design principles.

(3) Overview of an architecture for collaborative knowledge building. We present, in outline, our proposed new architecture. In this section we try to characterize a system for knowledge advancement that meets the new design principles.

New Design Principles

The idea of a knowledge-building community and of patterning schools after scientific research communities was not fully formed at the time CSILE 1.0 was designed. It took shape, in fact, as we began to see instances of CSILE students beginning to act like research groups--to be absorbed in working out an explanation of human evolution, for instance, rather than in the mastery of some body of subject matter or in performance of some task. As a result of thinking about and experimenting with ways to bring more of this collaborative knowledge-building activity to the fore, we have evolved a set of design principles. These are the principles we are using to guide design of the new version of CSILE, aimed more expressly at aiding the development of a knowledge-building community. The principles themselves are still undergoing change and elaboration. Presently, they may be summarized as follows:

1. Objectification. CSILE should help students treat knowledge as an object that can be criticized, modified, compared and related (one knowledge object to another), and regarded from different viewpoints, in different contexts.

Objectification of knowledge may have its dangers, but we follow Popper [8] in believing that treating knowledge (which we take to include theories, beliefs, empirical generalizations and the like) as an object is essential to the critical processes through which knowledge advances. The dangers--for instance, that knowledge may come to be treated as absolute, immune to human biases--are minimized in a community that values criticism, controversy, and belief revision. The problem in conventional classrooms is that knowledge receives little direct attention at all, being always subordinated to tasks and activities.

2. Progress. Knowledge-building activities should lead somewhere and progress should be perceptible to the students.

The outstanding achievement of the sciences, as Whitehead [16] observed, has been "disciplined progress"--progress toward increasingly powerful theories and technologies that does not depend on chance but that is a dependable result of the way scientific communities function. To whatever extent is possible, students ought to be able to have a similar sense that their quest for knowledge is yielding advances. In conventional schools, the course of events tends to be cyclical rather than progressive. Schemes like Mastery Learning aim to introduce progress through a sequence of learning objectives, but there are indications that students often do not see the sequence as progressive. Instead, they see the passing of a mastery test as merely marking the end of one task cycle and the onset of a new one much like the one before [6].

3. Synthesis. CSILE should encourage the construction of higher order representations and integrations of knowledge rather than the proliferation of loosely connected items.

School textbooks are often criticized for providing mere collections of facts under topical headings. Hypermedia have the potential to make matters worse, creating a sprawling network of items lacking even topical organization. Any system to promote knowledge building must counteract this tendency and must instead support the continual creating of higher level structures (such as theories, narratives, unifying descriptions) which help to make sense of and unite separate pieces of information. Knowledge synthesis is a continuous process, not a once-and-for-all accomplishment, and thus represents one aspect of progress. It seems sufficiently important and challenging, however, to be treated as a design principle in its own right.

4. Consequence. Something nice should happen to students as a result of knowledge-processing operations.

One nice thing that can happen is a perceptible advance in knowledge--a Eureka! experience or the settling of a pesky question. But more immediate consequences are needed to sustain the activities leading up to such advances. Consequences that make a difference in the scientific world include seeing one's work cited or commented on, one's ideas used or confirmed. CSILE should make provision for such consequences and ensure that, when they occur, the news gets back to the author.

5. Contribution. Students should see their entries in the communal database, not solely in terms of their independent merits, but also in terms of their contribution to advancement of the group's knowledge.

Only in the most backward disciplines are articles judged for publication solely on intrinsic merit. Ordinarily, a study must not only be good but must be judged a contribution to the discipline (i.e., to objective knowledge, in Popper's terms). Schoolwork, however, is usually judged solely by internal criteria (correctness, quality of thought, etc.), which reinforces the impression that it is mere exercise. CSILE environments should be designed to direct attention to contribution as well as to intrinsic characteristics of notes and should encourage students to see these as related--to realize that a reason to write clearly, check one's facts, and so on, is that it improves the chance for one's note to make a recognizable contribution. This principle should not be interpreted as discounting the value of private reflection and striving for personal excellence. The intent, rather, is to achieve the balance of personal and collective motives that characterizes any healthy learned discipline.

6. Cross-fertilization. CSILE should maximize chances for students to come into contact with related ideas, kindred spirits, and useful information, unrestricted by boundaries of space and category.

A computer can perform only shallow analyses of text content, but it can do so frequently and thoroughly. By analyzing lexical co-occurrences, for instance, it can identify notes that appear to have common themes and authors who appear to share an interest. In a networked system like CSILE, it can do this across subject-matter and classroom boundaries. Cross-linkages of many different kinds can help to create conditions for opportunistic learning and discovery.

7. Sociality. There should be no discontinuities between work in CSILE and other curricular activities. Instead, CSILE should help to integrate the social life of the classroom.

From being isolated in computer labs, microcomputers have been making their way inside elementary school classrooms, but it is still often the case that computer activities are sharply separated both from the rest of the curriculum and from the ordinary social life of the classroom. CSILE 1.0 has largely overcome the problem of separation from the curriculum, being used in all areas of the academic curriculum. To overcome separation from the social life of the classroom, however, future versions of CSILE should have ways of being relevant to classroom discussions, small group projects, laboratory experiments, field trips, and the like.

Evaluation of CSILE 1.0 in Light of New Design Principles

The seven design principles enunciated in the preceding section do not refer directly to technical features but rather to desired behavioral or psychological characteristics of the interaction between user and technology. Consequently they cannot be directly translated into technical implementations, and technical features cannot be directly evaluated according to them. There must always be an intervening consideration of the human-machine interaction. In evaluating design features of CSILE 1.0 we have the benefit of a great deal of information about how students actually used the system, although the linkage of behavior to specific features of the system remains speculative.

It is difficult to give a general characterization of what students do in CSILE because this varies considerably from subject to subject, from teacher to teacher, and from student to student, reflecting the extreme openness of the system. One constant is that students produce notes--text, graphic, or mixed--and enter them into the communal database. The nature and purpose of the notes can vary greatly, however. At one extreme are notes meeting a specific assignment in a specific subject--for instance, an assignment that instructs students to "tell what you concluded from the heating and cooling experiments." At the other extreme might be a cartoon created by a student, unrelated to any class project, and intended for the amusement of classmates. Notes can be searched and retrieved according to a variety of identifiers and for purposes that, again, may range widely--from carrying out a specific assignment to idle browsing. Such openness and diversity, however, make it all the more important to have principled ways of evaluating components and judging what features should be added, modified, or deleted. In the following paragraphs we discuss some of the main features of

CSILE 1.0 as they relate to the principles of objectification, progress, synthesis, consequence, contribution, cross-fertilization, and sociality.

Evaluation of individual features

Student-Generated Communal Database. The centerpiece of the system is the student-generated communal database, into which all notes in all curriculum areas are entered. Students write notes regarding their goals for learning, questions they would like to have answered, and discoveries they have made. Their notes are available for all users in the network (unless the author specifies otherwise) and readers are encouraged to engage the author in debate and refinement of ideas. We see the communal database as the major means for achieving the principle of objectification. It gives concrete form to the accumulating knowledge of the classroom group. It contributes to cross-fertilization in that notes are not segregated according to subject and searches may retrieve notes from unexpected domains and unexpected students. Because activities involving the database can serve to bring together information from off-CSILE activities (field trips, interviews, etc.) and to support off-CSILE activities (for instance, through having students scan CSILE notes for questions to be addressed in a class discussion), the communal database also serves purposes of sociality, integrating CSILE activity with the rest of classroom life. CSILE 1.0's database is weak, however, with respect to contribution and consequence. Notes of different students are often redundant. Unless notes receive comments, students have no way to know whether their notes have been read by others or if they have had any effect.

Keyword Assignment. Students are encouraged to summarize their ideas via labels that characterize the contents of their notes, making it easy for users to access their ideas, and for them to return to their own ideas for review and refinement. Although keywords have well-known limitations in information retrieval, we believe that the assigning of keywords is an important contributor to objectification. It encourages students to view their notes as knowledge objects and to find words to characterize and distinguish those objects. Keywords also serve cross-fertilization, since they allow connections to be made across subject-matter boundaries. Keyword assignment in CSILE 1.0 is weak in fulfilling these principles, however, because there is little to discourage students from assigning keywords in an arbitrary or indifferent manner. There is, in short, not enough consequence ensuing from keyword assignment.

Note Statuses. Students assign statuses to their notes--public (available to everyone) or private, anonymous or signed, and occasionally "Candidate for Publication." Notes designated as "Candidates for Publication" must be accompanied by a form which indicates, among other things, what the student feels to be the distinctive contribution of the note to the community knowledge base. "Candidate" notes are reviewed by teacher or committee. If accepted to "Published" status, the note appears with a different border. When searching, students may specify that they only want to see published notes. Publication, thus, provides some consequence for making a contribution to the community's knowledge. In practice, however, it has proved to take extensive teacher effort to make publication work and to keep students attending seriously to whether their candidate notes make a contribution to community knowledge.

Thinking Types. The system aims to support an additional metacognitive layer of objectification by encouraging students to identify the kind of knowledge activity that went into the construction of a note--stating a goal, identifying a problem in understanding, recording a new idea, etc. The major weakness of Thinking Types in CSILE 1.0 is a lack of consequence. Students are not required to assign Thinking Types to their notes, and the only consequence of doing so is that the Thinking Type becomes an attribute that others may use in retrieving the note. Where thinking types are operating effectively, teachers have built them into the structure of their lessons, thus honoring the sociality principle. In the new CSILE architecture we want to shift the focus from identifying kind of knowledge activity to identifying kind of knowledge contribution.

Searching and Reading. Students are provided with search parameters such as keywords, topics, authors, publication status, etc. that allow them to establish criteria for notes they wish to read. This provides them with ready access to their aggregated ideas, and gives them a kind of ownership over the growing body of knowledge. Although the searching done by elementary school students tends to be unsophisticated (few compound searches) it has nevertheless proved to be a valuable means of cross-fertilization at all grade levels [13].

Commenting. Students have facilities that make it easy for them to comment on each others' notes, with automatic notification to authors of notes receiving commentary. They are thus encouraged to engage each other in discussion and refinement of ideas. Comments provide an important consequence for the author. There are also instances where a series of notes and comments has shown progress toward more advanced understanding than any of the individual contributors had at the outset [11].

Linked Charts. Procedures for linking graphical notes enable students to add information via central coordinating structures (designed by students) that show how their charts are interrelated. For example, they create a central chart of a skeleton, with different students linking notes regarding various anatomical features. In this way they produce more complex knowledge structures than individual students are likely or able to produce. A set of interlinked charts provides a nice objectification of collaboratively-constructed knowledge. Linked charts may also serve the purpose of synthesis. A major limitation of CSILE 1.0 is that hierarchical linking is only possible with graphical notes and not with text notes.

Summary of evaluations

In one way or other, CSILE 1.0 provides support for all seven of the new design principles, but much more could be done on every count. Where we see CSILE 1.0 as weakest is in implementing the principles of progress and synthesis. It is not that CSILE 1.0 thwarts progress toward the solution of knowledge problems. On the contrary, we see instances of progress in CSILE that are exceedingly rare in regular schooling. But because this principle is so at variance with the way schools are structured--around cycles of activity, the learning of discrete bodies of material, and progress objectified only in individual performance--it requires powerful environmental changes to turn schooling around to where students will see knowledge building as a truly progressive enterprise. The main contribution to synthesis in CSILE 1.0 is keyword assignment, which serves to connect notes with shared topics. Like other hypermedia systems, it does not prevent but also does not support the creation of new, higher level representations which synthesize lower level organizations of knowledge. The new knowledge-building architecture to be implemented in CSILE 2.0 is intended to introduce powerful new elements into the structure of school activities that will promote progress toward integrative explanations and representations.

Overview of an Architecture for Collaborative Knowledge Building

CSILE 1.0 may be thought of as providing a single environment in which various operations may be carried out upon knowledge dealing with any of a variety of themes. The major architectural change in CSILE 2.0 is the creation of distinct environments for different kinds of knowledge operations and provision for distinct "thematic spaces" differentiated according to the topic or substantive problem (e.g., dinosaur extinction) that work is focused on. The concept of a single community database is preserved, however, as the basic means for objectification of the community's accumulating knowledge. All notes are accessible from any

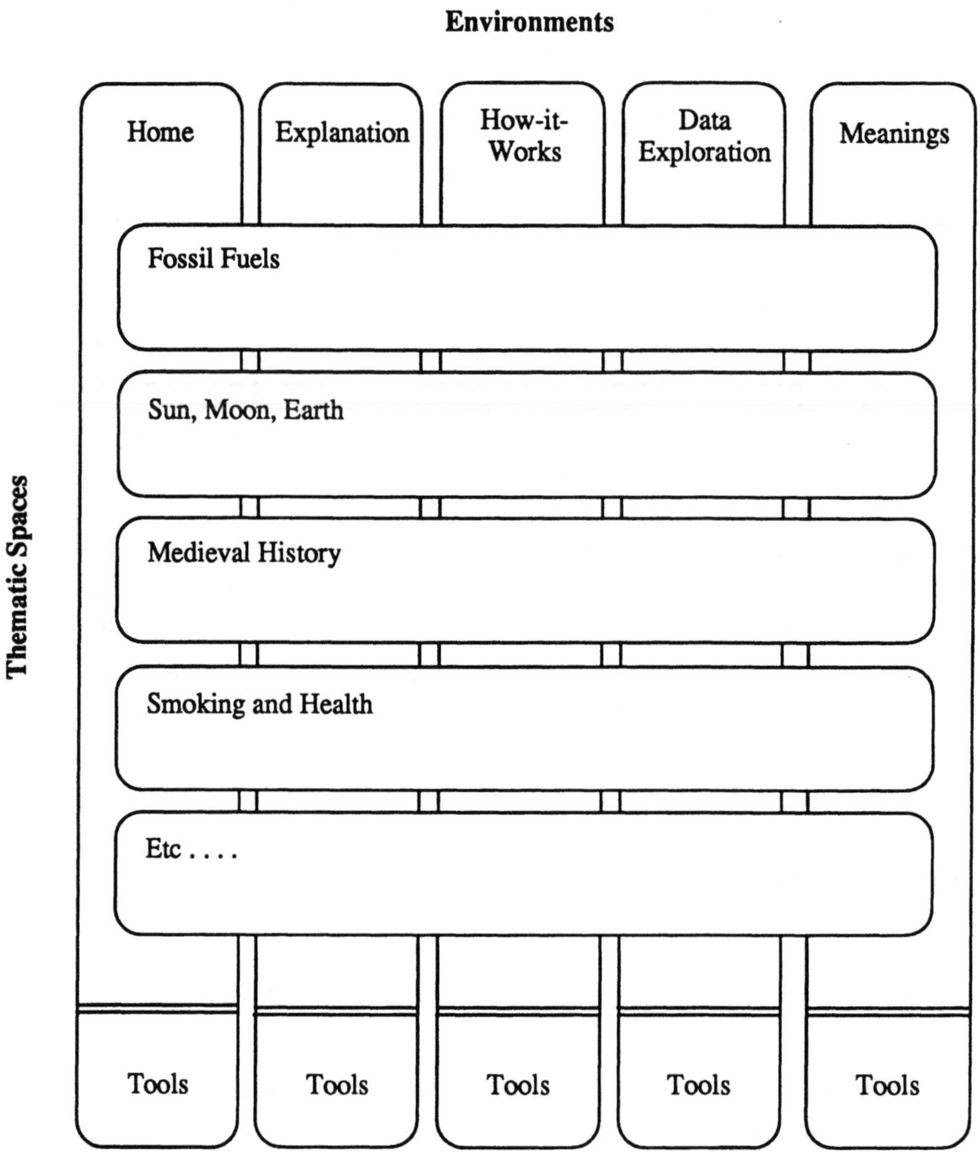

Figure 1. Environments and thematic spaces.

environment or thematic space, although the ways they are linked and the things that can be done with them vary. All of the previously discussed features of CSILE 1.0 will be preserved, with modifications to serve the more elaborate functions of the new architecture.

The relation between environments and thematic spaces may be visualized as a grid, as shown in Figure 1, with columns representing environments and rows representing different thematic spaces. A user may move horizontally from environment to environment while remaining in the same thematic space or may remain within a given environment and move vertically through different thematic spaces.

To illustrate how thematic spaces and knowledge-building environments are interrelated, we will develop an example based on material from an actual CSILE 1.0 database. One student produced several text and graphic notes attempting to show how the diaphragm works in coughing and hiccuping. (Examples are shown in Figure 2.) Another student tried to figure out what happens inside your body when you sneeze. This student also produced elaborated notes reporting information discovered through research into how the diaphragm works and its relation to hiccupping. Other notes in the database dealt with the relation between the diaphragm and laughing. To pursue the full set of related ideas, different students need to be brought together to elaborate their diagrams and notes. For example, the role of the epiglottis needs to be elaborated, as does the question of why a different sound is associated with a hiccup than with a sneeze. There seems little doubt that these students would explore these increasingly penetrating issues if they had these notes all conveniently available for joint consideration. That is the idea of a thematic space. It is analogous to a "conference" in the communications world except that it can be created in a bottom-up as well as a top-down manner, starting with unconnected notes that are brought together to form a meaningful group.

The thematic space gets students and notes concerned with a particular issue together, but then what do they do? They may just want to go on adding ideas and information related to the phenomena of hiccupping, sneezing, coughing, and laughing. In that case, they can stay in the HOME environment, which in CSILE 2.0 will correspond to the undifferentiated environment encompassing the whole of CSILE 1.0. But they may want to work more systematically toward a deeper and more explicit account of how the diaphragm creates these varying effects. For that purpose, they can move (while remaining within the same thematic space) to the HOW-IT-WORKS environment. In effect, they will have taken all their notes about the diaphragm with them, but the HOW-IT-WORKS environment will provide a framework for assembling notes and creating new notes that explain a process at increasing levels of specificity. It will also provide tools for producing slide-show sequences, simple animations, and simulations, which they may find useful in demonstrating their conceptions of the processes. They may

+a: Anatomaical Post
+b: hiccuping
+c: laughing
D: Diaphragm exhale
E: Diaphragm inhale
F: inhale
G: exhale

Diaphragm

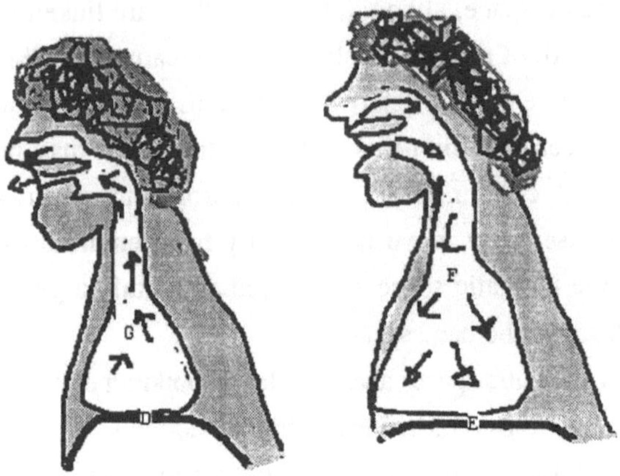

This is a picture of how the diaphragm works.

+a: Diaphragm
B: TRACHEA
C: EPIGLOTTIS
D: LUNG
E: DIAPHRAGM

hiccuping

Author:
Thinking Type: new learning
Note Level: SHARED
Topic: Biology
Keywords: BIOLOGY, diaphragm, hiccups.

HICCUPS

As you know from reading my note on the diaphragm the diaphragm has a certain rhythm. A hiccup is caused by a sudden spasm of the diaphragm. At the same time in a fraction of a second a small flap at the back of your throat called your epiglottis closes blocking off the trachea and when the air tries to to get in it bounces of the epiglottis and makes a "HIC" sound.

Figure 2. Sample of student notes on function of the diaphragm.

develop or read about competing theories, which they wish to compare and evaluate. For that purpose, they can move into an EXPLANATIONS environment, which provides ways of highlighting the explanatory aspects of notes and for investigating the coherence of complex explanations. Thus, thematic spaces and knowledge-building environments have complementary roles, which combine to support the focusing of cooperative knowledge-building effort on problems and issues and the pursuit of knowledge to deeper levels, in keeping with the principles of progress and synthesis.

Let us now consider how these new architectural components serve the seven design principles of objectification, progress, synthesis, consequence, contribution, cross-fertilization, and sociality. Thematic spaces are essentially ad hoc groups of notes within the communal database that provide thematic structure for this objectification, besides facilitating progress, through making knowledge problems to be solved more salient. The main goals for the special knowledge-building environments are (1) to serve the principle of progress by helping to sustain collaborative pursuit of explanations and understanding to deeper levels than students normally achieve, (2) to serve the principle of synthesis by providing means to combine information items into forms that facilitate the creation of integrative explanations or representations, (3) to provide task or problem contexts in which there are clear needs for students to make contributions to collective efforts rather than merely produce something meeting task requirements, and (4) to provide observable consequences for knowledge-processing operations that are directly relevant to the goals of the operations. In addition, the new architecture includes a number of background operations performed automatically by the computer. Some of these search out semantically related notes and suggest possible linkages to investigate and possible collaborators, thus serving the principle of cross-fertilization. Others monitor various aspects of system use--productivity, inquiry, and collaboration--and provide individual feedback serving the principle of consequence. Sociality is mainly served by the knowledge-building environments and thematic spaces in which notes are aggregated, leading to the natural formation of discussion and work groups. It is also served by a different component of the architecture, background operations, which will be described in a later section.

In the following sections we provide a more detailed account of the knowledge-building architecture, at this fairly early stage in its formulation, describing its five main components: the community database, knowledge-building environments, thematic spaces, tools and procedural facilitations, and background operations.

Community database

A single community database underlies all the thematic spaces and knowledge-building environments [14].

- All of the knowledge stored in the community database is stored in the form of user-generated notes, although these notes may be created in a number of ways.
- All notes are treated the same, regardless of whether they are text, graphic, copied from some external source, or generated as output of some knowledge-building tool.
- Information is brought into particular environments in the form of notes and the outcomes of knowledge-building operations carried out within environments are preserved in the form of other notes or note complexes. Thus the database at any time provides a representation of the totality of the public knowledge constructed by the user group.
- Material from other databases must be converted into notes before it can be stored in the community database. This requires that the user add keywords or other kinds of information that establish the contribution of the note to the knowledge base, so that a note is never simply a cut-and-pasted fragment.
- As in CSILE 1.0, students will have the option of designating notes as private. Private notes or note complexes may be taken into knowledge-building environments in the same way as public notes, thus putting the full resources of the system at the disposal of students who wish to work with knowledge in private or who wish to make their work public only after it has reached a certain state.

Knowledge-building environments

We aim in the design of these environments to foster knowledge processes such as the following: (a) testing the explanatory power of hypotheses; (b) exploring analogies; (c) identifying causal relationships; (d) illustrating and animating objects and thus identifying underlying mechanisms; (e) working out the historical relevance of ideas; (f) analyzing numerical relationships in terms of their relevance to mathematical principles; (g) extracting domain vocabulary from notes and creating background environments where these terms can be clustered and linked to create an overview of note contents; (h) engaging in research through integrated use of on- and off-line reference material.

The following are brief functional descriptions of some of the environments envisaged for CSILE 2.0.

The HOME environment. The HOME environment is the default environment that the user enters when signing on to CSILE. From the HOME environment, users can branch directly to any other environment or can branch to a thematic space. Being in the HOME environment and not being in any particular thematic space is equivalent to being in the CSILE environment in its current version. We would expect that for young and new-to-CSILE students this may be the only environment available initially, with other environments and thematic spaces available as they gain sophistication. This would also be the main environment for private notes of all kinds. All resources for creating, retrieving, and commenting on notes are available in this environment. The HOME environment is also the place for early drafts not ready to be made public.

The EXPLANATION environment. The motive for entering this environment should be a desire to understand some aspect of the world or some puzzling facts. The EXPLANATION environment's interface will require that notes created or brought into the environment be given short legends which summarize the fact or hypothesis they discuss. The interface also provides a means of linking these into a network of the kind that ECHO (a connectionist program that evaluates networks of propositions according to principles of explanatory coherence) can operate on [15]. This means that links indicating "explains" or "contradicts" can be created or removed. The interface will also provide procedural supports to help users think through the creation of links and will be designed with a view to incorporating other software related to explanation as it is developed.

The HOW-IT-WORKS environment. The focus in this environment is on explaining how things work--machines, physiological processes, geological processes, economic, political, etc. Like the EXPLANATION environment, it supports the building of coherent explanations. However, it is more suited to working out and elaborating causal mechanisms than with pitting one explanation against another.

The main challenge in creating this environment is to design an interface that facilitates a query-and-response process that leads to progressively deeper levels of process description. The linked charts feature of CSILE 1.0, described earlier, provides a framework for organizing such descriptions. A given chart could show, in comic strip fashion, three main steps in a process. Clicking on one of the steps would bring up a chart showing a finer breakdown of that step; clicking on one of those substeps would bring up a finer breakdown still, and so on. Other tools that might be available in the HOW-IT-WORKS environment would permit the construction of flowcharts, simulations, and animations. These would not only provide alternative descriptive media but also alternative modes of explanation--narrative, enactive, principled, systemic, and perhaps others.

The DATA EXPLORATION environment. This is an environment for exploratory analysis of data. It would include graphing tools, perhaps a spreadsheet program, and kinds of data exploration such as are available in Minitab. An important challenge in design of this environment is to maintain close connections between the data and what they refer to, so that users do not get lost in number-crunching.

The MEANINGS environment. The focus in this environment is on domain vocabulary and its interrelationships. From an educational standpoint, this is an important aspect of work within any thematic space.

A central feature being developed for the MEANINGS environment is a concept map that is automatically generated for any thematic space, based on co-occurrences of lexical terms in the sentences of notes in the thematic space. For any large thematic space, such an automatically generated map would likely be a monstrosity, containing irrelevant words and connections and redundancy resulting from synonyms and variant forms of words. Accordingly, the MEANINGS environment will also provide facilities for students to construct another, more intelligible map by selecting and moving nodes and links from the automatically generated map and assigning names and explanations to links. The automatically generated map remains important, however, because the only way new links can be created is by writing a note in which the to-be-linked words are used within the same sentence. This produces an automatically generated link which may then be transported to the student-constructed concept map. As the concept map grows in extent and coherence, it provides a representation of the growing collective knowledge of the students who have contributed to the thematic space. In this way the MEANINGS environment supports the progress principle. The sociality principle is supported in that the knowledge map provides a basis for live group discussions, in which names for various links can be proposed, discussed, and later actually assigned.

A further level of semantic investigation is foreseen, when CSILE 2.0 is linked to external (especially CD-ROM) information sources. Students may then designate external documents related to their thematic space and have CSILE extract the domain vocabulary that links to words on their own concept map. This additional vocabulary, appearing as shaded nodes and links in the background of their own map, provides clues to further relationships to explore.

The EXPERT environment. An analysis of questions by students in CSILE classrooms shows that approximately 35% [11] of their questions are ones that might require help from an outside expert to answer. The EXPERT environment will aim to help them connect with experts in their field of inquiry, while treating experts' time as a limited resource that must be used wisely. The challenge in creating the interface or educational layer will be to help students filter and upgrade material to be sent to experts. Questions should not be going to experts until users have taken advantage of the resources available to them, and in the course of doing this they should come to appreciate the fact that they can make progress without expert help even on very difficult questions.

The center of the EXPERT environment is, of course, an external communication facility. But knowledge-processing tools for outgoing communication will serve to synthesize and filter requests for information. For instance, when too many questions have accumulated for submitting to a particular external expert, the environment will provide means for students to comment on priorities and suggest revisions and ways to consolidate several questions into one. (Educationally, this pre-consultation work could prove as valuable as the actual consulting with an expert.) Responses from experts need to be converted into notes in order to become part of the community database. In accordance with the contribution principle, this requires that information be accompanied by some justification, elaboration, or interpretation by a student rather than just be pasted in as a note. The EXPERT environment will provide convenient means and procedural supports for transforming incoming communications into notes integrated into the communal database.

The DOCUMENTS environment. This environment serves two main functions: Preparing notes to become candidates for "published" status in the database, and preparing larger documents that are to be printed as reports, library entries, "take-homes," or whatever. Mainly this environment provides a range of editing and publishing tools beyond the basic ones available in all the other environments--e.g., grammar and style checkers, pagemakers, facilities for bringing separate notes into the environment and assembling them into larger documents.

Thematic spaces

When students sign on to CSILE they will enter the HOME environment. After completing such HOME environment activities as reading comments and checking their calendars, they have the choice of either going into a thematic space or going to another environment. Suppose they elect to go into a "dinosaur extinction" thematic space. If they then decide to search for notes, the note search screen will inform them that there are, say, 24 notes in the "dinosaur extinction" thematic space and they will have the option of calling up all of them or of carrying out an ordinary search. However, the search screen will only present them with keywords and other identifiers (such as authors) associated with those 24 notes. If they want to search more widely they will have to choose an option that lets them search "outside the thematic space."

Knowledge Organizers for Thematic Spaces. Knowledge organizers are visual frames, such as maps, timelines, and graphs that appear as the initial screen when one enters a thematic space. The knowledge organizers are labeled with keywords. Clicking on a keyword label permits attaching the keyword to a note in the write-a-note window and getting other information related to the keyword. CSILE needs to provide for these organizers to be created in several ways: (1) the teacher assigns a list of keywords, students produce the visual frame and place the keywords in it; (2) the organizer is built from a digitized image taken from a book, etc., and students attach keywords; (3) students build the whole organizer from scratch; (4) a previous organizer is used (retrieved from a library of these) and is edited for the current purpose; (5) the organizer is built using a template map, timeline, etc., drawn from a library of such templates.

The template option is especially desirable because it opens up the possibility of subsequently merging knowledge organizers. For instance, one group of students might develop a thematic space about artists, using a map template to situate the artist's names geographically. Another group might do the same with writers, political leaders, etc. A merging of these maps, with the different sets of keywords appearing in different colors, would allow for consideration of nations that appeared to be especially active culturally.

A more futuristic possibility involves knowledge-building tools that could retrieve and compile information on templates at the command of the user. Students could write simple programs to combine information in informative ways, such as, "Combine maps of Western Europe. Show names of poets in red and names of inventors in green." If students also used timelines when recording poets and inventors they could additionally specify that they'd like to see only those poets and inventors born between specified dates. With yet a bit more intelligence (in CSILE that is), they could do novel searches such as "Find all poets and inventors born in medieval times and combine on map of Western Europe."

Tools and procedural facilitations

Tools as Knowledge-Building Resources. The basic idea of a knowledge-building environment is that knowledge is brought into the environment and something is done to it that enhances its value. Tools available in the environment are means by which this value-adding work is done. Consequently, the design of environments and tools is guided by the goal of maximizing the value they add to knowledge--either the public knowledge represented in the community database or the private knowledge and skill of the user.

Some tools, such as basic text and graphics tools, are available in all environments. Others, however, are specific to certain knowledge-building environments and are integrated into the particular emphasis of the environment. We have mentioned ECHO as a possible tool in the EXPLANATIONS environment, providing a means for students to evaluate the coherence of linkages between explanations and facts. But it is integrated into the environment by means of an interface that gets students to identify and label explanations and "facts needing explaining" as notes are brought into this environment. These labels then become the tokens that ECHO operates on. STELLA [9], a dynamic system modeling tool, is a natural to integrate into the HOW-IT-WORKS environment; but again, we shall wish to design the environment's interface so that it provides a natural way for the information in notes to be translated into the data structures that STELLA works with.

As these examples illustrate, the architecture of CSILE 2.0 will be able to employ a variety of software tools designed by others. But we aim for CSILE to provide much more than an access route to these tools. CSILE's environments should provide a reason to use the tools, a context in which their use contributes to the overall development of the community's knowledge, and supportive means both for translation of information from notes into input to the tools and translation of output from the tools into notes that become a meaningful part of the community database.

Procedural Facilitations. The interface characteristics we have just described represent an important form of procedural facilitation. Procedural facilitation is a concept we have been promoting for some years and that is often incompletely or poorly understood [1, 10]. As applied to software, procedural facilitation refers to features that enable users to carry out more complex or sophisticated mental operations than they could otherwise carry out but that do not provide the substantive knowledge or skill required for those operations. Procedural facilitation is educationally significant because it does not substitute for the learner's competence but rather makes it possible for learners to engage more successfully in activity at the growing edge of their competence. Thus, an equation-solver is not a procedural facilitation as far as competence in solving equations is concerned, because the software, not the user, solves the equations. But an equation-solver could be a procedural facilitation with respect to developing competence in some type of empirical research where many equations need to be solved and where appropriate software would enable the student to investigate more complex relationships among variables.

Designing good procedural facilitations, even in such simple forms as hints that pop up on the screen, is not easy. The computer is intervening in complex and delicate cognitive processes, where it is easy to do more harm than good. That is why we have been very cautious about introducing procedural facilitations and why we are moving in CSILE 2.0 toward designing whole environments so that they will be supportive of certain kinds of mental activity, rather than depending on specific prompts or structuring devices that are triggered by actions of the user. There remains a role for specific procedural facilitations, however, and we will illustrate two that we intend to experiment with:

1. Procedural Facilitations for Commenting. The basic design of the main CSILE screen has proved to be supportive of commenting. A two-window design permits the student to have a note to be commented on in one window and to compose the comment in the other, thus aiding relevance and coherence. With no further supports, students produce many helpful comments, but these tend to consist of very global evaluations ("good") and low-level suggestions, such as pointing out spelling errors. We have been experimenting with off-line suggestions of sentence-openers that encourage more constructive comments: e.g., "My hypothesis is different from yours. I think ...," "I found this reference that explains ..." These, together with encouragement by teachers, have substantially increased both the quantity and quality of comments. CSILE 2.0 will provide menus of such sentence-openers, which students can select and pop into their comments whenever they wish. These menus will be teacher-authorable, so that the sentence-openers can be adapted to the age-level of students and to local needs and withdrawn when no longer needed.
2. Keyword Supports. As indicated earlier, we see keywords as having educational value both for the author who assigns them and for the students who use them in searching. But assigning relevant and useful keywords is notoriously difficult, therefore a good target for procedural facilitation. CSILE 2.0 will provide automatic support in checking the spelling of new keywords and screening for redundant variations (plurals, etc.). We also want to provide a kind of procedural facilitation that will help students think more effectively about the content of their notes. One simple support is the following: When a student highlights any word in a note and clicks the "keyword help" icon, CSILE searches for that word in the concept maps that have been created in the MEANINGS

environment (q.v.). If it finds the word there, it retrieves the portion of the map surrounding that word, thus displaying the other domain-relevant words that have already been associated with it. This may suggest: (a) that the word is already connected to other words relevant to the present note, and so it should be a good keyword to assign; (b) that the words already connected to this word are in quite a different domain (which could mean that the word is a poor choice or that it is a good keyword to assign because it will bring the note to the attention of students working on a different topic); or (c) that the word has presently little connection (or no connection if no map is retrieved) to other words in the database, so that if it is used as a keyword it should probably be accompanied by other keywords that have wider currency.

An Expanded Role for Thinking-Type Icons: From Thinking Types to Knowledge Contributions. CSILE is intended to make students more aware of the kinds of knowledge they are dealing with, and there is evidence--from increases in the use of epistemological terms, for instance--that it has that effect. We expect the number of knowledge concepts with which students actively work to increase as they deal more with "facts needing explaining," "hypotheses," "data," etc., in CSILE 2.0. Whereas CSILE 1.0 provided icons for students to label their notes according to "thinking types," CSILE 2.0 will provide for labeling "knowledge contributions." This change is in keeping with the focus on public knowledge and its growth. It is not so much what went on in the author's head when composing the note that needs to be recorded but what the note is believed to contribute to the community knowledge base. The list of possibilities could include standard types that are the same in every environment and special types relevant to particular environments (like "facts needing explaining" in the EXPLANATIONS environment). Whatever ones are chosen, the student is required to complete a brief label, which provides a short statement of the idea, question, fact, or whatever. These entries then become the objects that various tools can work with. There could, for instance, be tools that will display a list of the high-level questions associated with a certain keyword. The screen would show the 30-character-or-so question labels, and the user could then select from this screen any complete notes that looked interesting. Thus the assignment of knowledge types would play a useful role, and mislabeling would show up as notes appearing where they didn't belong.

Background operations

Background operations refer to computations that CSILE performs automatically rather than on the command of the user. Thus, for example, spelling checking is not a background operation when it is called for by the user (instead, we refer to the spelling checker as a tool available from all environments). But if CSILE automatically checks new keywords for spelling, that is a background operation (even though it may use the same program as the spelling-checking tool). Three classes of background operations may be distinguished. They differ in how closely they interact with the user, and consequently they differ in how fast they have to be able to run:

1. Automatic checking. This refers to operations such as checking new keywords for spelling and checking new keywords, propositions, or other distinguishing attributes for redundancy (for example, checking to see whether a new hypothesis in the EXPLANATION environment is semantically similar to one that already exists). These checks are ordinarily run at the time the user is storing a note, and therefore they have to be performed rapidly enough that they do not delay the storing process excessively. In general, automatic checking background operations involve some search and comparison and usually some uncertainty, so that the system must interact with the user before an action is taken.
2. Suggesting and reminding. An important class of background operations does automatic analyses of note attributes or of text content and then informs the user of such things as related notes that may be of interest, domain vocabulary relevant to note content, or uses that other people have made of the user's notes. Although speed is a virtue with these background operations, so that information can be provided to users at a time when it is most relevant, speed is not as important as with automatic checking, because no processes are held up waiting for the operation to be completed.
3. Monitoring. Included in monitoring are many kinds of data compiling and analysis that are currently being done off-line and used in preparing reports. The expectation is that in the new architecture these kinds of monitoring will be part of CSILE (although perhaps still carried on after hours) and will provide the basis for reports that can be called for by researchers, teachers, and students. A special kind of monitoring that does not exist presently will involve the assigning of activation values to notes. These values will not only be helpful for assessment reports but will also be used in the programs that carry out suggesting and reminding operations.

Through background operations running in CSILE, authors of related notes are notified of the new activity regarding their notes, or authors working on a similar topic are notified of each other's work. Background operations can collect information on individual students that can go into reports to or from teachers. They can also collect information on individual notes that can affect a note's status. For example, one design we are working on involves something similar to citation indices. Notes gain "credit" as they are used--read, copied, commented on, or attached to a larger structure in some knowledge-building environment. Additionally, if material is copied from a note, the copied information will be entered into the new note in a different font, with a bibliographic note recorded automatically. Such easy cross referencing and tabulating of sources should support the principles of synthesis and cross-fertilization.

In designing functions using background operations, everything possible needs to be done to defeat the "Big-Brother-is-watching-you" syndrome. Users need to be provided with an accurate model of the kinds of data used in background operations, so that they know what effect actions on their part can have. One might think of the design challenge as that of allowing teachers and students to track the information and kinds of educational activity they believe are valuable. When this is tracked with ease and consistency across users the information can serve to make explicit otherwise little known conditions for improvement.

Importing material from on-line reference material

In CSILE 2.0, on-line reference material will include CD-ROM, video, microworlds, and a broad array of commercially available software. The central educational problem to be addressed is how to allow students to take good advantage of these resources, and yet avoid mindless cutting and pasting of reference material.

To date, CSILE users have not had access to clip art, video, reference material on CD-ROM, or other resources that would allow wholesale importation of material into the student-generated database. However, they do take notes from reference books and then spend a good deal of their CSILE time copying their notes into CSILE. This is clearly wasteful of time on CSILE. But on the positive side they do add to the notes in interesting and oftentimes significant ways. First, they tend to add commentary, to elaborate ideas, to make inferences and generally to interpret ideas. That is, there is an interpretive layer to their notes, even though a good deal of the information might be taken directly from reference material. Second, they encode the note, selecting keywords so they can store and retrieve the note easily. Accordingly, there is value added to all notes that are in CSILE.

We want to ensure that future designs preserve the contribution principle, so that students not only actively process information they copy into CSILE but also consider what it contributes to the community's knowledge. The solution will be specially designed import protocols that place a layer of educational activity between imported material and the student-generated database itself. Accordingly, material from external databases will be required to be converted into notes before it can be stored in the community database. This minimally requires adding a keyword and a label indicating the nature of its knowledge contribution. We may want to require addition of some original text before it can be stored, or we may want to use background operations to assign low credit to a copied note, so that it needs to be enhanced or used in significant ways (e.g., needs to acquire "citation credits," as described in the preceding section).

Summary and Conclusions

Although the constructivist view of learning is well on its way to becoming a part of conventional educational wisdom, there are few circumstances, even in the most avant-garde schools, that actually encourage students to view what they are doing as the construction of knowledge. Conscious, cooperative development of shared knowledge is the focus of CSILE. Results to date, both from our own work and from similarly-motivated work [2, 4, 5] convince us that elementary school students can profitably make knowledge construction the focus of their efforts, although it is a novelty to them (and to their teachers) and requires a great deal of support.

Our development strategy has been to put a simple version of CSILE into full operation as quickly as possible and then to investigate experimentally the kinds of enhancements that best support the objective of cooperative knowledge construction. On the basis of what has been learned we are designing a much more powerful system, a knowledge-building architecture, which will be embodied in a second generation of CSILE. A major architectural change in CSILE 2.0 will be the addition of distinct environments for different kinds of knowledge operations. The basic idea of a knowledge-building environment is that knowledge is brought into the environment and something is done to it that enhances its value. Tools available in the environment are means by which this value-adding work is done. Consequently, the design of environments and tools is guided by the goal of maximizing the value added to knowledge--either the public knowledge represented in the community database or the private knowledge and skill of the user.

CSILE 2.0 will also permit the importing of information from electronic media such as CD-ROM and electronic mail, whereas in the current version of CSILE everything that enters the community database is authored by the students. We want to pursue the contribution principle here, as well, so that material is not mindlessly pasted into the CSILE database without having been justified and enhanced in some way by the student.

Knowledge-building environments of the kind proposed have no real precedent in educational practice, even though many of the tools included in the planned environments already exist and have proved useful in differently structured contexts. The environments must do much more than provide access to knowledge-processing tools. In terms of the seven principles set forth earlier, a knowledge-building environment should objectify knowledge, the operations performed on it, and the resulting changes in the knowledge. Thus the knowledge operations must have consequences and these in turn must be of a relevant enough kind that they help sustain progress beyond the usual limits of individual curiosity and zeal. The environments must be designed to encourage higher level synthesis of knowledge rather than mere accumulation and to focus students' attention on the contributions of their activities to the community's knowledge rather than solely on issues of individual merit. The bringing of information from different students and from varied thematic spaces should serve the principle of cross-fertilization. Finally, in keeping with the principle of sociality, the instructional program as a whole must be designed in such a way that systematic pursuit of problems of meaning, explanation, "how-it-works," etc., are integral to the curriculum.

Acknowledgments

This paper was previously presented at the NATO Advanced Research Workshop, "Computer-Based Learning Environments and Problem Solving," Leuven, Belgium, September 1990.

The authors wish to acknowledge the generous support of Apple Computer, Inc., External Research Division; IBM Corporation; and the James S. McDonnell Foundation. The work reported in this paper would not be possible without their support.

We collaborate with a team of teachers, elementary-school students, researchers, and systems analysts who have taught us a great deal. We are indebted to them for their contributions and challenges to the ideas presented. We also wish to thank Peter Goodyear for his careful reading of a previous version of this manuscript and for insightful commentary, and Denese Coulbeck for editing and manuscript preparation.

References

1. Bereiter, C., & Scardamalia, M. (1987). *The Psychology of written composition*. Hillsdale, NJ: Lawrence Erlbaum Associates.
2. Brown, A. L., & Campione, J. C. (1990). Communities of learning and thinking, or A context by any other name. In D. Kuhn (Ed.), *Contributions to Human Development, 21*, 108-126.
3. Collins, A., Brown, J. S., & Newman, S. E. (1989). Cognitive apprenticeship: Teaching the crafts of reading, writing and mathematics. In L. B. Resnick (Ed.), *Knowing, learning, and instruction: Essays in honor of Robert Glaser* (pp. 453-494). Hillsdale, NJ: Lawrence Erlbaum Associates.
4. Lampert, M. (1986). *Knowing, doing, and teaching multiplication. Cognition and Instruction, 3*(4), 305-342.
5. Lampert, M. (1989). *Teaching and learning division for understanding in school*. East Lansing, MI: Department of Education, University of Michigan, Mimeo.
6. Lancy, D. F. (1976). *The beliefs and behaviors of pupils in an experimental school: Introduction and overview* (Tech. Rep. No. LRDC-1976/3). Pittsburgh: University of Pittsburgh, Learning Research and Development Center.
7. Peters, T. (1987). *Thriving on chaos: Handbook for a management revolution*. New York: Knopf.
8. Popper, K. R. (1972). *Objective knowledge*. Oxford: Clarendon Press.
9. Richmond, B., & Peterson, S. (1988). *A user's guide to STELLAStack*. Lyme, NH: High Performance Systems, Inc.
10. Scardamalia, M., & Bereiter, C. (1985). Fostering the development of self-regulation in children's knowledge processing. In S. F. Chipman, J. W. Segal, & R. Glaser (Eds.), *Thinking and learning skills: Vol. 2. Research and open questions* (pp. 563-577). Hillsdale, NJ: Lawrence Erlbaum Associates.
11. Scardamalia, M., & Bereiter, C. (1991). Higher levels of agency for children in knowledge building: A challenge for the design of new knowledge media. *The Journal of the Learning Sciences, 1*(1), 37-68.
12. Scardamalia, M., & Bereiter, C. (in press). Schools as knowledge-building communities. In S. Strauss (Ed.), *Human development: The Tel Aviv annual workshop: Vol. 7. Development and learning environments*. Norwood, NJ: Ablex.
13. Scardamalia, M., Bereiter, C., Brett, C., Burtis, P. J., Calhoun, C., & Smith Lea, N. (1990). *Educational applications of a networked communal database*. Unpublished manuscript.
14. Scardamalia, M., Bereiter, C., McLean, R. S., Swallow, J., & Woodruff, E. (1989). Computer-supported intentional learning environments. *Journal of Educational Computing Research, 5*(1), 51-68.
15. Thagard, P. (1989). Explanatory coherence. *Behavioral and Brain Sciences, 12*(3), 435-502.
16. Whitehead, A. N. (1925/1948). *Science and the modern world* (Mentor edition). New York: New American Library.

How Do Lisp Programmers Draw on Previous Experience to Solve Novel Problems?[1]

Marcia C. Linn, Michael Katz, Michael J. Clancy, and Margaret Recker[2]

HyperMedia Case Studies in Computer Science Project, T-2241 College Avenue, University of California at Berkeley, Berkeley, CA 94720, U.S.A.

Abstract: In two exploratory studies, we looked at how expert Lisp programmers organize Lisp knowledge, and how they solve complex programming problems. First, using our previous work on Pascal *templates* as a framework, we generated a set of potential Lisp templates and asked experts to identify and to sort them. Second, we invented a set of challenging Lisp programming problems and asked experts to solve one or two of them while thinking aloud. The protocols from these sessions give a rich picture of some of the techniques experts use to design solutions, and the criteria they use to evaluate partial solutions and solutions by others. We discuss how the results of these exploratory studies suggest changes to current Lisp instruction, and describe some instructional materials we are currently developing, including a set of *Case Studies* and a hypermedia library of Lisp templates.

Keywords: Lisp, programming, templates.

How do Lisp programmers design solutions to problems? Which approaches are most effective? What are the implications for Lisp instruction? How can effective Lisp problem solving be taught?

In our investigations of Pascal programmers [16] we identified generalized, reusable programming building blocks referred to as "templates." Our hypothesis was that expert programmers organize their Pascal knowledge in templates to keep track of the details and manage the complexity of the language. Each template might integrate a general version of the

[1]This material is based upon research supported by the National Science Foundation under grants MDR-88-50552 and MDR-89-54793. Any opinions, findings, and conclusions or recommendations expressed in this publication are those of the authors and do not necessarily reflect the views of the National Science Foundation.

[2]The authors would like to thank Patricia Schank for contributions to the template sorting study and for helpful comments throughout the study. Thanks are also due to Chris Borton, Oliver Grillmeyer, Lydia Mann, and Annika Rogers for comments on the earlier draft.

Pascal code for a common action, such as sorting, with knowledge about implementation, testing, and debugging of the code. The first goal of the current exploratory study is to identify the building blocks used by Lisp programmers.

Our previous investigations also characterized the design skills used by Pascal programmers to solve complex problems. For example, we found that programmers recycled templates to streamline the programming task. In addition, we found evidence for more abstract problem solving processes such as "alternative generation" and "reflection." The second goal of this exploratory study is to examine the design skills used by Lisp programmers.

To motivate and investigate these questions we (a) discuss why more and more instructors teach Lisp as an introductory language, (b) describe existing methods for teaching Lisp problem solving and previous research on program design, (c) summarize our exploratory investigations of how Lisp programmers organize programming knowledge and solve moderately complex problems, and (d) discuss implications for teaching Lisp.

Why Teach Lisp to Novices?

Lisp allows instructors to focus on program design, both because the minimal amount of syntax can be taught quickly and because the language is remarkably flexible. Lisp supports a broad range of programming paradigms including functional, imperative (also called sequential or procedural), object-oriented, and rule-based styles. Lisp programmers must choose among these paradigms in solving programming problems and thus have considerable latitude in designing a problem solution. Ideally, programmers would match the programming paradigm with the characteristics of the problem.

The limited syntax and seemingly limitless flexibility of Lisp challenges instructors to prepare students who understand the versatility of Lisp and are able to use it productively. Just as Lisp program designers must deal with multiple, complex decisions, so are Lisp instructors faced with many possible alternatives.

Abelson and Sussman [1] offer one approach to Lisp instruction. With the publication of their text, *Structure and Interpretation of Computer Programs*, has come increasing use of Lisp to teach fundamental programming concepts. More and more universities, including M.I.T., U.C. Berkeley, St. Olaf's, Brandeis, and Columbia are basing their introductory course for computer science majors on this textbook. In addition, modern versions of Lisp are well-suited to introductory courses because they have rapid execution on workstations such as Sun,

Apollo, Digital, and Macintosh. Early versions of Lisp required specialized hardware or ran very slowly, but recent versions such as Allegro, MacScheme, Pearl, and others perform well.

Previous Work on Teaching and Studying Program Design

Our approach to Pascal instruction [16] has some similarities with the approach to Lisp instruction taken by Abelson and Sussman, as well as some differences. We compare and contrast these approaches, then summarize research on the behaviors of programmers, and finally discuss some ways of teaching tasks like program design which have proven successful.

Idioms and templates

Abelson and Sussman argue that programmers can reduce the overwhelming complexity inherent in the design and analysis of large programs through the process of abstraction. They discuss data abstraction, procedural abstraction, metalinguistic abstraction (building embedded languages), and other techniques that are used by programmers to hide masses of irrelevant details. Certain abstraction techniques are characteristic of particular programming styles, called "paradigms." For instance, emphasis on using generalized functions to capture very high-level recurring patterns is characteristic of functional programming. Emphasis on hiding the details of the computation by associating, with various procedure "objects," a store of local state information leads to object-oriented programming. By using metalinguistic abstraction, Abelson and Sussman show how to implement a rule-based logic programming language. By knowing various abstraction techniques and by being aware of the available programming paradigms, expert Lisp programmers learn to control complexity in large programs. They learn to write such programs, and also to read and to modify the well-written programs of others, retaining the style of the author.

However, knowing about these large-scale techniques is not enough. In order to assure the correctness of large programs, programmers need "an arsenal of standard program structures of whose correctness we have become sure" [1, p. xii]; these are called "idioms." Although similar in many respects, our templates (in Pascal, Lisp, or any language) are somewhat different from the idioms described by Abelson and Sussman both because we represent them in pseudocode and because we emphasize associating them with related information such as debugging strategies.

In previous and current research, our aim has been to teach program design in a way consistent with the approach of Abelson and Sussman and also consistent with our conception of templates. In particular, we have sought to help students reduce problem complexity by presenting them with a collection of standard templates and by modelling the design techniques used by experts.

Behavior of programmers

Analyzing the skills of expert program designers has proven difficult [e.g., 2, 12, 26, 28]. The difficulties are enhanced for Lisp since the programmers have such varied backgrounds and programming styles, as is further discussed below. Researchers have studied programmers by asking them to recall unfamiliar programs, to sort templates, and to solve problems while thinking out loud.

Recall tasks [e.g., 2] indicated that Pascal programmers, compared to non-programmers, could recognize and recall patterns of code. Soloway used this finding to support the idea that programmers develop what he calls "plans," which are summaries of how to carry out various programming goals represented as Pascal code with blanks for variable names. Soloway has developed successful instruction based on helping students acquire and reuse these plans. Plans and templates for Pascal differ in representation and associated knowledge but both perform the function of managing the complexity of a programming language.

Sorting tasks have been used to assess the way knowledge is organized in several fields [e.g., 4]. Marco [19] asked high school programming students learning Pascal to sort programs. She found that the students who solved programming problems better—and earned better grades—were also better at sorting programs by template. Less adept programmers sorted the programs by superficial criteria such as the names of the variables. Schank [22] investigated template use among college students in self-paced Pascal courses and among expert Pascal programmers. She found that experts tended to locate specific templates by using high level categories such as "processing procedure" (one-by-one or bulk), "data structure" (record, file, linked list, etc.), and "control structure" (recursion, iteration, etc.). Novices found this approach difficult and used other methods including searching the list of template names. These investigations show that programmers differ in the specificity of their templates. As they become more expert, they prefer more abstract representations. (In Pascal, however, most programmers retain certain templates which help them manage the complexity of syntax and other low-level details.) These results suggest that effective programmers organize knowledge (consciously or unconsciously) in large reusable chunks, regularly refine these chunks, and can access these chunks efficiently.

Studies of programmers solving problems while thinking out loud have helped researchers characterize the process of program design [7, 20, 23, 27]. These studies reveal that programmers often reuse patterns, algorithms, or code they have created in the past. For example, Soloway [27] reports that when experts are asked to solve problems similar to those they have solved before, they quickly design the whole solution and anticipate bugs. In contrast, when asked to solve unfamiliar problems these same experts fail to anticipate many common problems and start to implement solutions without completing their design. Soloway summarized this behavior as "design a little, do a little." Programmers seem to follow a given solution path until they encounter some obstacle and then revise the design to circumvent the problem. In addition, these studies indicate that programmers start the implementation process soon after they come up with a "kernel" idea, even if the idea is not guaranteed to succeed.

How do these investigation techniques apply to Lisp? Lisp has fewer syntax details for the programmer to keep track of and therefore may require fewer reusable chunks. The process of program design in Lisp may also differ from the process for Pascal since all Pascal programs are basically procedural, while Lisp supports a range of paradigms and an expert programmer would select among these paradigms as part of the design process. The nature of a template in Lisp and of design in Lisp are the questions this exploratory investigation is intended to address.

Teaching complex problem solving

Research on reading, writing, mathematics, and science suggests how instruction can help students learn to design solutions to complex problems. Taken together, these examples of successful instruction have been referred to as apprenticeship teaching [6]. For example, Ann Brown and her colleagues [3] have succeeded in helping students develop reading comprehension skill through a technique called "reciprocal teaching." Key features of this technique include (a) providing a new conceptual model of the reading task; (b) having teachers model the complex problem-solving processes that they employ; (c) reducing the cognitive load in reading comprehension by teaching specific skills such as formulating questions and predicting outcomes; (d) encouraging students to reflect on their knowledge by devising explanations and by evaluating explanations devised by others; and (e) encouraging students to take responsibility for their own learning by turning instruction over to groups of students.

Consistent with the results described for reading, Linn, Sloane, and Clancy [18] found that, in teaching program design, teachers who model their own design processes at a level accessible to students are far more effective than those who provide only abstract discussions of problems. Thus, teachers who talk through how they solve problems and include in their discussion some reinterpretation of the problem statement, some backtracking from initial conjectures, and some clear examples of how they use prior problem solutions to solve new problems, are more effective than those who provide a tightly-knit presentation of the subject matter but fail to communicate how the techniques discussed can be applied to a problem.

In addition to demonstrating these general skills, apprenticeship instruction has been used to model the process of applying particular templates to recurring problems in programming. Sloane and Linn [24] found that students benefit from instruction that identifies complex templates and specifies how these templates can be applied and tested. For example, students rarely test their programs, but skills for doing so in typical and extreme cases can be taught by emphasizing the debugging information associated with various templates. Other studies [17] have shown that modeling solutions to programming problems combined with providing opportunities to apply such models and get feedback from the technological environment is successful for a wide range of students.

However, not all teachers have the necessary skills to model program design. As a result, Clancy and Linn [5, 16] designed Pascal case studies to model the design process and to engage students in thinking about program design. These case studies were successful for teaching program design [15].

Skill in program design is often thought to result from discovery learning rather than to be the focus of teaching. Research on apprenticeship learning suggests that teaching these skills successfully requires an understanding of the problem solving process. The exploratory studies reported here sought to characterize the Lisp problem solving strategies that should be modelled for students.

Exploratory investigations

The problems of studying programmers are enhanced for Lisp due to the varied backgrounds of the programmers as well as the diversity of possible programming styles. Indeed, it became evident at the outset of our study that expert Lisp programmers form a heterogeneous group. Since instruction in Lisp has only recently been offered, most expert Lisp programmers learned their skills by unguided discovery and use personally-developed, idiosyncratic techniques for program design. Their techniques may be narrow in scope or unsuited to introductory instruction. Since Lisp has traditionally been used for a fairly constrained set of problems—in design of artificial intelligence programs, in the modelling of complex processes such as natural language, and in intelligent tutors—the techniques used by typical expert programmers may not generalize to a wider array of programs, and experts may have techniques that work for their type of problem but not for others.

For this reason, we studied a diverse group of Lisp programmers. We offer some generalizations about their behavior and some implications for instruction. The questions that guided our interviews included (a) How do programmers organize common templates or idioms? (b) When do programmers recycle aspects of previous solutions? (c) When do programmers seek alternatives to the solution they first design? (d) What criteria determine solution selection? and (e) What self-monitoring strategies govern programming behavior?

In our first investigation, we asked programmers to recognize and organize templates. In our second investigation, we observed programmers solving reasonably complex problems.

Study I: Recognizing and Organizing Templates

Based on the success of sorting tasks for inferring knowledge organization in other domains, we designed a sorting task to determine whether programmers recognized certain chunks of code as templates and to assess how they would organize the templates they recognized. In our Pascal work, templates included some (perhaps abstract) form of code and the knowledge required to implement the code in a problem solution [14]. In this study we defined a template simply as any reusable chunk of Lisp code. We did not attempt to determine what additional information was associated with Lisp templates. We did, however, distinguish three categories of potential Lisp templates and selected a range of examples from among the three for the task. We also probed for additional examples during the interviews and added a few after our first interviews to ensure that we had a representative assortment.

In deciding which pieces of code in Lisp might be general and useful enough to be called templates, we considered analogies to the Pascal templates. As noted above, some Pascal templates are necessary to manage the detailed syntax of the language. For instance, Pascal has three kinds of loops, as well as other flow of control mechanisms, many forms of data structures, and numerous types of variables. The novice is immediately challenged to manage all this syntax and templates are a natural mechanism for this task [16]. Although the core of Lisp has very little syntax, most Lisp dialects contain many special forms that each have their own arbitrary syntactic structure. So, it is possible that Lisp programmers have templates for organizing syntax, although they are likely to play a much smaller role than their Pascal counterparts.

It would seem likely that many of the more high-level Pascal templates would have useful counterparts in Lisp. An investigation of this question led us to focus on the difference between the imperative programming style typical in Pascal and the functional style typical in Lisp. It became apparent that many of the Pascal templates were useful because they helped the programmer remember the *order* in which a complex sequence of events needed to take place. On the other hand, one of the great advantages of strict functional programming is that, by avoiding assignment statements, the programmer need not worry about the sequences of events in a program (see Abelson & Sussman [1]). Thus it seemed that many Pascal templates would apply in only a limited way to Lisp. In fact, at one point, we considered the possibility that templates, at least in the sense that they exist in Pascal, are not necessary in Lisp.

Managing complexity, however, remains a challenge for the Lisp programmer, even with the limited syntax and even when using a purely functional style. It seemed that the place to look for Lisp templates would be in the ways expert programmers developed building blocks for writing large, complicated programs. We examined textbooks, reviewed course syllabi, and interviewed programming instructors to develop potential candidates for templates in Lisp. From this research, three categories of potential templates emerged.

First, focusing on functional style, we identified as potential templates the class of Lisp functions called "applicative operators." These "higher order" functions take other functions as input and apply the given function in some way, usually to each member in a list. In a sense, they can be thought of as the functional counterpart of looping operations in Pascal. For instance, `map`[3] is the most common applicative operator, and applies a given function to each element in a list, returning the list of resulting values. Another example is `filter`, which takes a predicate as its input function and includes in its output list only those elements of its input list which pass the test of the predicate. Other examples are named `accumulate` and `repeated`. It turns out that a very small number of these applicative operators (the four mentioned and perhaps a few others) capture the majority of the recursive patterns that turn up in typical Lisp programs. Abelson and Sussman [1] and Touretzky [29] stress the use of these functions as a way to write concise, elegant programs by hiding the details of recursion in the applicative operators. Some programmers believe that problem solutions composed primarily of applicative operators are easier to debug, to test, and to verify than are solutions which make heavy use of explicit recursion. (These same programmers admit, however, that programs composed

[3] Throughout the paper, we distinguish the names of Lisp functions, special forms, and data with this type.

primarily of applicative operators are sometimes hard to read and inefficient.) As a result, applicative operators are good candidates for templates.

A second type of template summarizes various forms of recursion. Lisp programmers typically break recursion into various types such as linear recursion, tree recursion, and other variations. These might be associated with information about implementation. For example, some programmers initially implement recursive routines that work "backwards" and then alter them so the action is done from the beginning of the list. These recursion templates are not unique to Lisp and apply, for instance, to Pascal programming as well.

Finally, we were interested in seeing if Lisp programmers did indeed keep track of certain details about the syntax of special forms using templates. Special forms, unlike regular functions, break from the standard pattern of evaluation in that their arguments are passed unevaluated. For instance, conditional structures like `if` and `cond` need to be special forms since only one of their consequent arguments is meant to be evaluated (see Abelson & Sussman, [1]). Since the evaluation rule for each special form is arbitrary, programmers must remember the syntax of each. Thus it is common to hear beginners complain about the (arbitrarily) complicated nesting of parentheses typical in `cond` and `let` expressions. It seemed that templates might be useful for remembering how these special forms are used.

Methods

The subjects in this investigation were volunteers including 1 undergraduate in computer science, 4 graduate students in computer science or related fields, and 3 Lisp instructors. The students had all served as teaching assistants for Lisp courses. One instructor was self-taught and the others had largely developed Lisp expertise on their own.

We identified twenty templates from the three categories discussed above. Five of our seven recursion templates were labeled "basic recursion," "tree recursion," "nested recursion," "recursion with an iterative helper function" and "recursion with a helper function to transform input" (there were two other recursions with helper functions). Our eight applicative operator templates were labeled "`map`," "`map` with multiple lists" (sometimes called "`zip`"), "`accumulate`" (or "`reduce`"), "`accumulate` with a helper function," "`accumulate` with multiple lists," "`filter`," "`find`" (or "`lookup`"), and "operate on pairs of consecutive list elements." The five remaining templates included one complex but non-applicative function called "merge," the syntax template for "let" constructs, and a group of set operations including "set-difference," "member," and "intersection." An example of a template which was presented to the subjects is given in Figure 1.

We represented the templates in three forms. "Blackboard" Lisp captures the abstract pattern of a template, using English-like representations of Lisp expressions and leaving unspecified such things as the actual predicates in `cond` clauses and the number of these clauses. The pseudocode gives an English translation that is more restricted than the blackboard Lisp but more general than the specific example which follows. Although we hypothesize that programmers represent their templates abstractly so they can apply them widely, we included the concrete examples because programmers may abstract the same general template in quite different ways. Thus, having both the abstract and concrete representations increased the probability that a respondent would recognize a template. Also, we realized that recognizing templates from isolated examples may be more difficult than recognizing them in the context of a big program. To test this hypothesis, we conducted one interview where we asked therespondent to analyze a complete, complex program and comment on templates as they arose in the solution.

Your Title: ____________________________

Blackboard Lisp:

```
(define (f L)
      (cond          ((null? L) nil)
            ((base-case-2 (car L) (cdr L)) base-result-2)
            ...
            ((base-case-n (car L) (cdr L)) base-result-n)
            ((recursive-case (car L) (cdr L))
                  computation-on (f (car L) (cdr L)))) ))
```

Pseudocode:

if the list is empty, then
    return `nil`
else return a value computed from a recursive call, which itself is
    based on a computation on the first element and the
    remaining elements of the list.

Example:

```
(define (first-duplicate L)
      (cond          ((null? L) nil)
            ((member (car L) (cdr L)) (car L))
            (else (first-duplicate (cdr L)))))
```

Title: nested recursion

Figure 1. One of the 20 templates used in the interviews.

Each interview had three parts. In the *naming* part, respondents were required to invent a name for each template *before* looking at the name we provided. Their name was then compared to the name we had generated. In the *critiquing* part, respondents were asked to indicate whether they would use the template, whether it was too general, or whether it was too specific. In the *sorting* part, respondents were asked to group the templates they would use into categories.

Three response categories were identified. To get a broad picture of the way the templates broke down, we determined for each respondent the number of templates which that subject categorized (at least roughly) as applicative operators, recursive functions, and iterative functions.

Results and discussion

Respondents had considerable difficulty with this task. Many did not recognize the templates in the form we provided. As a result, some respondents could not come up with names for the templates. In addition, respondents gave names at varied levels of abstraction, sometimes focusing on the example and sometimes focusing on the abstract versions of the code. The names we supplied sometimes interfered because the respondent had a different name for the same template. For example, some people prefer the name (and imagery of) "`reduce`" to "`accumulate`."

On the critiquing task, several respondents reported that the recursive templates were too general to be used in the design of problem solutions. On the other hand, some respondents found a subset of the applicative operator functions too specific. That is, they recognized functions like `map` and `accumulate`, but found variations like `zip` and "`accumulate` with a helper function" too specific to be counted as separate templates.

On the sorting task, six of the eight respondents differentiated applicative operators from recursive functions, one differentiated iterative functions from recursive functions, and one singled out only applicative operators. The seven respondents who identified applicative operators put them in one or several categories. Among those subjects who recognized the basic versions of the applicative operators, some misunderstood the special case variations of those operators, some did not see the connection between the general and special case versions, and some grouped the two versions together. The seven respondents who grouped recursive templates into a category identified one or several types of recursion, including tree, basic, tail, and/or linear recursion. Several singled out recursion that used "helper functions" or "preprocessors" as a separate category.

The Lisp instructors recognized more of the functions and grouped the functions in more abstract categories than the students. We hypothesized that instructors were motivated to come up with abstract categories for elements of the language in order to help students recognize patterns.

As we hypothesized, some respondents found the examples to be contrived and difficult to process. They commented that the templates might be easier to recognize in a larger program. This was among the motivations for the one interview where we presented a long program and asked the respondent to comment on the templates it contained. This respondent really seemed to be identifying familiar patterns. Interesting comments such as "Oh, that's just a `map`," or "I see, it's `accumulate`" arose regularly. Both recursion and applicative operator templates were recognized. In addition, the respondent volunteered associated information about implementation and selection of the templates. He regularly offered criticism of the template choices and of the mode of implementation. The respondent also regularly indicated alternative approaches. In retrospect, this type of interview seems effective for getting at the templates that programmers use and for eliciting information about implementation and selection associated with each template. This approach deserves further investigation.

Implications

In summary, most of the programmers interviewed recognized both recursive and applicative operator templates. The two who did not use these categories, however, had interesting reasons for their choices.

The one who recognized only iterative and recursive functions did so based on recent experience in a course based on the Abelson and Sussman textbook. The grouping contrasted the main control structure of Lisp with the main control structure of Pascal. Our observations suggest that many students have this perspective after their first Lisp course. They generally have not adopted the applicative operators as personal templates, so they group them in the recursive category. This point is reinforced in the next exploratory study.

One Lisp instructor refused to create a recursive category, saying that such a category is too general since recursion is such a basic feature of the language. Several other respondents fundamentally agreed with this point. They commented that it was possible to differentiate recursion but that the recursive templates were less useful than the applicative operator templates. This group of respondents organize their knowledge around applicative operators and think of recursion as a super-category. The other respondents essentially organize their Lisp knowledge in terms of recursion. They pay attention to the types of recursion, such as linear or tree recursion. They consider whether a function does most of the work as it goes down the list or as it returns.

Respondents differ in how they organize applicative operators. Some have established categories for applicative operators, referring to the "five basic operators." Others group versions of `map`, for example, but do not have a specific set of operators as a part of their repertoire. Others view even the various types of `map` as separate categories, while some have only a general category of applicative operators. We suspected that these differences would be reflected in programming style and found some evidence for that in our next study.

Templates are difficult to recognize out of context, especially for students. We could improve recognition by refining the code, pseudocode, examples, and type of templates selected. Even if the task were refined, however, it is likely that novices would have difficulty. Our one interview involving templates in the context of a complex program elicited quite useful information and might be a preferred mode of investigation.

Study II: Designing Solutions to Lisp Problems

To get a sense of the process of designing solutions to problems in Lisp, we observed programmers solving complex problems. We asked programmers to "think out loud" so we could understand how they solved the problems. This technique has proven successful for many tasks. Erikson and Simon [8] argue that respondents' mental operations when engaged in this form of problem solving can be inferred from analysis of their verbal protocols. Others [9, 13] mention drawbacks. Certainly, some respondents find talking and problem solving difficult and may refuse to participate. In addition, the technique of gathering and analyzing protocols is time-consuming, so only a few data points are generally included. Furthermore, if researchers follow the usual approach of selecting experts and novices, then it is difficult to infer the intermediate steps on the path to expertise. In the case of Lisp, as mentioned earlier, there is no one type of expertise. Thus, to get a general picture of the process, we selected programmers with a variety of backgrounds and asked them to solve problems appropriate to their expertise. In general, we sought solutions by at least two different subjects for each of our problems.

A useful framework for understanding the process of program design is to consider the nature of design in general. Design is the process of developing a solution to a problem [10, 15, 25]. It involves successively reformulating the problem statement, clarifying features of the solution, generating kernel ideas, determining and testing partial solutions, implementing and revising ideas, debugging implementations, and testing the approach.

We abstracted a set of principles, described in Figure 2 (these are discussed in detail elsewhere, see Linn & Clancy, [15]), that seemed to govern successful design in Pascal programming. These principles are somewhat general since the specifics of the design process are unique for each problem. We used the principles to guide the design of Pascal case studies.

A case study is what we call a description of an expert's solution to a "complex" problem. By complex, we mean a problem that seems complex to the students learning from it. Case studies are used to illustrate decision making in many complex domains including finance and psychiatry. They offer promise for teaching computer programming as well.

We have developed a number of Pascal case studies. Each includes (a) a statement of the programming problem; (b) a narrative description of an expert's solution written so a student can understand the expert's approach; (c) a listing of the expert's code; (d) study questions to provide practice analyzing the program; and (e) test questions to assess students' understanding of the program solution. Our case studies explain and defend the decisions made in producing a

---

<u>The Recycling Principle</u>: Reuse ideas, specifically design templates, that worked before. Elaborate on a known template to solve a new problem rather than attempting to design new code from scratch for every problem encountered.

<u>The Multiple Representation Principle</u>: Consider multiple representations of each design template to attain a robust understanding of programming templates. Promising representations include natural language, pseudocode, diagrams and dynamic illustrations.

<u>The Alternative Paths Principle</u>: Generate and evaluate alternative designs for programming problems. Develop processes for finding alternatives and criteria for selecting among alternatives.

<u>The Reflection Principle</u>: Reflect on alternative designs for programming problems and on problem solving processes. Develop skill in monitoring progress and in diagnosing unproductive activities. Anticipate personal weaknesses and guard against them.

<u>The Fingerprint Principle</u>: Develop effective debugging skills by associating symptoms of common bugs (their "fingerprints") with the appropriate bug.

<u>The Divide and Conquer Principle</u>: Code and test complex programs a piece at a time to isolate program bugs and reduce the time spent searching for problems. Use this process of "incremental development" to reduce the cognitive demands of programming by concentrating on one component at a time.

<u>The Persecution Complex Principle</u>: Adopt a "persecution complex" to test computer programs. Test for all possible weaknesses in the program using typical and extreme cases, and spend time determining the full range of cases.

<u>The Literacy Principle</u>: Produce code that is self-documenting so that it is easy to understand, modify, and debug. Use self-documented code to make the purpose of each section clear and to automatically update documentation while modifying the program.

Figure 2. Principles of Pascal program design.

solution to the problem at hand, and contrast effective and ineffective strategies for solving related problems. They also describe how expert programmers organize their programming knowledge and illustrate the application of program design skills. Our case studies have been successful in teaching Pascal program design.

The principles in Figure 2 serve as a way to tie together the templates and the design process. For instance, the "fingerprint" principle is a general design strategy, and is also directly related to the debugging components of our Pascal templates. In this study we explore whether the same principles that govern design and give a context for the templates in Pascal might do the same for Lisp. We also speculate about additional principles that might be especially appropriate for Lisp. At the outset, it should be made clear that there are some obvious differences between design in Pascal and Lisp. In particular, design in Pascal tends to be much more structured than in Lisp.

Whether or not the principles turn out to be similar or drastically different, the case study approach (and apprenticeship in general) has been an excellent approach for communicating the principles and templates for Pascal [15], and we are exploring case studies in Lisp.

In summary, improved understanding of the process of program design seems likely to offer some promising directions for teaching Lisp. Problem solving while "thinking out loud" has proven useful for understanding the process of program design in the past. Our exploratory investigations used this methodology to examine how a variety of Lisp programmers solved problems. We then reflected on how the most promising design techniques might be taught using case studies and other instructional techniques.

Methods

We asked eight programmers to solve problems while thinking out loud. We selected problems that we expected would challenge the respondents and require them to use an extensive range of design skills.

Of the eight programmers who participated in this study, six had also participated in the previous study and two were new to this investigation. The group included four programming instructors, three graduate students in computer science or related fields and one undergraduate in computer science. As before, all students had served as teaching assistants in Lisp courses. One instructor was completely self-taught and the others had largely developed Lisp expertise on their own.

Programmers solved one to three problems selected to match their expertise. We only analyzed responses if the programmer made some reasonable progress towards a solution. In this paper, results from eight problems are discussed. The problems are briefly described in Figure 3.

Triangle Graph: Suppose the vertices of a directed graph are represented by positive integers, and the edges in the graph are represented by two-element lists of integers. A graph is a list of such edges. Write a function `(missing-edges graph)` that returns a list of all missing edges of triangles in the graph. That is, your function will return a list of all two-element lists `(a c)` such that `(a b)` and `(b c)` are in the argument list, but `(a c)` is not.

File Directory: A file is represented by a list called a file entry. The file entry for a non-directory file is a two-element list whose first element is the word `FILE` and whose second element is the name of the file. The file entry for a directory is a list whose first two elements are the word `DIRECTORY` and the name of the directory, and whose remaining elements are file entries for the files within the directory (which may be directories themselves). Write a Lisp function `(file-list directory)` that returns a list of the names of the non-directory files that appear anywhere in the given directory tree.

Cyclic Graph: Write a Lisp function `(convert adj-list)` that takes an adjacency list representation of a (possibly cyclic) directed graph and returns a (possibly cyclic) list structure that corresponds to the adjacency list. For instance, the adjacency list `((a b) (b c) (c a))` can be implemented as a list structure with the two commands `(define cycle '(a b c))`, `(set-cdr! (cdr (cdr cycle)) cycle)`. Write the inverse procedure `(unconvert structure)` that returns the adjacency list corresponding to the given (possibly cyclic) structure.

Recursion: Write a Lisp procedure that takes as input a list of lists where each sublist is the text of a Lisp procedure definition. The list of procedures constitute a Lisp program. Your procedure should return a set of statistics about how recursion is used in the given Lisp program. In particular, report how many times tail, linear, tree and mutual recursion are used in the given program.

Intervals: Write a Lisp procedure that takes a list of intervals (start-time/end-time pairs) as an argument and returns an equivalent list of intervals in which no intervals overlap or adjoin. For instance, given the interval list `((5 10) (17 20) (3 7) (10 15))`, your program should return the equivalent interval list `((17 20) (3 15))`. The order of the intervals in both the input and output lists is unspecified.

Addition Tutor: Write a program that drills students in addition problems. The lesson proceeds in a sequence of levels, moving from simple one-digit additions without carry to several-digit additions with multiple carries. Students can only move from one level to the next when they get five problems in a row correct at the old level. If students are doing sufficiently poorly at a given level, they should be dropped back a level.

Poker: Write a Lisp procedure `(evaluate hand)` that takes as input a representation of a poker hand and returns a description of the value of the hand (e.g. that it's a full house with kings and fives). The output from the evaluate procedure would be used in the context of a complete poker program.

Stable Marriage: Let W be a set of n women and M be a set of n men. Each woman ranks the men from 1 to n in the order she would like to be married to them. Each man, likewise, ranks the women from 1 to n in the order he would like to marry them. A ***pairing*** is a one-to-one correspondence of women to men. A pairing is ***stable*** if for each two women w_1 and w_2 and two men m_1 and m_2 with whom they have been paired respectively, the following two conditions are both satisfied:
1) either w_1 ranks m_1 higher than m_2, or m_2 ranks w_2 higher than w_1
2) either w_2 ranks m_2 higher than m_1, or m_1 ranks w_1 higher than w_2

(a) Explain intuitively the meaning of the two conditions.
(b) Prove that a stable pairing always exists.
(c) Write a Lisp procedure `(match-make women men)` that produces a stable pairing.

Figure 3. Problems used in the interviews.

The interviews had three phases. In the first phase, respondents were introduced to the "think out loud" procedure, asked about their background and, in six of the eight cases, given the template sorting task. In the second phase, respondents were given one or more problem statements and asked to solve problems. One to four members of the project staff observed each problem solving session. In addition, the sessions were audiotaped and all notes made by the problem solver were gathered. The interviewer answered clarification questions and prompted the respondent to keep describing the problem solving process. In the third phase, the respondent returned several days to a week later and reflected on the problem solving process. Most of the project staff members participated in these reflection interviews. Programmers were asked whether they had come up with different solutions since the interview, were queried on their reuse of familiar programming patterns, and were encouraged to comment on specific aspects of their solution. We also asked programmers to comment on alternative designs for the problem that had been generated by others. We were particularly interested in programmers' reflection on their problem solving processes in this interview.

To analyze the protocols for the problem solutions, we focused on three main activities. First, to complement our investigation of the nature of templates, we looked for evidence of recycling parts of problem solutions. Second, to assess the sequence of program design, we characterized the conditions under which programmers considered alternative designs or modified their designs. Third, to understand programmers' view of the design process, we analyzed the "self-monitoring" statements made during problem solving. By "self-monitoring" we mean statements that assess behavior or performance with the goal of checking on its accuracy, appropriateness, or completeness. For example, programmers say things like "Did I get all the base cases?," or "I wonder if that's true, let's check." Often these self-monitoring behaviors can be viewed as "rules of thumb" used by the programmer to manage the task.

Each interview was carefully analyzed by one member of the research team, generally someone who was present at the interview. The analysis was then reviewed by the remaining members of the team. If there were differences of opinion these were resolved by discussion, by consulting the respondent, or by eliminating the interpretation. In most cases the analyst and the research team agreed on the interpretation of the respondent's behavior.

Results

The protocols give a rich picture of the process of program design. There are substantial differences between programmers in how they approach problems, in the amount of ambiguity they will tolerate, and in their personal criteria for judging alternatives and completed solutions.

The protocol analysis technique yielded the most information about self-monitoring and the least information about reuse of code. Programmers tended to describe the things they wanted to keep in mind as they planned a problem solution. They appeared to rehearse those ideas they wanted to remember. They seemed to remind themselves of important ideas. As illustrated below, their "thinking out loud" included a great deal of self-monitoring.

Although many respondents reported considering alternatives for pieces of their problem solutions, it was rare for programmers to generate alternatives which differed from one another in radical ways. In the reflection interviews many respondents reported that the protocol analysis approach we used inhibited them from abandoning the course of action they originally selected and going in a new direction. During the problem solving session they said things like, "this is brute force but it will solve the problem" or, "I could be more elegant but I know this will work." They felt pressed to get to a quick solution rather than considering different possibilities and commented that they might have explored other alternatives if they were working on their own. It may be that the mental demands of talking out loud interfered with generating completely different approaches. It may be that the unnatural conditions of the problem solving interview—programmers do not usually do their work in front of observers—altered the nature of the task the programmers set for themselves. This is an empirical question that deserves further study.

The third phase of the interview, where programmers were asked to reflect on their solutions and comment on solutions generated by others, motivated some programmers to generate alternatives. Those programmers whose personal criteria had not been met by their initial solution tended to come up with alternatives. For example, programmers who preferred to use applicative operators and had not yet generated a solution featuring this approach continued to generate alternatives and discussed them in the reflection interview. When asked to comment on alternative designs, programmers were more likely to criticize than to embrace solutions generated by others.

Thus, the interviews shed light on the process of designing Lisp problem solutions. We describe some specific examples of programmer behavior. Then, in the discussion, we compare these results to our Pascal investigations.

Self-monitoring

Since self-monitoring was the most prevalent activity observed, we report results in that category first. We found that many programmers showed evidence of "rules of thumb" which surfaced in the form of instructions to themselves. For example they would say things like, "I should do the hard things first, then the easy things will fall into place."

Specifically, we observed two main categories of self-monitoring. One involved managing the complexity of the situation. The other involved keeping track of progress.

Programmers manage complexity by solving hard problems first, by picking off easy parts first, by postponing details, by choosing descriptive names early in the design process, and by using multiple representations of their problem decompositions. They choose among these (sometimes conflicting) approaches, depending on the nature of the problem and the nature of their prior knowledge.

Four subjects used the strategy of managing complexity by concentrating first on the hard part of the problem and only later filling in the details of the other parts. For example, in solving the "poker" problem, one subject concentrated first on the issue of the best form for the output. Once he had decided on an acceptable form, the program structure came together rather easily for him. In solving the "cyclic-graph" problem, another subject started by inventing an intermediate form which was about half-way between the adjacency list and the cyclic structure. He decided that the trickiest part of the problem was turning the intermediate form into the cyclic structure and worked on this first. Indeed, once he had figured that out, the conversion from the adjacency list to the intermediate form proved relatively easy for him. Two subjects who were solving the "addition tutor" problem together spent more than the first 80% of the session trying to work out a convenient and general way to represent the different types of addition problems. Although some of this time was spent unproductively, when they did come up with a satisfactory representation, the rest of the problem, including a somewhat complicated outer control structure, came together very quickly. Finally, in solving the "file directory" problem, one subject postponed worrying about the details of the actual directory structure given and worked out a more general procedure that could be used for a whole class of directory types. It was then trivial to use this general tool in solving the given problem. It is interesting to note not only that in each case a particular part of the problem turned out to be the key to the rest of the problem, but that all four subjects singled out this key section very early on in the problem solving session.

An opposite strategy for controlling complexity is to "trim away" the easy part of the problem before turning to the parts which seem troublesome at the outset. For example, a subject solving the "recursion" problem quickly broke the problem down into cases, some of which seemed fairly straightforward to him, and some of which seemed quite difficult. Whenever he became bogged down with a section that seemed to be leading to more and more unanswered questions, he would turn his attention back to an easier case. This not only allowed him to make progress on the easier parts, but his work on those parts gave him some insight into the more difficult cases. Thus we see that the strategy of concentrating on the difficult parts first and the strategy of postponing the hard parts in order to make progress on and gain insight from the easier sections, can both be useful.

The well-known strategy of choosing descriptive names for variables and procedures was seen in many of our interviews. Perhaps the clearest example was seen in the two subjects mentioned above who solved the "addition tutor" problem together. They each had their own ideas about what names would be best, and in the process of resolving their disagreements they came up with an elaborate system for naming the many functions they had invented (one for each type of problem in the arithmetic drill).

Finally, a pervasive method for controlling complexity was to use alternative and multiple representations for procedures and data. For example, in solving the "cyclic graph" problem, one subject quickly translated the problem statement into box and pointer notation. This notation shows the details of how list structures are stored in memory. Since this problem involved a tricky manipulation of those structures, this representation proved fruitful for the subject. The same subject, in solving the "recursion" problem, chose to write most of his procedures in pseudocode instead of actual Lisp. This allowed him to concentrate on the important high-level issues without becoming overwhelmed by some tricky but relatively minor details. Similarly, a subject solving the "stable marriage" problem came up with a representation for the data that was simpler than that specified in the problem. In particular, at some points he treated the list of people as a list of symbols, even though the problem statement implied that each person was represented by a sublist of preferences. The subject recognized that his simplification helped him analyze the problem without causing him to overlook anything relevant to the solution.

Lisp programmers use self-monitoring to check on their own progress and to assess the accuracy of their solutions. This sort of self-monitoring is very similar to the self-monitoring we observed in Pascal and summarized in the principles of design for Pascal. For example, one common method of checking the partial solution to a programming problem is to simulate the partial program on some sample input. A subject solving the "poker" problem did this very close to the start of the session to make sure his overall design would work out. A subject solving the "file directory" problem did this close to the end of the session to check his final answer. We also observed some specific criteria used by subjects in coming up with sample data for their procedures. In analyzing the "triangle graph" problem, one subject spent considerable time coming up with a very general graph example to make sure that the algorithm he had in mind would work in all cases. On the other hand, another subject solving the "cyclic graph" problem was careful to pick a small example because he knew that tracing a large example by hand would be error-prone and time-consuming.

Judging alternatives

Our interviews suggest a process of program design that proceeds from the generation of a kernel idea, to the refinement of that idea by applying certain criteria, to the testing of partial implementation of the idea, to the complete implementation of the idea using both newly generated code and previously designed code. This process is cyclical and proceeds until a solution that meets the programmer's criteria is met. We described above how programmers pick off easy or hard parts of the problem to be completed first. In this section, we look at examples of places where programmers used various criteria to judge their ideas.

One of the clearest cases involved a subject solving the "triangle graph" problem. His first solution attempt resulted in a somewhat complicated scheme of translating the list of edges into what he called a "maximal normal form." His plan was to add several redundant edges to the list so that he could then operate on the list without having to worry about whether certain edges existed. Simplicity was a definite criterion for this subject, and he spent some time rationalizing his first solution, claiming that there was something "fundamentally arbitrary and complicated" about finding triangles in graphs. Therefore he thought his initial complicated solution was justified. In the reflection interview, however, this subject described a new solution he had come up with that was extremely short, using only a single function built out of several applicative operators. In general, we found that those subjects for whom simplicity of code was an important criterion tended to like using the applicative operators as much as possible.

A similar case of using simplicity as a criterion in the design process occurred with a subject solving the "cyclic graph" problem (the same subject described above solving this problem). In thinking about how to convert from his intermediate form, the subject first devised a multi-step scheme which would have involved the creation of yet another intermediate form. But he backed away and realized that his first intermediate form was very close to the desired output (he only need to reset a few pointers to already existing structures). It seems likely that this programmer is constantly on the lookout for any unwarranted complexity, since the change in his design plan came quickly after he started considering the complicated alternative.

Another criterion for judging partial solutions has to do with the level of abstractness of the code generated. For instance, there was considerable variation in the level of abstractness of solutions to the "file directory" problem. One subject solved the problem by writing two procedures which were very specialized to the structure described in the problem statement. For instance, this subject's solution relied on the fact that a directory consists of the word `DIRECTORY`, followed by the name of the directory, followed by a list of files and sub-directories. A different subject decided to write a solution that depended as little as possible on the details of the structure given in the problem. He came up with a solution that relied only on the fact that file names came after the word `FILE` in some sublist. In the reflection interview, this subject explained that it seemed better to keep the procedure as general as possible, since it would allow for changing the representation of directories without changing the procedure. Also, the subject felt that his solution was less prone to errors since it included so few details. Finally, thinking of the problem in these terms, the subject was able to give a solution that was only a slight variation on a standard applicative operator that was well-known to the subject, called `tree-filter`. On the other hand, subjects who gave "specific" solutions to the problem commented that the structure of this problem was unique enough that it warranted a solution which was not in terms of a general tool function. We discovered that there is a subtle balance between the criterion of writing abstract, general code and the criterion of making the purpose and function of the code clear.

Finally, although efficiency of code is often de-emphasized in Lisp texts, we found that for many expert Lisp programmers efficiency was an important issue. For instance, in solving the "poker" problem, one programmer noticed that his initial solution had some redundancy in checking for a straight twice, once in checking for a straight-flush and again in checking for a regular straight. He sacrificed a bit of simplicity in his code to avoid this redundancy. Similarly, the programmer who used an intermediate form in the "cyclic graph" problem analyzed his solution in terms of the number of passes that needed to be made over the data. Finally, during their reflection interviews, many subject criticized another subject's solution to the "intervals" problem because it had a running time proportional to the cube of the length of the input list.

Recycling

When implementing an idea, programmers reuse techniques used in the past. The problems we used in this investigation were not chosen to facilitate reuse of code and the typical reuse we observed was for built-in functions. We suspect that problems designed to detect this process would shed more light on this issue.

Of course, all of Lisp programming involves constructing functions from built-in functions. However, we think it is useful to single out certain built-in functions which are powerful enough to be a key feature to a problem solution. For instance we have already suggested that the (built-in) applicative operators are among the candidates for templates in

Lisp. In our problem solving interviews we saw the use of several of these functions. For instance, both `map` and `filter` were used repeatedly by a subject solving the "cyclic graph" problem. We mentioned above that one subject based his solution to the file directory problem on the `tree-filter` applicative operator. The clearest example, however, came from the subject who gave a solution to the "triangle graph" problem that consisted of a single function which included several uses of applicative operators. In addition, several standard list processing primitives played key roles in various solutions, especially when they were used in conjunction with an applicative operator. For instance, a common pattern was to use `append` with the `accumulate` (`reduce`) applicative operator to achieve a one-level flattening of a list. Another technique was to use `flatten` in conjunction with `filter` to achieve the results of doing a `tree-filter`.

Discussion

We found evidence for the principles of design we had previously generated for Pascal. We discuss how each of these principles might apply to Lisp. We also found some Lisp-specific design techniques having to do with abstraction.

Recycling principle

In Pascal we found recycling of previous solutions to be beneficial and to permeate the behavior of both experts and novices. On the other hand, Lisp programmers vary in their use of previous solutions. In addition, the results of our sorting study indicate that Lisp programmers differ in the potential templates they hold. Some programmers prefer applicative operators for elegance, for ease of debugging, or for ease of understanding, and use these as templates. Others prefer to design recursive programs and cannot even recall how to implement some of the common applicative operators. Those who rely on recursion vary as to whether they view their knowledge of recursion as organized in reusable patterns or as a single, general pattern. Preferences may be due to course background, experience, or personal style.

Applicative operator templates are taught at the beginning of courses using the Abelson and Sussman book and later or not at all in other courses. Applicative operators may be difficult to learn or use depending on how they are taught. For example, applicative operators learned in one context may be hard to apply in new situations. If applicative operators are taught to hide the details of recursion, then programmers might abandon them once they master recursion. Combining applicative operators to solve problems improves with experience. The techniques for making such combinations are rarely taught. Students may benefit from instruction in this area, rather than learning to combine applicative operators by unguided discovery.

Some experts prefer to use applicative operators in solving most problems. Applicative operators are preferred for clarity, for easy of debugging, and for ease of determining whether a program is accurate. Recursive solutions are preferred by some for efficiency, since the generality and power gained by using applicative operators usually comes at some cost in speed. Those preferring applicative operators report that efficiency is rarely an issue for the first design of a program.

As our interviews reveal, even experts may take additional time to figure out solutions using applicative operators. For example, the respondent who solved the triangle graph problem completely in terms of applicative operators did so only after struggling with a recursive version first.

Lisp programmers report that unusual uses of applicative operators emerge as they gain experience, much as Pascal programmers report gaining understanding of broader applications of their repertoire of templates with experience. Those committed to applicative operators sometimes push this approach to the extreme and end up with "wacky" or bizarre solutions which can be extremely difficult to read. For example, one respondent described how to implement a selection sort completely in terms of applicative operators, even though a recursive solution would have been much less complicated. These uses of applicative operators seem bizarre because they attack a simple problem with a complicated solution. Programmers committed to this approach justify this complication on the grounds that solutions made up of a very small set of simple tools seem fundamentally more elegant than solutions that use all the features of the language.

Having emphasized the applicative operators so strongly as a potential candidate for templates, we should reiterate that there was not general agreement among subjects as to how to classify the operators, or even as to which functions counted as applicative operators. An example of the first issue is that some programmers have a somewhat elaborate classification of applicative operators and distinguish clearly between things like `map` and `pairwise-map`, while other programmers think of all applicative operators as slight variations on a single idea. Then there is the tricky issue of what should count as an applicative operator. On the one hand we might call any function that accepts another function as input and applies that function an applicative operator (and hence a template). For instance, a version of a sorting function that takes a comparison predicate as input is more general than one that can only sort certain types of data. Thus the first sorting function might be called a template because it captures a general sorting pattern that is applicable to different types of data. Some respondents were happy with this rationale for considering `sort` to be a template. Some were even happy to consider certain powerful standard functions such as `flatten` (which do not take other functions as input) to be templates. On the other hand, others were adamant that the only functions that could rightly be called templates were the standard applicative operators such as `map` and `filter`. And even within this greatly restricted set there was some disagreement. For instance, there are two common versions of `accumulate` that are quite similar. Yet we had one subject who was familiar with one but didn't recognize the other in the context of the template sorting task.

In summary, we found evidence for two potential categories of templates: recursive functions and applicative operators. Our third potential category, functions with complex syntax, was not common in these investigations. Further study is needed to determine when or whether this is a category of template. However, we found that the details of applying applicative operators were not always recalled by programmers either because they had not been taught or because they were not learned. We also found that programmers did not always generalize applicative operators to all appropriate problems. As a result, students may benefit from a "library of templates" to help recall details and to help determine novel applications.

Multiple representation principle

As in Pascal, Lisp programmers use many different representations to design problem solutions. We observed many representations for functions and data including pseudocode, "blackboard" Lisp, "plumbing diagrams" (showing inputs and outputs of black-box functions) and box and pointer diagrams. These representations seemed to help programmers to understand problems, to recall appropriate templates, and to find ways to reuse templates.

Alternative paths principle

Since many respondents felt inhibited from considering large reorganizations of their programs, we cannot reach definite conclusions about the process of alternative generation. No respondent generated several paths before starting a more detailed design as we advocated for Pascal. The one respondent who made major reorganizations did so as the result of reflection, not during the problem solving interview.

On the other hand, programmers did have clear criteria for generating single alternatives to pieces of their designs. For example, programmers came up with alternatives if they thought their initial approach was too complex, too specific or too error-prone. Many programmers also generated alternatives to increase the efficiency of their solutions.

Divide and conquer principle

As in Pascal, we found that Lisp programmers use the divide and conquer strategy to simplify the implementation of their design. Programmers started on parts of the problem at different points in the design process, either as the result of personal style or due to the nature of the problem. Some picked off small parts of the problem before generating a complete design in order to simplify the design of the remainder of the solution. Others completed the design in a top-down way, dividing the problem into several big problems and then dividing each of these into smaller problems. Some problems were divided into a series of smaller problems immediately, possibly because the programmer had templates for most of the pieces. Still others immediately started on a "kernel" idea, and assumed that the details would fall into place.

Fingerprint principle

As in Pascal, we found that expert programmers recognized the symptoms of common bugs and designed their solutions to avoid situations likely to generate bugs. Student programmers in our study were more likely to continue with designs that were prone to bugs. In addition, even when they were shown a less complex solution, student programmers did not always agree that the less complex solution was preferable. Since most respondents did not fully implement their solutions, we did not investigate how they detected bugs in their completed programs. These preliminary investigations suggest that programmers would benefit from instruction that emphasized guarding against bugs.

Persecution complex principle

Strategies for testing all the potential weaknesses in a program were not fully investigated in this study since few of the programs were completed. During design, experts working on complex problems such as the "recursion" problem or the "arithmetic tutor" problem generated good test cases to evaluate their preliminary designs. Some of the student respondents also generated test cases to clarify the problem statements. Others persevered without generating cases. In general, those generating test cases early ended up with better solutions. These exploratory investigations suggest the value of emphasizing test cases in instruction.

Literacy principle

As in Pascal, Lisp programmers found self-documenting code to help in the design process. Programmers assigned informative names to their functions and parameters early in the design process. They used selector functions [1] to give informative names to the parts of data structures. These techniques were more common and more extensive among experts and could be effectively taught to students.

Reflection principle

Our reflection interviews illustrated the potential benefits of reflection for effective program design. Expert programmers engaged in more reflection than novices. Experts seemed to engage in reflection to apply criteria to their solutions that they had not considered during the initial problem solving session. Experts also used reflection to evaluate their personal programming style. For example, some programmers reported that they preferred the "wacky" applicative operator solutions even when they were overly complex, or indicated that they neglected to go back and recheck the problem statement after deciding on the appropriate design. In both cases, the programmers recognized drawbacks to their personal styles.

Limitations of the principles

We found evidence that the principles generated for Pascal also governed the behavior of Lisp programmers. There are aspects of Lisp programmer behavior, however, that are not fully captured in these principles.

Lisp permits a degree of abstraction that is not easily achieved in Pascal. An "abstraction principle" seems to inform much of the design activities of expert Lisp programmers. Whenever possible, Lisp programmers seek to make their functions apply to a broader range of situations. Functions that take functions as parameters make this approach ideal. For example, a Lisp designer would prefer a general calculation function that took the operation as a parameter to separate functions for addition, division, etc.

Lisp offers choices in programming style that are not available in Pascal. Lisp programmers might prefer applicative operators or might avoid them yet still design equally satisfactory solutions. Communicating the alternative styles and teaching students to modify programs while maintaining the original programmer's style is a crucial part of Lisp instruction. The alternative generation principle and the recycling principle can be modified to emphasize differences in style. Certain other aspects of style, including the importance of maintaining the style of a program when modifying it, are suggested by the literacy principle.

Finally, Lisp has much less syntax than Pascal so the recycling principle takes on a different meaning, having more to do with reuse of things like applicative operators and common recursive patterns and less to do with remembering syntactic details.

Communication of principles

Much of the programming knowledge observed in respondents during "think out loud" studies is implicit. Our investigations identified behaviors that programmers find effective but might not think to communicate to others. Textbooks and instructors often fail to make these ideas explicit as well [11, 24]. Research on problem solving suggests that many students benefit from instruction that makes the steps of problem solving explicit. By conducting these investigations we hoped to identify information that would help students. We hope to communicate this information in case studies.

Implications for Teaching Lisp

In conclusion, these exploratory investigations suggest criteria for the design of instructional materials, and on-line tools. We have preliminary designs for an on-line library of Lisp templates and for Lisp case studies. In this section we describe these tools and explain how they were informed by our investigations.

The "Perspective Library" is a hypertext database of Lisp programming examples. The library is organized by including each example under any number of "perspectives." Many of the components of our previous Pascal Template Library [21] such as student notes, information about debugging and verbal descriptions, are just instances of perspectives in the new library. Students can reorganize the library to fit their personal viewpoint by creating new perspectives, creating new examples and including or excluding an example in a given perspective. A screen from the Perspective Library is shown in Figure 4.

We have also implemented a Lisp interpreter in the form of a Hypercard external command (XCMD). When this XCMD is used within the Perspective Library, the result will be an integrated environment where students can not only examine and reorganize programs, but also run, test and debug those programs.

The basic motivations for the design of the Perspective Library are similar to those for the Pascal Template Library and are described elsewhere [21]. Here we concentrate on the motivation for the new "perspective" feature. After debating for some time the question of the best way to organize the Lisp templates in an on-line library, we realized (a) that many of the different suggestions were equally reasonable and we had no basis for deciding between them, and (b) that most of the suggested organizations were relatively "flat" in the sense that they divided the set of templates into categories and perhaps divided those categories into sub-categories, but deeper subdivisions were rare. So instead of thinking of the set of templates as being arranged in a fixed, deeply nested tree, we wanted to think of them as falling into a number of weakly structured categories. We also wanted to take into account that some categories, such as applicative operators, had enormous importance to some expert programmers, but were only mildly interesting to others.

The idea of a "perspective" is that each category can impose the main framework for understanding the language. In addition to allowing for a difference of opinions among experts, we are using the perspective feature as a way to trace the evolution of students' categorization of templates.

For instance, we might suppose that initially students would classify functions based on somewhat superficial criteria, such as the types of data the functions operate on or return. They might start with categories such as "lists," "arithmetic" and "predicates." Later, they would add more subtle categories such as "applicative operators" or "tree recursion."

Our exploratory studies of Lisp problem solving provided the beginnings of a conceptual model of Lisp programming—the first step in devising an apprenticeship approach to instruction. This conceptual model is summarized in the principles of design initially described for Pascal and now extended to Lisp. For instance, programmers monitor their own progress, use a variety of "rules of thumb" to make the task easier, and differ in their preferences for recycling applicative operators. Students must study a large and complex problem (from their perspective) to see the advantages of these problem solving procedures. Space, time and other constraints mean that these aspects of programming are not addressed in current textbooks nor in most Lisp course lectures. Based on the success of our Pascal case studies, we devised Lisp case studies to teach these problem solving techniques. We targeted these materials for an introductory self-paced Lisp programming course.

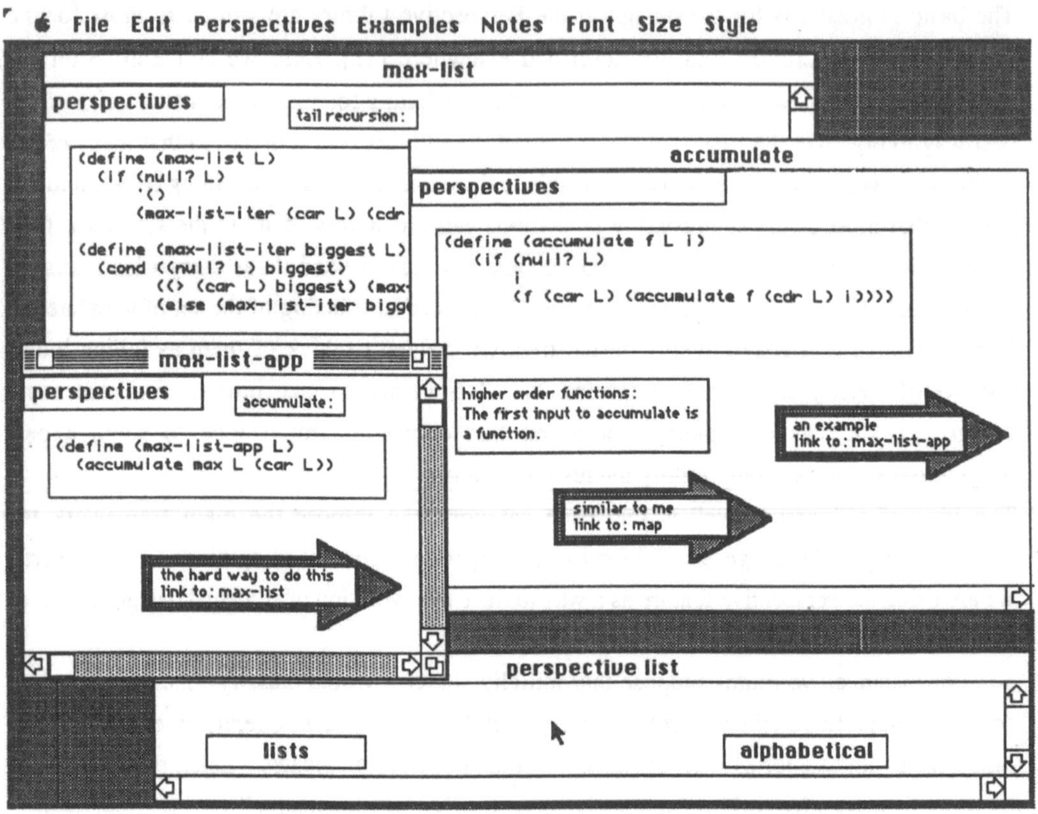

Figure 4. A screen from the Perspective Library.

Currently, we have designed three Lisp case studies. An introductory case study illustrates basic techniques of composing functions in the context of a program to find the difference in the number of days between two given dates. A more advanced case study makes heavy use of the applicative operators to write a set of statistics functions. Finally, a program to evaluate a poker hand (represented as a list of card descriptions) illustrates some advanced recursion techniques. Selections from the "Difference Between Dates" case study appear in Figure 5.

Again, the motivations for these case studies are largely the same as those for the Pascal case studies [16] and reflects research advocating an apprenticeship approach to instruction [6]. The case studies implement this approach to instruction by involving students in the processes used by experts to solve complex problems. In solving the study questions (see examples in Figure 5), students second-guess the expert problem solver, assess potential blind alleys and incorrect decisions made by experts, and practice self-monitoring procedures. Students contrast methods of solving a complex problem, for example the difference between doing the hard part first so that the easy parts will fall into place, or picking off the easy parts in hopes of gaining insight into the hard parts. They then compare their experience to that of the experts. In addition, we designed the case studies to gradually turn responsibility for learning over to the student. As students progress through the series of case studies they are asked to perform increasingly more complex and complete tasks.

These exploratory investigations, especially when combined with our investigations of Pascal learning and with other research on learning complex tasks, have implications for Lisp instruction in general. Our preliminary conceptual model of Lisp programming suggests that instructors should address topics such as self-monitoring that are often left to unguided discovery. Instructors might model their own problem solving procedures to help students recognize the subtle but important skills used to manage complexity and evaluate options. The individual differences our studies revealed in how experts organize Lisp knowledge also suggest instructional implications. The categories used by experts such as "applicative operators" or "linear recursion" should be introduced and discussed so that students can make a choice rather than end up with an organization by default. Students might also be encouraged to categorize their templates and to seek progressively more powerful and useful ways to organize their understanding. The Perspective Library and Lisp case studies are designed to complement instruction that builds on the preliminary conceptual model and that focuses on the organization of templates.

"Difference Between Dates" Case Study
by M. J. Clancy and M. C. Linn

Problem

Write and test a Lisp program to compute the length of intervals between two dates. The program will include a function called days-between that returns the number of days between and including its two arguments. The arguments to days-between represent dates in 1990 (a non-leap year); each argument is a two-element list whose first element is a month name and whose second element is a legal date in the month. The first argument date should be assumed to precede the second argument date. ....

Part I: A solution using only functions, lists, and conditional expressions

The reader should have experience with defining his or her own Lisp functions and constants, with the use of conditional expressions in Lisp (if and cond), and with the use of lists and the built-in functions first, second, member, and assoc.

A design based on a procedure for solving the problem by hand

How do programmers design problem solutions?

Figuring out how to solve a problem by hand usually provides ideas for the design of Lisp functions to solve the problem.

...

We consider three cases:

First, to compute the interval between two dates in the same month, for instance, between January 3 and January 9, we merely subtract the date in the first month from the date in the second month and add 1. We add 1 because *both* the days are to be included in the returned value. Thus the interval from January 3 through January 9 spans 9 – 3 + 1 = 7 days.

...

How is adjacent-months? coded in Lisp?

Next we work on adjacent-months? One way to code this is as an eleven-way test:

...

Study question: Complete the definition of adjacent-months?
Study question: Test your completed version of adjacent-months? Did you make any mistakes?
Study question: What errors would you expect someone to make when typing the completed version of adjacent-months? into the computer?
Study question: What are other ways to code this function?

Ugh. This code is much too complicated. A better way in Lisp is to represent the months in *data*—a list—rather than in the *program* code of the cond, and to use built-in functions that return information about a list.

(continued on next page)

Figure 5. Excerpts from the "Difference Between Dates" Case Study.

How can a list of month names be used to simplify the code?

We can code adjacent-months? more simply, using a list of month names, along with the member function. Member takes two arguments, an item to look for and a list. Member returns a sublist of the list. The sublist starts with the item being looked for and includes all subsequent items. If the item is not in the list it returns the empty list (nil). Normally the value returned by member is used merely in a yes-or-no fashion to determine whether or not an element occurs within a list, but in this case we are interested in the contents of the list returned as well.

...

This is much nicer.

Study question: Why is this a better solution?
Study question: Rewrite this function so that it compares single months instead of comparing lists.

...

What went wrong, and what can be done about it?

In retrospect, we see that computing the number of days between nonconsecutive months is not significantly simpler than the original problem. List-handling functions alone are not powerful enough to allow implementation of the procedure for computing the answer by hand.

To design an acceptable solution, we must back up and find a more suitable algorithm. We may, however, be able to take advantage of some of the functions we already designed in the new solution.

...

Study question: Show that days-preceding's argument must be one of the twelve month names if the arguments to days-between are legal dates.
Study question: What happens if the dates provided as arguments to days-between are legal dates, but out of order (i.e. the first date follows the second)?
Study question: What happens if a date with an illegal date-of-month is provided as an argument to days-between?
Study question: What happens if a date with an illegal month name is provided as an argument to days-between?
Study question: What happens if a date with more than three elements is provided as an argument to days-between?
Study question: What test cases would be good to determine if the program works correctly?
Study question: What bugs are likely in this program?
Study question: How could the table of dates be checked for accuracy?

Figure 5 (cont). Excerpts from the "Difference Between Dates" Case Study.

References

1. Abelson, H., & Sussman, G. (1985). *Structure and interpretation of computer programs.* Cambridge: MIT Press.
2. Adelson, B., & Soloway, E. (1985). The role of domain experience in software design. *IEEE Transactions on Software Engineering, SE-11*(11), 1351-1360.
3. Brown, A. L., Campione, J. C., Reeve, R., & Palincsar, A. (in press). Interactive learning and individual understanding: The case of reading and mathematics. In L. T. Landsmann (Ed.), *Culture, schooling and psychological development.* Hillsdale, NJ: Erlbaum.
4. Chi, M. T. H., Feltovich, P., & Glaser, R. (1981). Categorization and representation of physics problems by experts and novices. *Cognitive Science, 5*(2), 121-152.
5. Clancy, M. J., & Linn, M. C. (1990). *"Difference Between Dates" Case Study.* Hypermedia Case Studies in Computer Science Project, University of California at Berkeley.
6. Collins, A., Brown, J. S., & Newman, S. E. (1989). Cognitive apprenticeship: Teaching the craft of reading, writing, and mathematics. In L. B. Resnick (Ed.), *Cognition and instruction: Issues and agendas* (pp. 453-494). Hillsdale, NJ: Lawrence Erlbaum Associates.
7. Curtis, B., Krasner, H., & Iscoe, N. (1988). A field study of the software design process for large systems. *Communications of the ACM, 31*(11), 1268-1287.
8. Erickson, K. A., & Simon, H. A. (1985). *Verbal reports as data.* Cambridge: MIT Press.
9. Eylon, B., & Linn, M. C. (1988). Learning and instruction: An examination of four research perspectives in science education. *Review of Educational Research, 58*(3), 251-301.
10. Goel, V., & Pirolli, P. (1989). Motivating the notion of generic design within information processing theory: The design problem space. *AI Magazine, 10,* 18-36.
11. Husic, F., Linn, M. C., & Sloane, K. D. (1989). Adapting instruction to the cognitive demands of learning to program. *Journal of Educational Psychology, 81*(4), 570-582.
12. Jeffries, R., Turner, A. A., Polson, P. G., & Atwood, M. E. (1981). The processes involved in designing software. In J. R. Anderson (Ed.), *Cognitive skills and their acquisition* (pp. 255-283). Hillsdale, NJ: Lawrence Erlbaum Associates.
13. Linn, M. C. (1986). Science. In R. Dillon & R. J. Sternberg (Eds.), *Cognition and instruction* (pp. 155-204). New York: Academic Press.
14. Linn, M. C., & Clancy, M. J. (1990). Designing instruction to take advantage of recent advances in understanding cognition. *Academic Computing*(April, 1990), 20-41.
15. Linn, M. C., & Clancy, M. J. (in press-a). Can experts' explanations help students develop program design skills? *International Journal of Man-Machine Studies.*
16. Linn, M. C., & Clancy, M. J. (in press-b). The case for case studies in programming instruction. *Communications of the ACM.*
17. Linn, M. C., & Dalbey, J. (1989). Cognitive consequences of programming instruction: Instruction, access, and ability. [Reprinted]. In E. Soloway & J. C. Spohrer (Eds.), *Studying the Novice Programmer* (pp. 57-83). Hillsdale, NJ: Lawrence Erlbaum Associates.
18. Linn, M. C., Sloane, K. D., & Clancy, M. J. (1987). Ideal and actual outcomes from precollege Pascal instruction. *Journal of Research in Science Teaching, 25*(5), 467-490.
19. Marco, R. (1988). Knowledge organization in novice student programmers: Relation to skill and instruction in programming classes. Unpublished dissertation, University of California, Department of Educational Psychology, Berkeley.
20. Rosson, M. B., Maass, S., & Kellogg, W. A. (1988). The designer as user: Building requirements for design tools from design practice. *Communications of the ACM, 31*(11), 1288-1298.
21. Schank, P. (1989). *A Pascal template library.* Unpublished master's thesis, University of California at Berkeley.
22. Schank, P., Linn, M. C., & Clancy, M. J. (1990, April). *How does an on-line template library help students learn Pascal?* Paper presented at the American Educational Research Association Annual Meeting, Boston, MA.
23. Shneiderman, B., & Carroll, J. M. (1988). Ecological studies of professional programmers. *Communications of the ACM, 31*(11), 1256-1258.
24. Sloane, K., & Linn, M. C. (1988). Instructional conditions in Pascal programming classes. In R. Mayer (Ed.), *Teaching and learning computer programming: Multiple research perspectives* (pp. 207-235). Hillsdale, NJ: Lawrence Erlbaum Associates.
25. Soloway, E. (1985). From problems to programs via plans: The content and structure of knowledge for introductory LISP programming. *Journal of Educational Computing Research, 1*(2), 157-172.

26. Soloway, E., & Ehrlich, K. (1984). Empirical Studies of Programming Knowledge. *IEEE Transactions on Software Engineering*, *10*(5), 595-609.
27. Soloway, E., Pinto, J., Letovsky, S., Littman, D., & Lampert, R. (1988). Designing documentation to compensate for delocalized plans. *Communications of the ACM*, *31*(11), 1259-1267.
28. Spohrer, J. C., & Soloway, E. (1986). Analyzing the high frequency bugs in novice programs. In E. Soloway & S. Iyengar (Eds.), *Empirical studies of programmers* (pp. 230-251). Norwood, NJ: Ablex Publishing Corp.
29. Touretzky, D. S. (1990). *Common LISP: A Gentle Introduction to Symbolic Computation*. Redwood City: Benjamin-Cummings.

Analysis-based Learning on Multiple Levels of Mental Domain Representation[1]

Rolf Plötzner and Hans Spada

University of Freiburg, Department of Psychology, Niemensstr. 10, D-7800 Freiburg, FRG.

Abstract: In this paper we present the computerized learning model KAGE (Knowledge Acquisition Governed by Experimentation) and MULEDS, a multi-level diagnosis system. Both systems are based on a framework which describes learning in physics domains as a succession of multiple levels of mental domain representation ordered along the dimension qualitative/quantitative. The emphasis is on the mental representation of functional relationships between physics variables. We exemplify our methods in the domain of elastic impacts, a subtopic of classical mechanics. KAGE is a cognitive model of how students learn physics by means of a computerized learning environment. It reconstructs the acquisition of knowledge about functional relationships between physics variables by analysis-based learning mechanisms. KAGE accounts for the question which knowledge states have to be expected when specified learning mechanisms are applied to given instructional information. MULEDS assesses the students' knowledge about a certain physics domain. It incorporates mechanisms for tailored testing as well as for active adaptation of instruction to diagnosed misconceptions. A study with 132 subjects has been conducted in order to test the empirical validity of our approach.

Keywords: Multi-level domain representation, qualitative and quantitative reasoning, knowledge diagnosis, analysis-based learning

Introduction

Our objective has been to uncover some basic learning principles students utilize when they seek to develop expertise within a physics domain. We started out from the conviction that knowledge in physics domains can be acquired easiest if learning takes place on *multiple levels of mental domain representation* ordered along the dimension *qualitative/quantitative*. The basic instructional idea is that different levels of mental domain representation have to be ad-

1. The research reported herein was supported by the German Research Foundation (Deutsche Forschungsgemeinschaft), Grant Sp 251/2-x to the second author.

dressed by corresponding levels of information presentation.

From a psychological point of view students should be enabled to start reasoning on that level of mental domain representation which is in accordance with their intuitive reasoning about the considered physical phenomena. Very often such intuitive reasoning can be described as a kind of *qualitative reasoning*. Hence, students should first be provided with an accurate qualitative conception of the domain under study before *quantitative reasoning* is taught. Quantitative information should then extend the qualitative knowledge in order to enable more precise problem solving. According to this view, teaching has to support not only the knowledge acquisition within one such level of domain representation, but also the transitions from one level to the next. This well-established instructional claim has been discussed recently by several authors with respect to the development of tutorial systems [22, 25, 36].

We introduce the computerized simulation model KAGE (Knowledge Acquisition Governed by Experimentation) which reconstructs learning on multiple levels of mental domain representation. KAGE models learning as it might occur in the context of computerized learning environments which simulate a physics domain and allow students to perform (simulated) experiments.

We intend not only to give a theoretically well-founded characterization of learning as it takes place within the investigated class of learning situations, but also to take advantage of our findings for improving computerized learning environments. From a psychological point of view, the foremost advantage of a computerized intelligent learning environment is that it can adapt to the students' needs in a knowledge-based way. In order to provide this adaptability, an intelligent learning environment has to be developed in a way such that the instructional knowledge which is implicit in a given learning model such as KAGE is utilized as a prescriptive theory of knowledge acquisition.

We consider *adaptability* as a feature which contributes heavily to the acceptance and effectivity of a tutorial system. Very often, early computer aided instruction systems provided a simple page turner to the student, realizing a rigid flow of information which frequently led to poor learning results. In contrast, an intelligent learning environment should enable flexible and individualized access to different sources of information supporting self-guided learning. However, to provide unsupervised access to different sources of information often has the effect that the student gets lost, resulting again in insufficient learning progress. What is required is knowledge-based adaptability of the learning environment to the student's needs. Adaptability can be related to different aspects:

- The student's domain-relevant everyday knowledge;
- The student's preferred learning strategies and presentation modes;
- The individual student's knowledge states and their changes during the interaction with the

learning environment;

- A generic theory of knowledge acquisition in order to enable an appropriate communication of knowledge depending on the student's present knowledge state.

The research reported in this paper is concerned with the latter two topics. We aim at a computerized diagnosis of the knowledge states of individual students, and a generic theory of knowledge acquisition. Together, *knowledge diagnosis* and *prototypical learner modeling* make possible a wide range of instructional interventions which realize adaptability.

We distinguish between passive adaptability and active adaptability. *Passive adaptability* is concerned with the idea that the system should allow the student to work on that level of instructional presentation which is in accord with his[2] level of reasoning. *Active adaptability* presupposes that the teaching system itself changes the domain presentation. This has to be based on an automated assessment of the student's knowledge states as well as on a model of knowledge acquisition processes which specify how certain knowledge states can be acquired or modified.

Generally spoken, passive adaptability enables interactions between learning environment and student which build on a domain presentation understandable to the student. We consider passive adaptability especially important in the case that a theory of knowledge acquisition claims that learning takes place on multiple levels of mental domain representation. In such a case the tutorial system should be equipped with a user interface that is flexible enough to make it possible for the student to interact at any time on that level of instructional domain presentation which corresponds to his actual mental conceptualization of the domain. A theory of knowledge acquisition will thus serve as a basis for a theoretically founded design of the user interface.

Active adaptability is concerned with the feature of a learning environment to perform goal-directed instructional interventions. A satisfying realization of this feature has to take into account the student's present knowledge state. Interventions have then to provide the instructional information which makes it possible for the student to master the next learning step.

An intelligent learning environment equipped with the features mentioned above will support learning in various ways. A student might pursue self-guided discovery learning having access to that level of instructional presentation which corresponds to his level of reasoning. If the system's student model reflects the assumption that the student has acquired an incorrect piece of knowledge this should be pointed out to him. One possibility is the automated generation of instructional information which addresses the incorrect knowledge, thereby realizing a kind of *guided* discovery learning. Furthermore, the learning environment might also present tutorial demonstrations to the student by articulating its own reasoning steps while deriving a problem solution. An intelligent learning environment might thus not only be utilized in order to teach a

2. Masculine expressions are used as generic terms. No bias is intended.

certain physics domain, but also to teach different learning strategies and to improve students' own information search behavior.

The domain of instruction we have selected in order to exemplify the methods described in this paper is the physics domain of elastic impacts, a subtopic of classical mechanics. However, these methods are applicable to any subject area with a comparable structure.

The rest of the paper is organized as follows. Section 2 describes a computerized learning environment which simulates the physics domain of elastic impacts by means of a kind of disk billiard. It realizes an experimental setting which exemplifies the class of learning situations the analysis-based learning model KAGE has been tailored to. Section 3 introduces the framework for a multi-level domain representation and presentation of physics domains. This framework serves as the basis for the reconstruction of qualitative and quantitative reasoning in these domains. The utilization of this framework for a computerized diagnosis of qualitative and quantitative knowledge will be delineated in Section 4. Section 5 describes the cognitive model KAGE which simulates the analysis-based acquisition of qualitative and quantitative knowledge.

A Learning Environment for the Physics Domain of Elastic Impacts

Stumpf et al. [28, 29] have implemented the *computerized learning environment* DiBi[3] (disk billiard) for the physics domain of *elastic impacts*, a subtopic of classical mechanics. DiBi illustrates the domain under study as a kind of disk billiard whereby the phenomena of central and oblique impacts upon disks, disks rebounding off a cushion and collisions of disks can be explored by a student in a rather self-guided way. The quantitative treatment of these different phenomena is based on the general laws of particle kinematics in two dimensions. In particular, the impacts are assumed to be completely elastic, friction and rotation are ignored, and the extension of a body is considered only for the impact event. The domain representation is based on the laws of conservation of momentum and energy.

DiBi sets up a learning environment wherein a student learns by arranging simulated experiments as well as by predicting and watching their outcomes. DiBi puts exactly those physical objects at a student's disposal which are necessary and sufficient for realizing the simulation of a certain phenomenon. Each physical object can be varied by the student within a certain determined range of values ensuring that the resulting experiment exemplifies the intended phenomenon. A typical interaction cycle in DiBi comprises four steps termed SPFP sequence [26]:

3. DiBi has been implemented in Interlisp-D and runs on XEROX Lisp-machines.

- The student selects a certain type of phenomenon and designs an experimental arrangement. If necessary, such an arrangement might also be designed by the system itself.
- The student predicts the outcome of the simulated experiment based on hypotheses formulation.
- The student receives feedback from the system about the correct outcome. Feedback might be provided in a static and/or dynamic mode. In the dynamic mode, DiBi displays the actual movements of the disks. In the static mode, the disks are presented in a certain position along with their velocity vectors. DiBi is also able to display the movements of disks according to incorrect predictions made by a student, thus visualizing the consequences of his own misconceptions.
- During a processing phase the student might revise and refine his hypotheses based on the comparison of his predictions with the system-generated feedback.

SPFP sequences emphasize the hypotheses-guided and incremental process of knowledge acquisition. They give prominence to the assumption that in order to understand complex relations between multiple concepts, it is more important to support a student in learning how to develop and debug his own theories rather than to teach him directly. We are especially concerned with learning situations in which knowledge acquisition is structured by repeated cycles of search for information and its subsequent processing. For instance, such a cycle might include changes of states of certain (independent) variables, the prediction of the consequences of these changes on other (dependent) variables, and the subsequent processing of the system's feedback.

The knowledge a student has to acquire in such a learning situation comprises domain-specific knowledge about functional relationships between certain variables as well as knowledge about how to integrate the effects different independent variables have on a certain dependent variable. Thereby a student has to utilize not only his knowledge about experimental designs, but also his mathematical knowledge about functional relationships and various arithmetical procedures.

If DiBi made possible only a quantitative-numerical treatment of the simulated phenomena, this would often not correspond to a student's actual understanding of the domain. What is required is the consideration of multiple levels of mental domain representation which reflect the different knowledge structures and reasoning procedures that can be observed on the long way from the novice to the expert. In the next section we discuss a framework for multiple levels of domain representation ordered along the dimension qualitative/quantitative. This framework can be utilized to support understanding of elementary functional relationships between physics variables such as proportional and inversely proportional relationships. It provides the ground for a reconstruction and encouragement of qualitative and quantitative reasoning in physics domains.

A Framework for a Multi-level Domain Representation

Proportional and *inversely proportional relationships between physics variables* are the most important types of functional relationships students have to acquire during physics lessons in secondary school. For example, several early taught laws in physics exclusively involve these relationships, such as the law of uniformly moving objects, Hook's law and Boyle's law.

A well-established method for introducing functional relationships consists in the construction of data rows by systematically varying and observing the variables followed by a search for invariants. This method is well-suited to be used by advanced students in order to test their hypotheses about relationships between physics variables. Several computerized models exist which are concerned with the discovery of regularities in data based on this method. For example, BACON [15, 16] denotes a whole series of computerized models of the discovery process of quantitative and qualitative laws. Based on data-driven mechanisms and guided by heuristic search processes these models are capable of discovering a considerable range of physics laws like Ohm's law, Kepler's laws and the law of uniform acceleration. Falkenhainer and Michalski [10] have developed ABACUS, a computerized model which is capable not only of discovering multiple equations describing disjoint subsets of data, but also of stating the conditions under which these equations are valid. Though both models do not aim at being cognitive models of how humans formulate laws in physics, they provide sufficient mechanisms to account for discovery processes.

While such a quantitative method is appropriate for a precise identification of functional relationships between physics variables, we argue that it is inappropriate for acquiring understanding of them. Pure quantitative treatment of the phenomena combined with a rather abstract introduction of these relationships often even prevents students from gaining a deeper understanding of those equations which produce the numerical solutions. This corresponds to the observation that students often do not reconstruct numerical equations on the basis of their understanding of the underlying functional relationships, but tend to simply memorize them with all its negative consequences. With regard to learning mathematical procedures Ohlsson and Rees [22, p. 4] have pointed out that:

> Memorized procedures are brittle. They lack the flexibility required to transfer to unfamiliar problems or even minor variations of familiar problems. ... Memorized procedures are also prone to nonsensical errors. ... Finally, memorized procedures resist being incorporated as subprocedures into higher-order procedures.

Chi, Feltovich and Glaser [4] have found corresponding differences between experts and novices with regard to reasoning about formal problems in physics domains. Experts frequently structure a given problem domain by analyzing it qualitatively before they proceed to quantitative reasoning. The subsequent quantitative problem analysis is then guided by the results of qualitative reasoning. In contrast, novices who had introductory courses in physics often seem to resort directly to quantitative reasoning by searching for equations which match the givens in a current problem. Such an unreflected search for equations frequently leads to incorrect problem solutions, or no solution at all.

On the other side, research on students' conceptual frameworks in physics [7, 12, 19, 34] suggests that novices employ, though often incomplete and incorrect, qualitative reasoning when confronted with conceptual physics problems. An analysis by Chi and VanLehn [5] revealed that the knowledge which students express while explaining worked out physics examples to themselves is exclusively qualitative. Together, these findings provide evidence for the assumption that often students' numerical/algebraic and qualitative/conceptual knowledge is not properly connected.

Plötzner, Spada, Stumpf and Opwis [25] have developed a *framework for a multi-level domain presentation and representation* within the context of tutorial systems. Within this framework the acquisition of knowledge about functional relationships is described as the evolution of a sequence of consistent mental domain representations ordered along the dimension qualitative/quantitative. This view is comparable with regarding learning as a progression of qualitative, causal models as introduced by White and Frederiksen [36]. The assumption of multiple levels of mental domain representation helps to understand why one has to expect structural changes in the problem solving performance of a student while he is developing expertise in a physics domain. Such changes will always take place when aside from acquiring additional knowledge on a certain level of mental domain representation a further, more powerful level of domain representation becomes available. However, it also has to be expected that the reasoning processes on such a new level of mental domain representation are still rather unpracticed and thus are prone to errors.

In the following the phenomenon of a *central elastic impact upon a resting disk* will serve as an example to illustrate the progression through the different levels of domain presentation and representation. Reasoning about functional relationships which hold between the impacting force, a disk's mass and the resulting velocity of the disk might take place on three different representational levels:

- A *qualitative-relational* (QlR) *level*;
- A *quantitative-relational* (QnR) *level*;
- A *quantitative-numerical* (QnN) *level*.

Relational reasoning of a student on the QlR and QnR level of mental domain representation is provoked by the corresponding instructional domain presentation: A student is always requested to make a prediction on the basis of comparing a certain experiment with a reference experiment.

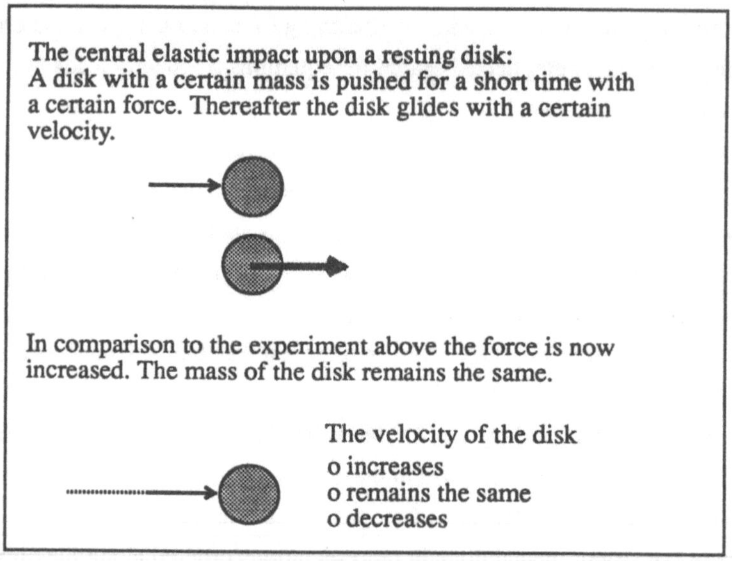

Figure 1. A problem addressing reasoning on the qualitative-relational level.

A novice will most likely start reasoning on the *qualitative-relational level* based on his pre-instructional everyday knowledge. Figure 1 presents a problem which addresses reasoning on this level of domain representation. The upper part describes a reference experiment where a resting disk is pushed with some force. Thereupon the disk glides across the surface with a certain velocity. Then the impacting force is increased. The question to be answered by the student refers to the effect of the force's increase on the velocity of the disk. Does the velocity remain the same, increase or decrease? Correct reasoning on the QlR level is capable of solving such a problem guided by knowledge which takes the form of hypotheses and can be expressed as scope-phenomenon pairs:

IF the force increases

THEN the velocity increases, too.

Such a hypothesis correctly describes the monotonously increasing relationship between the involved variables. Examples of wrong pieces of knowledge on the QlR level would be to assume either no relationship, or a monotonously increasing relationship between the mass of the disk and its resulting velocity. Reasoning on the QlR level allows the formulation of domain-specific hypotheses which are very close to the pre-instructional intuitive reasoning of a novice. However, it has its clear-cut limitations. If more than one independent variable is changed, resulting contradicting effects on the dependent variable cannot be integrated. Correct reasoning on the QlR level in this case leads to the prediction that the resulting effect is not determinable.

This corresponds to findings of research in developmental psychology stating that in early phases of cognitive development existing knowledge structures do not enable the integration of contradicting effects of two variables [13, 37]. For example, Klahr and Siegler [14] coined the term "conflict problems" within the domain of balance scale tasks for those experiments in which the effects of the two independent variables weight and lever length contradict each other.

Instruction should help students understand such level-specific limitations in order to motivate the progression to another representational level on which they can overcome them on the basis of more information. Such understanding is a prerequisite for the ability to use knowledge acquired on multiple levels of mental domain representation in a flexible way in order to cope with the various problems posed in the domain.

The presentation of simulated experiments which offer more information allows the processing of more differentiated hypotheses on the next level of reasoning, the *quantitative-relational level*. Instead of specifying only that a certain variable increases or decreases, the addition of the rate of changes enables the formulation of more precise hypotheses such as

IF the force increases by some factor

THEN the velocity increases by the same factor

and

IF the mass increases by some factor

THEN the velocity decreases by the same factor

These hypotheses reflect the proportional relationship between the impacting force and the velocity as well as the inversely proportional relationship between the mass and the velocity. The QnR level of reasoning is of outstanding importance. First, it enables the restriction of monotonous relationships to proportional or inversely proportional relationships, for example, and thus permits more precise predictions. It still does not require the determination of the proportionality constants themselves. Second, it supports the learner in inferring the quantitative equation during the next step where he is provided with numerical information, and helps him understand and reconstruct this equation later on.

With respect to elastic impact phenomena, correct reasoning on the *quantitative-numerical level* leads to an equation which is derived from Newton's second law **F**=ma. The introduction of the simplifying conditions that the impacting force **F** is constant during a particular push and the duration t of a push is (very short and) defined as 1 results in the equation v=**F**/m which describes the functional relationship between impacting force F, the mass m of the disk and the disk's velocity v. Since the duration t has been omitted in this equation, units are not considered. This equation has to be used by the students in order to predict experimental outcomes on the quantitative-numerical level. Although from the viewpoint of physics instruction it might be argued that elastic impacts are not a suitable domain to introduce proportional and inversely proportional relationships, it has to be emphasized that the introduced framework of multiple levels of domain presentation and representation is applicable to many other physics domains such as balance scale tasks, uniformly moving objects, etc..

The framework described above bears several implications on passive as well as on active adaptability of a learning environment. With regard to passive adaptability, the learning environment has to make it possible for a student to interact on that level of instructional domain presentation which corresponds to his current level of understanding of the domain. Active adaptability has to be based on a diagnostic assessment of the student's present knowledge state as well as on a model of the knowledge acquisition processes which specifies how the different knowledge states can be acquired.

A first step in order to meet the requirements of active adaptability is to determine the space of all possible domain-specific knowledge states on the different levels of mental domain representation. The determination of such a space of knowledge states demands a detailed analysis of the domain as well as thorough empirical studies. Correct as well as incorrect and fragmentary domain-specific knowledge on each representational level has to be defined. The remaining diagnostic task is to identify that knowledge state which accounts best for an individual student's behavior. In the next section we discuss a computerized diagnostic model which relies on this approach.

Computerized Diagnosis of Qualitative and Quantitative Knowledge

One method to construct a diagnosis system which incorporates correct as well as incorrect and fragmentary knowledge states by means of explicit enumeration is known as the bug library approach which can be traced back to the work of Brown and Burton [3, see also 35]. We have developed MULEDS [25, 27], *a multi-level diagnosis system for the domain of elastic impacts* which is based on the bug library approach. It has been implemented as a rule-based system by means of the production system interpreter PRISM [21]. Figure 2 presents the different components of MULEDS and their interplay with the learning environment DiBi. The two grey-shaded components in Figure 7 denote the domain-specific knowledge as well as the mathematical knowledge represented in MULEDS. The domain-specific knowledge constitutes a *hypotheses library* which contains *correct* and *fragmentary* as well as *incorrect domain-specific rules* with regard to the phenomena of central and oblique impacts upon disks. These rules express various hypotheses about the functional relationships between the independent variables impacting

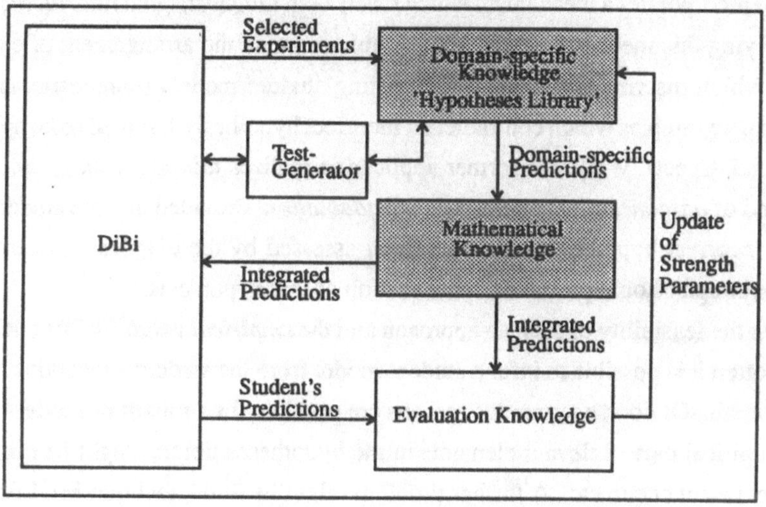

Figure 2. The components of MULEDS.

force, impact angle, and mass of a disk, and the dependent variable disk velocity. For each of the representational levels described above, a different set of hypotheses has been formulated, making possible *knowledge diagnosis on different levels of mental domain representation.* The mathematical rules represent the knowledge required for the integration of the effects of two or more independent variables on the dependent variable on different levels of domain representation. Within MULEDS this knowledge is introduced only as correct knowledge and is not yet part of the knowledge to be diagnosed.

The domain-specific rules of the hypotheses library along with the rules which represent the mathematical/experimental knowledge make up a space of potential student models. The diagnostic task is to identify the subset of rules which allows the best possible reconstruction of the predictions provided by a particular student. Once such a set of rules has been identified, it can be regarded as a *model of the student's knowledge* about the instructional domain. It is predictive and executable via interpretation, that is, the assessed model is capable of predicting the predictions a student will make on the basis of particular experimental situations. This diagnostic task is realized by a comparison of predictions. The predictions yielded by the interpretation of all applicable hypotheses out of the hypotheses library are compared to the student's predictions regarding the same sequence of experimental situations. For those hypotheses whose predictions match the student's predictions, an associated strength parameter is increased which results in tagging the appropriate hypotheses.

MULEDS also incorporates a mechanism which meets both *diagnostic and tutorial requirements* [9]. By applying this mechanism MULEDS is able to guide the arrangement of experimental situations which discriminate between competing student models (diagnostic aspect) and to provide counterexamples which contradict an incorrect hypothesis assumed to be applied by a student (tutorial aspect). While the former application realizes *tailored testing*, the latter accomplishes a kind of *misconception-oriented active adaptation*. Provided that the student actually applies the incorrect hypothesis which has been assessed by the diagnosis system, the generated counterexample should provoke a modification of this hypothesis.

One way to test the feasibility of such an approach and the *empirical validity* of its outcome is to analyze how often it is possible to infer a student model from the student's reactions to the problems of the domain. Of course numerous reasons are possible for a mis-fit of model-based predictions and empirical data: Relevant elements in the hypotheses library might be missing, unsystematic errors might occur, etc.. A further problem might lie in the fact that MULEDS is not able to assess a knowledge state if it changes over a sequence of test items.

In order to determine the validity of our diagnostic approach, we investigated[4] 132 students of the grades 5, 7, 9, and 11 of a secondary school. The subjects were about 11, 13, 15, and 17 years old. 30 problems were presented. 12 problems addressed the central elastic impact upon a resting disk. The remaining 18 problems were concerned with the oblique elastic impact. All problems were presented without any feedback. In this paper we will discuss only the results concerning the central elastic impact.

The collected data have been analyzed with the goal to identify for each student that model which matches his responses to the problems best. In 18 cases (14%) the subjects were characterized by correct domain-specific knowledge on all three levels of reasoning with respect to the functional relationships between the impacting force and the disk's resulting velocity as well as the disk's mass and the resulting velocity of the disk. If more than one response of a student could not be predicted by the model, the knowledge state on the corresponding representational level has been considered as not identifiable. Table 1 shows[5] that on the qualitative-relational level only 5%, and on the quantitative-relational level only 9% of the knowledge states were not identifiable[6]. These results are encouraging although the number of students with correct knowledge has to be taken into account. The picture changes drastically on the quantitative-numerical level. On this level, 47% of the knowledge states were not identifiable. We suspect that this result is largely due to the fact that the hypotheses library of MULEDS does not yet com-

4. We would like to thank Johannes Bellert, Bernhard Bührlen, Christoph Herderich, and Petra Schweizer who have been involved in designing, realizing and analyzing this empirical study.
5. The 101 subjects who were characterized by correct domain-specific knowledge on the qualitative-relational level include 61 subjects who were not able to integrate the effects of the two independent variables correctly.
6. The considered knowledge states might even have not existed in these cases.

Table 1
Empirical results of 132 subjects: The diagnosis of qualitative and quantitative knowledge for the domain of elastic impacts.

	Knowledge states		
	correct	incorrect or fragmentary	not identifiable
Qualitative-relational level	101 (77%)	24 (18%)	7 (5%)
Quantitative-relational level	103 (78%)	17 (13%)	12 (9%)
Quantitative-numerical level	53 (40%)	17 (13%)	62 (47%)

prise any assumptions about incorrect versions of arithmetical procedures. Arithmetical procedures are required on this representational level in order to integrate the effects of independent variables on the dependent variable. For instance, a rule which integrates the effects of the impacting force and the disk's mass by means of addition would correspond to an incorrect arithmetic procedure in the considered domain. Furthermore, since the problems given without feedback were beyond the subjects' level of training on the quantitative-numerical level, impasses during problem solving were a frequent phenomenon. As a consequence many different errors have to be expected which in our case often rely on the application of nonsensical arithmetic procedures.

The most frequent misconception has been observed on the qualitative-relational level. 50% of the subjects reasoned in the following way. If both the impacting force and the mass of the disk increase or decrease, the effects of these two variables on the dependent variable compensate, and hence there is no change in the disk's velocity. This misconception as part of the mathematical knowledge has been assessed by a hand analysis since MULEDS is in general not designed to diagnose incorrect knowledge of this type. Other misconceptions diagnosed by MULEDS were the neglection of the disk's mass and even more frequently (in 8-10% of the cases on the different representational levels) the assumption of a monotonous/proportional relationship between the mass and the velocity of a disk. A further interesting finding was that subjects (5%) applied the correct numerical equation, but were characterized by incorrect, fragmentary, or not identifiable knowledge on the qualitative-relational level. However, the overall consistency of misconceptions over the different representational levels and between the considered phenomena of central and oblique impacts was not as high as we expected.

A diagnostic model which is equipped exclusively with an enumerative library of knowledge states, and which contains no model of how these knowledge states evolve is rather insufficient. It does not provide an adequate basis for diagnosing the dynamically changing knowledge structures of a student while interacting with a learning environment, and for addressing and supporting those learning mechanisms which produce the different knowledge states to be acquired. What is needed are models of the learning processes which specify how the different knowledge states evolve on the basis of which instructional input and by means of which learning mechanisms. Such learning models should not only describe how knowledge within one representational level can be acquired, but should also reconstruct the transitions from one level to the next. In the following section we introduce such a learning model which simulates the acquisition of qualitative and quantitative knowledge in a prototypical way while progressing through the different levels of mental domain representation. Such a model of prototypical learning processes can then be utilized in a prescriptive way in order to enhance active adaptability of a learning environment.

Analysis-based Acquisition of Qualitative and Quantitative Knowledge

In this section we delineate the *computerized generic learning model* KAGE [24]. We start with a brief characterization of the kind of learning situations the model has been tailored to. Thereafter, we describe the basic learning principles which have been implemented in KAGE in order to model learning in the considered instructional domain.

KAGE proposes a cognitive model of how *knowledge about functional relationships between physics variables* develops step by step while progressing through different levels of mental domain representation. It reconstructs learning as it might take place within a computerized learning environment such as DiBi, that is within an instructional setting structured by SPFP sequences. From such a learning environment, KAGE expects experiment descriptions which require relational reasoning. Relational reasoning is provoked if experiments on a certain level of domain representation have to be compared to a reference experiment in order to enable correct predictions. The knowledge to be acquired by the model consists of *domain-specific knowledge as well as mathematical integration knowledge on different levels of domain representation*. Domain-specific knowledge denotes knowledge about the functional relationships which hold between the various physics variables of the domain under study. Mathematical integration knowledge is necessary for the integration of the effects of different independent variables on the dependent variable. Being provided successively with experiment descriptions and the corresponding feedbacks, KAGE simulates the acquisition of both kinds of knowledge. It accomplishes the learning task by means of *analysis-based learning mechanisms* which produce those pieces of knowledge necessary to solve the posed problems. KAGE emphasizes the use of knowledge acquired on a qualitative level of mental domain representation for the formulation of hypotheses on a quantitative level.

A learning environment making possible self-guided learning structured by SPFP sequences suggests a kind of learning which relies heavily on the analysis of examples. This assumption is supported by empirical observations[7] which reveal that students who are requested to formulate a prediction that could not be derived by applying already acquired domain-specific knowledge often resort to past examples. They attempt to identify the functional relationships underlying the past example, and to apply them to the new problem in order to derive a prediction. Subsequently provided system-generated feedback confirms or contradicts a student's prediction. In the latter case, students frequently try to explain the actual outcome to themselves in order to make their current hypotheses consistent with the observed results.

The importance of examples for learning and problem solving has often been noted in the literature. For instance, Pirolli and Anderson [23] observed that students solved initial problems of recursive programming by the modification of examples. Learning mechanisms then transformed solutions of these initial problems into new problem solving operators. A detailed psychological analysis of how students employ 'self-explanations' during problem solving and learning with respect to the domain of mechanics is provided by Chi, Bassok, Lewis, Reimann and Glaser [6]. They found that successful students flexibly generated many explanations during the study of known examples in order to understand how they instantiate the general principles they exemplify.

The analysis of examples as well as their subsequent generalization in many cases requires the utilization of domain-relevant pre-knowledge. Lewis [18] suggested the term 'analysis-based generalization' for those methods in which generalizations of examples are based on an analysis of single instances. The rationale behind these methods is to model how already acquired knowledge can be used to understand an example in order to provide an analysis-based justification for generalizations. Pure realizations of this approach are known as 'explanation-based generalization' (EBG) [20] and 'explanation-based learning' (EBL) [8]. These methods accomplish the analysis of an example by an explanation of how it satisfies the definition of the concept under study. This explanation relies on a formal and complete theory of the example

7. In the respective study 9 ninth and eleventh graders of a secondary school participated as well as 21 college students.

domain and serves as the basis for the generalization of the example. Since such an explanation constitutes a proof that an example is an instance of a particular concept, these methods realize a kind of reformulation of already available, though abstract domain knowledge guided by concrete examples.

Analogical generalization denotes a different analysis-based method. Since it emphasizes the elaboration and extrapolation of the internal structures of examples without requiring a comprehensive domain theory, KAGE's learning mechanisms have been developed around this learning strategy. However, analogical reasoning is important with regard to both problem solving and learning.

Many existing approaches to analogical reasoning (for an overview of computational approaches see [11]) focus on single conceptual steps of the investigated reasoning process. It has not been our goal to design yet another approach, but rather to tailor some basic well-known principles of analogical reasoning to the considered learning situation and content domain in order to model learning as it takes place under these circumstances. The mechanisms implemented[8] in KAGE comprise:

- A set of principles for identifying that a certain variable affects another one;
- A set of analogical mechanisms for elaborating and extrapolating the kind of functional relationship between variables in order to enable predictions in new problem situations;
- A proceduralization mechanism for building abstract rules which explicate the generalizations implicitly performed in these extrapolations;
- A problem solver for applying acquired domain-specific rules as well as mathematical integration rules to future problems.

The first three mechanisms form the learning mechanisms of KAGE. The last mechanism is concerned with the problem solver implemented in the model which makes use of the knowledge the system has already acquired at a certain point of time. Table 2 lists the various learning mechanisms employed in KAGE along with their input and output. In particular, KAGE is applied to experiment descriptions which specify how the different independent and dependent variables vary.

If KAGE is applied to examples comprising variables about which no knowledge is available, it tries in a first step to *identify* which of these variables affect each other. This is accomplished by making use of experimental and heuristic knowledge. Experimental knowledge enables the interpretation of experimental data resulting from factorial designs. The application of this knowledge leads to the identification of variables in the considered example which affect each other. However, students often do not make use of factorial designs. Instead, they vary two or even more independent variables at once and hypothesize about their effects. In many situa-

8. KAGE has been implemented in Prolog. For details of the implementation, see [24].

tions this will lead to wrong hypotheses. In order to be able to model such frequent mislearning, KAGE has also been provided with heuristic knowledge. Heuristic knowledge enables the interpretation of experimental data resulting from non-factorial designs. The employment of heuristics has been inspired by the work of Lewis [18] and in our case results in a systematic preference for tying together independent and dependent variables which change in the same way.

If KAGE has access to known examples analyzed in the way described above, and does not yet know the functional relationships between the variables in the present problem, it tries to predict the outcome on the basis of analogical reasoning. *Analogical reasoning* in KAGE comprises four steps: *retrieval*, *elaboration*, *extrapolation*, and *evaluation*.

During the *retrieval* step, KAGE searches for an analogous known example on which analogical reasoning could be based. Two experiments are considered analogous if both instantiate the same experimental design. Note that this definition of analogous experiments is independent of the representational level. KAGE always first searches for examples formulated on the same level as the present problem. Only if no analogous example can be found on this level, KAGE searches on a representational level which has been processed previously. Currently, KAGE implements the assumption that retrieved examples are well-remembered. Although this is in general an idealized assumption, it is justified with respect to the learning environment DiBi. DiBi makes the retrieval of already processed examples easy for students by means of a graphically presented history list.

The goal of the *elaboration* step is to hypothesize which functional relationships hold between those pairs of variables in the known example which have been identified as affecting each other. In order to perform such an elaboration, KAGE has been provided with knowledge about various mathematical functional relationships. During the elaboration simple relationships are tried before more complex ones.

The functional relationships hypothesized during the elaboration of the base are extrapolated to the present problem within the next step. This is realized by applying them to the corresponding variables in the present problem. By making use of mathematical knowledge for integrating the effects of different independent variables on the dependent variable, the model is capable of predicting how it will change in the present problem. Such a prediction is then compared to the system-generated feedback during the *evaluation* step. This results either in a confirmation or in a rejection of the above reasoning steps.

While analogical reasoning is a powerful learning and problem solving strategy, it has the drawback that it is always tied to concrete examples. However, with increasing proficiency in a certain domain, human problem solving becomes more and more independent from past examples. It will rely on rules which have been abstracted from concrete examples and which can

Table 2

The various learning mechanisms implemented in KAGE.

INPUT	MECHANISM	OUTPUT
example description; experimental pre-knowledge; heuristic pre-knowledge;	*Identification* of variables which affect each other:	assumptions about which variables affect each other;
	Analogical Reasoning:	
present problem description which requires a prediction; known examples; definition of analogous examples;	*Retrieval* of a known analogous example;	known example which is considered as analogous to the present problem description;
known analogous example; mathematical pre-knowledge about functional relationships;	*Elaboration* of the functional relationships in the known example;	assumptions about the functional relationships between the variables in the retrieved example;
assumptions about functional relationships in the retrieved example; present problem description which requires a prediction; mathematical pre-knowledge about consistent extrapolations;	*Extrapolation* of the elaborated functional relationships to the present problem;	prediction of variable changes in the present problem;
system generated-feedback;	*Evaluation* of the correctness of the prediction;	confirmation or contradiction of the predicted variable changes;
traces of successful analogical reasoning about functional relationships between the variables;	*Abstraction:* inference of domain-specific knowledge about functional relationships between variables;	abstract and experiment-independent rules which state the functional relationships between the variables;

directly be applied to future problem situations. The process of knowledge *abstraction* is often described as *proceduralization* [1, 2]. KAGE possesses such a proceduralization mechanism capable of explicating the generalizations carried out during the elaboration and extrapolation steps of analogical reasoning. This mechanism is invoked in all those cases in which a certain elaboration and extrapolation step has proven to be useful a certain, prespecified number of times. Since the objects of elaboration and extrapolation are functional relationships between variables, these relationships will be built into abstract rules in the proceduralization step. These rules are domain-specific in that they are related to certain independent and dependent variables out of the physics domain under study. They are directly applicable to future problems by the problem solver implemented in KAGE without the need to refer to past examples.

In the following we illustrate KAGE's different learning mechanisms with a few experiments concerned with the central elastic impact upon a resting disk. We begin with the description of how KAGE, starting without any knowledge about the domain, acquires domain-specific knowledge on the qualitative-relational level. Thereafter, it will be shown how the model uses the knowledge acquired on this level for the formulation of hypotheses on the next level, the quantitative-relational level. KAGE expects experimental data to be encoded in schema-like structures. For a certain experiment, these structures specify the considered variables and their modifications on a certain representational level.

Each sequence of experiment descriptions starts with a reference experiment. On the qualitative-relational and the quantitative-relational level, a reference experiment specifies that a given force applied to a disk with a given mass results in a certain speed[9] of the disk. In the following experiments the independent variables are systematically varied. For instance, the mass of the disk is increased while the force is held constant. The question to be answered is how the speed of the disk changes as a result of these modifications. The problem given below corresponds to such an experiment on the qualitative-relational level.

experiment_1:
    level: QlR
    independent_var: (FORCE =)
    independent_var: (MASS >)
    dependent_var: (SPEED ?)

If KAGE is initially applied to such a problem, it is not able to formulate a prediction since the necessary domain-specific knowledge is missing. In the next step, the learning environment provides the correct feedback that the disk's speed is decreasing.

9. Since velocity is a vector quantity, its magnitude and direction are considered independently. In the case of central elastic impacts, the velocity's direction is always constant, and is therefore neglected in the provided examples. The velocity's magnitude corresponds to the speed.

dependent_var: (SPEED <)

On the basis of its experimental and heuristic knowledge, KAGE now tries to identify which independent variable is responsible for the observable change in the dependent variable. Since experiment_1 is an instance of an experiment which satisfies a factorial design, KAGE only makes use of its experimental knowledge and infers that the variables mass and speed affect each other:

affect_each_other: (MASS, SPEED)

If KAGE is subsequently faced with a new problem like

experiment_2:
level: QlR
independent_var: (FORCE =)
independent_var: (MASS <)
dependent_var: (SPEED ?)

it tries to infer a prediction on the basis of analogical reasoning. Therefore an analogous example has to be retrieved. Since experiment_1 instantiates the same experimental design as experiment_2, it is retrieved as the base of analogical reasoning.

During the elaboration of this known example, KAGE tries to infer which kinds of functional relationships hold between those pairs of variables which have previously been identified as affecting each other. In order to accomplish this, KAGE utilizes pre-specified knowledge about functional relationships. In this case it is inferred that the variables mass and speed vary in opposite ways. This elaboration corresponds to the correct assumption that the relationship between these two variables is monotonously decreasing. During the extrapolation step, the application of the inferred relationship to the variable mass in the current problem yields the prediction that the speed will increase. Since this prediction is correct, it is confirmed in the subsequent evaluation step. If such analogical inference processes are repeatedly successful, the generalization implicitly performed within them will be explicated by creating an abstract rule like the one presented below.

$\forall x \in \{=,<,>\}$: (MASS x) --> (SPEED y) $\wedge$ opposite(x,y)

Such a rule is domain-specific, since it is related to certain independent and dependent variables. It is abstract, because it can be applied to future problem descriptions without the need to go back to any concrete examples.

After having acquired knowledge about the relationship between the mass of a centrally impacted disk and its resulting speed, subsequent experiment descriptions provided to KAGE might exemplify the effect of the impacting force on the disk's speed. Experiment descriptions with more than one varying independent variable additionally require mathematical knowledge for the integration of the different effects of the independent variables on the dependent variable. KAGE acquires this mathematical integration knowledge on the basis of the same analogical inference processes which it employs for the acquisition of domain-specific knowledge.

If KAGE has acquired knowledge about the relationships between the variables on the qualitative-relational level of domain representation, it takes advantage of this knowledge when reasoning about problems on the quantitative-relational level. If the model is applied to a problem description[10] on the QnR level such as

experiment_3:
 level: QnR
 independent_var: (FORCE [✱,1])
 independent_var: (MASS [✱,2])
 dependent_var: (SPEED ?)

and possesses no knowledge about the functional relationships between the variables on this representational level, it searches for an analogous example. Because no other example on the QnR level is available yet, KAGE considers examples on the already processed QlR level. Since experiment_1 and experiment_2 instantiate the same experimental design as experiment_3, both are candidates for retrieval. KAGE always selects the one which has been processed more recently and thus retrieves experiment_2.

The monotonously decreasing relationship between the variables mass and speed which has been successfully applied to experiment_2 is now extrapolated to experiment_3. However, the information provided by experiment_3 enables a more precise prediction than just stating that the speed will decrease. In order to exploit the additional information, KAGE searches for a functional relationship which meets the condition of being monotonously decreasing. This search takes place within a pre-specified library of functional relationships and their properties. The inversely proportional relationship is the most simple relationship which meets the condition stated above and which is applicable to experiment_3. Thus KAGE applies this relationship to experiment_3 which results in the correct prediction that the disk's speed will divide by two. Like in the previous case, if such extrapolations are repeatedly successful, the functional relationship will be built into an abstract rule. The result of this process is a newly acquired rule which states that the relationship between the mass of an impacted disk and its resulting speed is inversely proportional.

Analogical reasoning which utilizes knowledge already acquired on a certain level of domain representation in order to predict on another representational level involves a kind of *transfer of knowledge between different levels of domain representation*. The central feature of this transfer is that knowledge which has been acquired on one representational level serves as

---

10. The modification operators '[✱,q]' provided by such a description on the QnR level in this case denote the factors by which the corresponding variables vary.

a constraint for the search of appropriate hypotheses on the next level. Currently, we design an empirical study in order to test the validity of some of KAGE's assumptions. A further test could be based on an analysis of the efficiency of instructional measures which have been derived from the model.

Given a model of knowledge acquisition such as KAGE, it is possible to determine those *sequences of instructional input* which are *optimal, suboptimal,* or even *misleading* with respect to the implemented learning mechanisms. In order to be able to optimally utilize the learning mechanisms described above, KAGE has to be provided with sequences of examples and problems which are designed according to specific conditions. These conditions define conventions the instructional material has to obey in order to be felicitous for the learning model [30, 32]. For each piece of knowledge that has to be acquired by the model, an example is presented that introduces exactly one additional aspect exemplifying this knowledge. In addition, for every submitted example a problem has to be designed which presupposes the piece of knowledge for a correct prediction which has been addressed by the preceding example. Hence, the basic felicity condition that governs the design of instructional material derived from KAGE is that each piece of knowledge which has to be acquired will be taught by its own examples and problems, or its own sublesson [31]. This felicity condition enables a learner to concentrate on only one new aspect in each experiment and hence reduces the space of hypotheses drastically. Whether such pre-determined sequences of instructional input are optimal for students is a question of the model's validity which still has to be empirically tested.

Discussion

In this paper we described the computerized diagnosis system MULEDS and the learning model KAGE. They have been exemplified in a learning environment which simulates experiments in the physics domain of elastic impacts. The development of both systems is rooted in questions such as which knowledge structures are applied by students in physics domains, how these knowledge structures are utilized, and how students learn in these domains. MULEDS and KAGE model knowledge structures about functional relationships between physics variables as well as about mathematical procedures. Knowledge about the functional relationship between variables makes it possible to predict how a dependent variable will change as a consequence of a varying independent variable. The application of mathematical procedures is required when the effects of different independent variables on a dependent variable have to be integrated.

MULEDS and KAGE are based on a framework for a multi-level domain representation ordered along the dimension qualitative/quantitative. This framework evolved during our attempts to model the knowledge structures students utilize when they try to solve the problems in the considered domain. The conceptualization of physics domains on multiple levels of domain representation supports flexible and successful problem solving. It enables the qualitative interpretation of physics equations as well as their successive (re-) construction. These abilities play an important role in the problem solving behavior of experts. Typically, they analyze a problem situation qualitatively before they proceed with a quantitative analysis [4, 17]. The qualitative analysis is often guided by the interpretation of abstract domain principles which are expressed as equations. Later, when the relevant physics concepts and their qualitative relationships have been established in the problem situation, these relationships are verified by a (re-) construction of those equations which involve the concepts.

The framework can be utilized in a descriptive way in order to delineate the knowledge structures students possess on the different levels of mental domain representation. MULEDS implements a diagnostic model which assesses the knowledge applied by students on these different representational levels in the domain of elastic impacts. It relies on an enumerative library of correct, fragmentary, and incorrect pieces of knowledge about the functional relationships between the considered variables on the different levels. Within a tutorial system, knowledge states diagnosed by MULEDS can serve as a basis for active adaptability. For instance, MULEDS guides the arrangement of counterexamples in order to contradict incorrect hypotheses, thus provoking their modification.

If the design of instructional material is guided by the framework, it is utilized in a prescriptive way. The acquisition of knowledge about functional relationships between physics variables is then described as the evolution of a succession of consistent mental domain representations. On the basis of correspondingly designed sequences of examples and problems, KAGE simulates how this knowledge develops step by step on the different representational levels. Especially, it emphasizes how knowledge which has been acquired on a qualitative level of mental domain representation can be used in order to formulate hypotheses on a quantitative level. A simulation model like KAGE constitutes a "pseudo-student" [33] which enables the investigation of instructional material. For instance, it might be applied to a certain kind of instruction in order to test whether it leads to the desired outcome. If the knowledge acquired by the model is incorrect, one can suggest changes to the instructional design on the basis of how the model has acquired the incorrect knowledge. Within a tutorial system, the knowledge states produced by a pseudo-student might be used as a basis for knowledge diagnosis, and during a tutorial demonstration the different reasoning steps which are simulated by the model might be explicitly carried out, thus teaching a human student a strategy for solving a certain class of problems.

Although MULEDS and KAGE help to understand the knowledge structures used by students on the different levels of mental domain representation, and how they are acquired, there are still many open questions. For instance, learning on multiple levels of mental domain representation requires not only the acquisition of domain-specific knowledge. It also demands the acquisition of domain-independent knowledge about which classes of problems can be successfully processed by applying domain-specific knowledge on a certain representational level. Poor understanding of functional relationships between physics variables is only one reason for the often missing relatedness of students' qualitative and quantitative knowledge. Inconsistent and incorrect conceptual frameworks in physics is another one. Although there exists a considerable documentation of the content of students' naive concepts, only little is known about how they interact with formal, quantitative physics knowledge.

References

1. Anderson, J. R. (1983). *The architecture of cognition*. Cambridge, MA: Harvard University Press.
2. Anderson, J. R. (1986). Knowledge compilation: The general learning mechanism. In R. S. Michalski, J. G. Carbonell, & T. M. Mitchell (Eds.), *Machine learning: An artificial intelligence approach. Vol. 2* (pp. 289-310). Los Altos, CA: Morgan Kaufmann.
3. Brown, J. S., & Burton, R. R. (1978). Diagnostic models for procedural bugs in basic mathematical skills. *Cognitive Science, 2*, 155-191.
4. Chi, M. T. H., Feltovich, P., & Glaser, R. (1981). Categorization and representation of physics problems by experts and novices. *Cognitive Science, 5*, 121-152.
5. Chi, M. T. H., & VanLehn, K. (1991). The content of physics self-explanations. *The Journal of the Learning Sciences, 1* (1), 69-105.
6. Chi, M. T. H., Bassok, M., Lewis, M. W., Reimann, P., & Glaser, R. (1989). Self-explanations: How students study and use examples in learning to solve problems. *Cognitive Science, 13*, 145-182.
7. Clement, J. (1982). Students' preconceptions in introductory mechanics. *American Journal of Physics, 50* (1), 66-71.
8. DeJong, G., & Mooney, R. (1986). Explanation-based learning: An alternative view. *Machine Learning, 1*, 145-176.
9. Evertsz, R. (1989). Refining the student's procedural knowledge through abstract interpretations. In D. Bierman, J. Breuker, & J. Sandberg (Eds.), *Proceedings of the 4th International Conference on AI and Education* (pp. 101-106). Amsterdam: IOS.
10. Falkenhainer, B. C., & Michalski, R. S. (1986). Integrating quantitative and qualitative discovery: The ABA-CUS system. *Machine Learning, 1*, 367-402.
11. Hall, R. P. (1989). Computational approaches to analogical reasoning: A comparative analysis. *Artificial Intelligence, 39*, 39-120.
12. Halloun, I. A., & Hestenes, D. (1985). Common sense concepts about motion. *American Journal of Physics, 53* (11), 1056-1065.
13. Inhelder, B., & Piaget, J. (1958). *The growth of logical thinking from childhood to adolescence*. New York, NY: Basic Books.
14. Klahr, D., & Siegler, R. (1978). The representation of children's knowledge. In H. W. Reese, & L. P. Lipsett (Eds.), *Advances in child development*. New York, NY: Academic Press.
15. Langley, P. (1981). Data-driven discovery of physical laws. *Cognitive Science, 5*, 31-54.
16. Langley. P., Simon, H. A., Bradshaw, G. L., & Zytkow, J. M. (1987). *Scientific discovery*. Cambridge, MA: MIT Press.

17. Larkin, J. H. (1983). The role of problem representation in physics. In D. Gentner, & A. L. Stevens (Eds.), *Mental models* (pp. 75-98). Hillsdale, NJ: Lawrence Erlbaum Associates.
18. Lewis, C. (1988). Why and how to learn why: Analysis-based generalization of procedures. *Cognitive Science, 12,* 211-256.
19. McCloskey, M. (1983). Naive theories of motion. In D. Gentner, & A. L. Stevens (Eds.), *Mental models* (pp. 299-324). Hillsdale, NJ: Lawrence Erlbaum Associates.
20. Mitchell, T. M., Keller, R. M., & Kedar-Cabelli, S. T. (1986). Explanation-based generalization: A unifying view. *Machine Learning, 1,* 47-80.
21. Ohlsson, S., & Langley, P. (1984). *PRISM: Tutorial, manual, and documentation* (Technical Report). Pittsburgh, PA: The Robotics Institute, Carnegie-Mellon University.
22. Ohlsson, S., & Rees, E. (1988). *An information processing analysis of the function of conceptual understanding in the learning of arithmetic procedures* (Technical Report No. KUL-88-03). Pittsburgh, PA: Learning Research and Development Center.
23. Pirolli, P. L., & Anderson, J. R. (1985). The role of learning from examples in the acquisition of recursive programming skills. *Canadian Journal of Psychology, 39,* 240-272.
24. Plötzner, R. (1990). *Analysis-based induction of elementary functional relationships* (Research Report No. 65). Freiburg: Department of Psychology.
25. Plötzner, R., Spada, H., Stumpf, M., & Opwis, K. (1990). Learning qualitative and quantitative reasoning in a microworld for elastic impacts. *European Journal of Psychology of Education, 4,* 501-516.
26. Spada, H., Reimann, P., & Hausler, B. (1983). Hypothesenerarbeitung und Wissensaufbau beim Schuler. In L. Kotter & H. Mandl (Hrsg.), *Jahrbuch fur Empirische Erziehungswissenschaft* (S. 139-167). Dusseldorf: Schwann.
27. Spada, H., Stumpf, M., & Opwis, K. (1989). The constructive process of knowledge acquisition: Student modeling. In H. Maurer (Ed.), *Proceedings of the 2nd International Conference on Computer-assisted Learning* (pp. 486-499). Berlin: Springer.
28. Stumpf, M. (1990). Adaptivitat durch Flexibilitat in Reprasentationsform und Kontrollstruktur im System DiBi-MR. In A. Reuter (Hrsg.), *GI - 20. Jahrestagung* (S. 377-385). Berlin: Springer.
29. Stumpf, M., Branskat, S., Herderich, C., Newen, A., Opwis, K., Plötzner, R., Schult, T., & Spada, H. (1988). *The graphical user interface of DiBi, a microworld for collision phenomena* (Research Report No. 44). Freiburg: Department of Psychology.
30. VanLehn, K. (1983). *Felicity conditions for human skill acquisition: Validating an AI-based theory* (Technical Report CIS-21). Palo Alto, CA: Xerox Palo Alto Research Center.
31. VanLehn, K. (1987). Learning one subprocedure per lesson. *Artificial Intelligence, 31,* 1-40.
32. VanLehn, K. (1990). Mind bugs: The origins of procedural misconceptions. Cambridge, MA: MIT Press.
33. VanLehn, K. (1991). Two pseudo-students: Applications of machine learning to formative evaluation. In R. Lewis, & S. Otsuki (Eds.), *Advanced research on computers in education* (pp. 17-26). New York, NY: North-Holland.
34. Viennot, L. (1979). Spontaneous reasoning in elementary dynamics. *European Journal of Science Education, 1* (2), 205-221.
35. Wenger, E. (1987). *Artificial intelligence and tutoring systems.* Los Altos, CA: Morgan Kaufmann.
36. White, B. Y., & Frederiksen, J. R. (1990). Causal model progressions as a foundation for intelligent learning environments. *Artificial Intelligence, 42,* 99-157.
37. Wilkening, F., Becker, J., & Trabasso, T. (1980). *Information integration by children.* Hillsdale, NJ: Lawrence Erlbaum Associates.

Modeling Active, Hypothesis-Driven Learning From Worked-Out Examples[1]

Peter Reimann

University of Freiburg, Department of Psychology, Niemensstr. 10, 7800 Freiburg, FRG.

Abstract: Starting from the observation that worked-out examples play an important role in the early phases of knowledge acquisition, a strategy is described that learns "optimally" from examples in the sense that it allows for intelligent transfer from an example solution to a new problem. The components of this strategy as well as the knowledge they process are delineated and it is indicated how these components can be realized on a computer. Implications for the design of an Intelligent Tutoring System that helps students to improve their example-processing skills are outlined.

Keywords: Physics learning, problem solving, intelligent tutoring systems.

The long term goal of the research reported here is to develop a computer program that helps students with learning from worked-out examples. This computer program should demonstrate and eventually tutor a student on how to get the most out of an example. The computer receives the same example as a human student and tries to understand this example based on a domain specific knowledge base and further knowledge sources outlined below. The process as well as the outcome of the example understanding attempts form the basis for a dialogue between the machine and the human student. The nature of the dialogue can range from a mode where the student simply observes what the program is doing with the example up to more sophisticated shared-initiative and tutorial dialogues during which the program attempts more actively to teach to the student specific skills for learning from examples. These tutorial activities should be adapted to the specific knowledge and needs of the particular student, that is, the computer should not only elaborate the example for itself, but should watch the elaboration activities of the student, this way building a diagnostic model of the student's elaboration skills.

---

1. The research reported here is supported by a grant from the Deutsche Forschungsgemeinschaft (Re 814/1-1). I would like to thank Thomas J. Schult and Hans Spada for comments on earlier drafts of the text.

Another form of intervention that requires information about the individual student is to display to the student a particular example in a form where parts that demand further elaborations are highlighted or where the student is even prompted with a specific question. The decision which parts of the example demand elaborations is made by the program as the result of processing the example, looking for statements that should be difficult for the student to comprehend. For the initial phase of system development I assume that the dialogue will be a very limited one, a demonstration of how the program works on the example. The student can control the part of the example she wants to see elaborated by the machine and can control the verbosity of the program's output.

Both to generate output that is potentially meaningful for the student and to enable the program to follow the student through her elaboration activities, the program needs information about elaboration strategies as put to use by human learners. Therefore, as a prerequisite for the above tutorial scenario we need computational models of elaboration strategies and the knowledge they process. Work on this component is the focus of our current research activities.

On the following pages, I first describe some psychological research findings concerning the use of examples for problem solving and learning. As we will see, worked-out examples are very important during the initial learning phase and are therefore often used by students to guide their problem solving attempts. But learning from examples is not without problems. In particular, novices seem often to be at a loss when it comes to decide how to generalize from individual examples. These two findings motivate my intention to improve learning from example skills in students.To accomplish this, an "ideal" strategy for learning from examples is described that is based on general insights into the nature of acquiring knowledge from text and on specific empirical observations. I will sketch out how this strategy can be realized as a computer program and close with a comparison to other attempts that identified and taught strategic knowledge for learning.

The Function of Examples in Learning

When solving new or complex problems, people often rely on examples of former solutions to similar problems. This is true for experts [1, 18], but is particularly often found in novices, i.e., during the early stages of knowledge and skill acquisition [7, 23, 27, 28]. Novices tend to rely extensively on worked-out examples, even in domains where the solution knowledge is explicated (or even formalized) and the solution principle is known to them [28]. For example, Pirolli and Anderson [23] found in the protocol of a subject solving a recursion programming problem clear evidence that he couldn't have solved the problem without close reference to an example solution. Chi, Bassok, Lewis, Reimann and Glaser [7] observed that students rarely used a general problem solving procedure described in a physics textbook even after intensive study, but instead relied frequently on worked-out examples when solving problems. Ross [28, 30] reports similar observations for the case of mathematical problem solving. LeFevre and Dixon [22] found that students actually ignore written instruction and focus exclusively on example information when reading a technical text.

The specific kind of examples I am considering appear typically in science textbooks after the verbal description of a law or principle and a general description of how the principle or law is to be applied. Later, the student is usually requested to solve a set of problems, building on the general declarative knowledge and the information from the examples presented in the text. Figure 1 shows a prototypical worked-out example, taken from a widely used textbook on physics [17]. It illustrates how Newton's Second Law is applied to solve a specific mechanics problem.

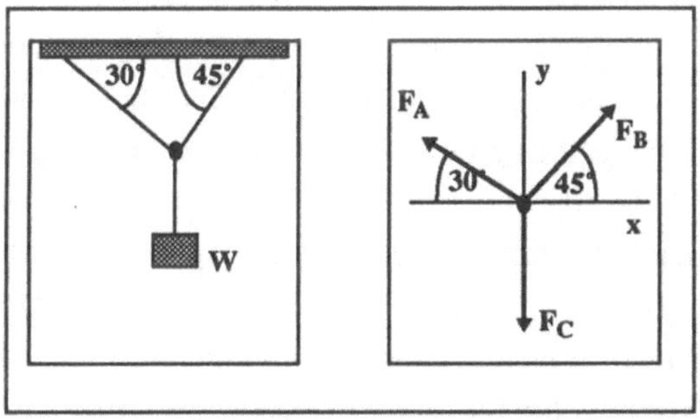

(1) The left figure shows an object of weight W hung by massless strings. (2) Consider the knot at the junction of the three strings to be "the body". (3) The body remains at rest under the action of the three forces shown in the right figure. (4) Suppose we are given the magnitude of one of the three forces. (5) How can we find the magnitude of the other forces? (6) F_A, F_B and F_C are all the forces acting on the body. (7) Since the body is unaccelerated, $F_A + F_B + F_C = 0$. (8) Choosing the x- and y-axes as shown, we can write this vector equation as three scalar equations: (9) $F_{Ax} + F_{Bx} = 0$, and (10) $F_{Ay} + F_{By} + F_{Cy} = 0$. (11) The third scalar equation for the z-axis is simply: (12) $F_{Az} = F_{Bz} = F_{Cz} = 0$. (13) That is, the vectors all lie in the x-y plane so that they have no z component. (14) From the figure we see that (15) $F_{Ax} = F_A \cos 30 = 0.86\, F_A$, (16) $F_{Ay} = F_A \sin 30 = 0.5\, F_A$, (17) and $F_{Bx} = F_B \cos 45 = 0.707\, F_B$, (18) $F_{By} = F_B \sin 45 = 0.707\, F_B$.

Figure 1. A worked-out example for a mechanics problem.

What is the function of this kind of examples? In general, they are intended to illustrate how problem solving procedures described in abstract terms in the textbook are to be applied to specific problems. The example in Figure 1 illustrates the general procedure to solve mechanics problems as described in the textbook by Halliday and Resnick [17, p. 69]:

1. Identify the "body";
2. Identify the objects in the problem situation which exert forces on the body;

3. Draw a free body diagram for the body by first selecting a suitable reference frame and then drawing all the forces acting on the body;
4. Apply Newton's Second Law to each component of force and acceleration.

To be more precise, in formal domains such as physics worked-out examples have two main functions for the novice when solving problems: They provide her with control information and with hints how to interpret concepts appearing in the formal principles and laws. First of all, novices use examples to guide their search for a solution. A worked-out example serves as a sort of plan which prescribes a sequence of operators. It is well known from research on problem solving in formal domains that novices lack domain-specific control knowledge and are therefore doomed to apply weak problem solving strategies or to rely on examples when they are available [21, 26]

A second function of examples is to show students how to interpret abstract concepts as they typically appear in the definition of principles and laws in formal domains. In semantically rich domains such as physics, general descriptions of procedures such as the one provided above cannot be used directly for problem solving since they include references to physical concepts (such as *force*) and principles (e.g., Newton's Second Law) which have to be instantiated by the problem solver to be applied [6]. For example, solving a mechanics problem requires often to find all the forces acting on a particular particle or system. These forces, however, cannot be found by syntactically manipulating the information given in a problem statement. When solving physics problems, the student has to generate a theoretical description of the problem situation which includes concepts and relations not found in the original problem statement. Put differently, the function of examples is primarily to provide students with what Reif [25] calls *ancillary knowledge*, knowledge that allows for reliable, easy and flexible application of knowledge about concepts and principles. Ancillary knowledge about concepts and principles *sensu* Reif includes information such as specification of the applicability conditions, of permissible values, and of the various symbolic representations. Ancillary knowledge includes further information about likely errors in applying a concept or principle and how these errors can be avoided.

Other arguments for the value of learning from examples are based not so much on the kind of information they convey but *how* they support learning. Sweller [33] argues that learning from examples is under certain circumstances superior to learning from problem solving since it focuses the learner on schema acquisition instead of just reaching a problem solving goal: "Goal attainment and schema acquisition may be two largely unrelated and even incompatible processes" (p. 283). Problem solving (meaning means-ends analysis in this context) can be ineffective to acquire knowledge because it is a cognitively demanding process and leaves not sufficient cognitive capacity free for further analysis of the problem in terms of its general structure, relation to similar problems, and so forth. This argument is supported by Sweller with a computer simulation. He demonstrates also in training studies that learning from examples is at least as effective as problem solving [34, 35].

All this said, we can conclude that examples should foster both problem solving and knowledge acquisition. Yet, using examples for these purposes is in itself not without problems.

Obstacles in Learning From Examples

The main problem with learning from examples is that students do not generalize over specific example solutions (corresponding to the observation that they often don't generalize over their problem solving experience). That leads to the inability to transfer from the worked solution to problems that are not quite similar [13, 24, 35]. To analyze possible reasons for limited transfer in more detail we have to look into the processes that are involved in solution transfer. I take the model developed by Ross [30], which is in accordance with other accounts of analogical reasoning [15, 36] as an orienting framework. He distinguishes four phases concerning the use of examples for problem solving: Noticing, reconstruction, mapping, and generalization. During *noticing*, a problem solver is reminded of one or more earlier examples. Both superficial and structural similarities[2] are used to determine the similarity between the current problem and earlier examples. The common observation is that novices rely more on superficial similarities than do more advanced problem solvers [8]. As for the problem of too narrow transfer, we can speculate that problem solvers do not have stored the example in terms of structurally relevant features or that during memory access these features are not attended to sufficiently. It is not enough to notice the similarity to an example, the example has also to be retrieved. To the extent that retrieval involves *reconstruction*, many problems can occur during problem reconstruction which can contribute to a less than perfect use of the example solution. In the next step, *mapping*, solution steps from the retrieved example have to be applied in the context of the new problem and may have to be modified to account for differences between the new and the retrieved problem. Ross' work [28, 29] is particularly interesting here since he studied how subjects use examples even so the formal principles (formulas) required to solve the problems (simple probability theory) had been explicitly given to them. Ross observed that subjects used extensively details from the example solution in the new solutions, even details which were not required to solve the new problem. Superficial similarities thus do not only play a role in noticing, but also during mapping.

2. Ross distinguishes structural and superficial similarity according to whether a change of a problem feature requires to modify the solution procedure (structural similarity) or not (superficial similarity)

Whereas noticing, reconstruction, and mapping are processes required to solve a new problem given a former problem solution, ***generalization*** influences how an example solution is stored into memory and accessed. Generalizations can occur at several occasions; for instance, as a side-effect of mapping from the example to the new solution. In this situation, the problem solver may be forced to apply a solution step from the example to the new problem even so not all the conditions for applying the step are met in the problem context [31]. If this leads to a successful solution, the conditions not met are not necessary for applying the operator and can therefore be dropped. Generalizations also can occur when a solution to a new problem or an example solution is stored into episodic memory that indexes solutions in terms of differences to solutions already in memory. Schank [32] describes the organizing principles of such a memory. Kolodner [19] reveals the details of generalization and specialization processes that operate over such episodic structures.

The quality of generalizing - be it during mapping or during storing - depends on how the student mentally represents the example. Generalization processes work over mental representations and depend therefore on the content and the accessibility of those representations. How an example is represented in the students' mind is to a large degree determined by the form of the example in the textbook. This is particularly true for novices who often do not possess much domain specific knowledge that could be used to deeply elaborate on the example so that the mental representation would be substantially different from the information presented in the text.

An important observation is that examples often do not contain the information necessary to come up with an example representation that is useful for generalizing. In particular, examples do often not contain the reasons, the rationale for *why* a certain step in the solution was performed. Consider again the example solution provided in Figure 1. In statement 2, the example says "consider the knot at the junction of the three strings to be the body". Although this is an important decision that determines many aspects of the solution, no reasons are given why the knot and not another object in the problem situation could be the body. In statement 6, it says that the forces shown in the force diagram are "all the forces acting on the body". However, no procedure is specified that describes how this could be determined by the student.

Providing reasons for a problem solving step would ideally comprise: (a) the goal worked on, (b) how the goal was generated, (c) reasons for selecting the one operator that is actually applied. Since this or similar information is usually not given in an example, students will have considerable difficulties to learn from a single example. Recall that we consider students during early knowledge acquisition, students that have read more or less quickly through some pages of difficult text and have not necessarily a clear understanding of the domain principles. Based on such a fragile knowledge base, it is hard for them to on their own come up with the above mentioned pieces of additional information required to understand an example.

Given these problems, it is hardly surprising that so often little is learned from examples. How can we help students? One solution would be to improve the examples themselves, enriching them with the kind of information indicated above. But, as Chi and Bassok [6] observe, this strategy might not lead to much success. The main problem is that this kind of information would not be tailored to the individual student, thus being redundant for one student, not comprehensible for another. Information that does not fit the individual learner tends to be ignored by the student or, worse, confuses her. A second, related argument against pre- elaborated examples is that self-generated elaborations may be more effective than elaborations provided by somebody else [2]. The consequence can only be to teach the student to herself become better in using information that is contained in examples instead of trying to optimize the examples.

I want now concentrate on the question of how to develop effective learning-from-example skills in students. As a starting point, I refer to observations made in a study by Chi et al. [7] capturing characteristics of successful and less successful example study behavior.

How Human Learners Overcome Problems With Learning From Examples

Chi et al. [7] (from now on called "self-explanation study") made some interesting observations concerning differences in learning from examples. Their subjects - male and female undergraduates with no formal education in physics - had first to read several chapters of a physics textbook that dealt with the foundations of mechanics (part 1 of the study). As part of reading the chapter, subjects had to demonstrate after each chapter their understanding of the main concepts taught in the text by a series of tests, and had to study parts of the text again in case they failed the tests. The goal was to have each subject acquire an at least rudimentary understanding of the physics of motion and force. The primary goal was, however, to analyze how subjects studied worked-out examples that were contained in the chapter on Particle Dynamics (part 2 of the study), and how they later solved problems on their own (part 3). Some problems had been carefully designed to vary superficial and structural features, resulting in "close", "medium" and "far" transfer problems with respect to the examples. The examples dealt with the application of Newton's motion and force laws; one of the examples is displayed in Figure 1. Both during example study and problem solving, subjects had to "think aloud", to verbalize what they were thinking of while studying the examples and solving problems. Based on these (transcribed) verbal protocols, a group of successful subjects was compared with a less successful group, distinguishing between the two groups according to frequency of correct solutions during problem solving (median split). The two contrasting groups were compared with respect to example study behavior and problem solving behavior.

For the learning from examples phase, the following differences between the two groups were found. Successful students mentioned more often that they didn't understand a certain part of the worked-out example. Technically speaking, they reported that they didn't understand why a specific problem solving operator appeared at a certain stage of the example solution. Besides this difference in "monitoring" understanding of the example text, successful students also engaged in a series of activities to overcome their problems: They elaborated on the relations between a particular step in the example and the goals behind that step. They further attempted to come up with a specification of conditions that could explain why the operator under question was applied. Finally, they elaborated on the effects the application of an operator had beyond those mentioned in the example. The less successful subjects displayed either none or considerably less of these elaborative inferences.

It is interesting to note that the successful students tried to improve the weak points of usual textbook examples. In worked-out examples, it is often not shown clearly how the specific actions taken relate to the principles they are assumed to illustrate; further, neither the conditions for when to apply a problem solving step nor the effects of the application are specified in detail and exhaustively.

The differences with respect to learning had effects on the problem solving behavior[3]. Not only did those subjects who elaborated examples solve more problems, they also showed more often "far transfer" in that they were able to solve problems that only remotely resembled the ones presented to them as examples. The not successful students were less often able to solve problems that did not share surface features with the example problems. It also turned out that the successful students seemed to have acquired more generalized problem solving schemata since they less frequently turned back to the example pages while solving new problems, and when they turned back used the example as a source of specific information and as a check on the correctness of their solution, whereas the less successful students turned frequently back to the examples, studied again the complete example, and tried to map each step of the example to the new problem.

From this study it can be concluded that in order to foster learning from examples one should teach students to do two things: To actively monitor the understanding of each part of an example solution and to try to explain steps in the solution that are not well understood. As a result, the example representation will be of a kind that allows for better transfer to other problems and builds the basis for generalization processes. This effect is primarily due to a better understanding on side of the learner as to which features of the example solution are essential for problem solving.

---

3. It should be noted that we tried to assess carefully whether other factors such as general intelligence or more specific skills in physical reasoning could have accounted for these differences. For arguments against these and other possibly confounding influences see [7, 9].

The differences between the two types of students with respect to learning from worked-out examples can help us to understand how one can overcome problems with learning from procedural examples in general. The successful students seem to gain a better understanding of the example and, based on that, of the domain. After having studied an example the successful students have acquired a representation that is different from the one built up by the less successful students. Their representation is more coherent in the sense that the particular example is embedded within the domain's concepts and principles; their representation of problem solving operators has more (and the right) conditions attached (describing both the legal application of the operators and control heuristics); and their representation of operators includes more information on the effects of applying an operator. Finally, both the conditions and effects of operator application seem to be represented in a somewhat generalized form, not bound to all the specificities of the example. In other words, the successful learners acquire the *ancillary knowledge* [25] or the *utilizational competence* [16] that is necessary to apply abstract domain principles during problem solving.

Modeling the Processing of Example Solutions

To learn more about elaboration skills, I use the same materials (text, examples, problems) as were used in the self-explanation study [7]. Further, the observations about successful and less successful example study behavior from that study are taken as points of reference. The goal is to develop computational models of learning from examples that can generate some of the central characteristics of effective and less effective learners. This requires to capture the learning mechanism in a more precise manner than it was possible in our purely descriptive attempts. Besides reproducing the elaboration behavior on some level of abstraction, the models should also be sufficient to explain the observed differences in students' problem solving performance. The outcome of this endeavour should be a learning program that takes as input worked-out examples as they were given to the subjects in the self-elaboration study and returns as output a representation of the examples and the domain knowledge which can in turn be given to a problem solving program that takes as input a problem description and returns as output a solution attempt (see Figure 2).

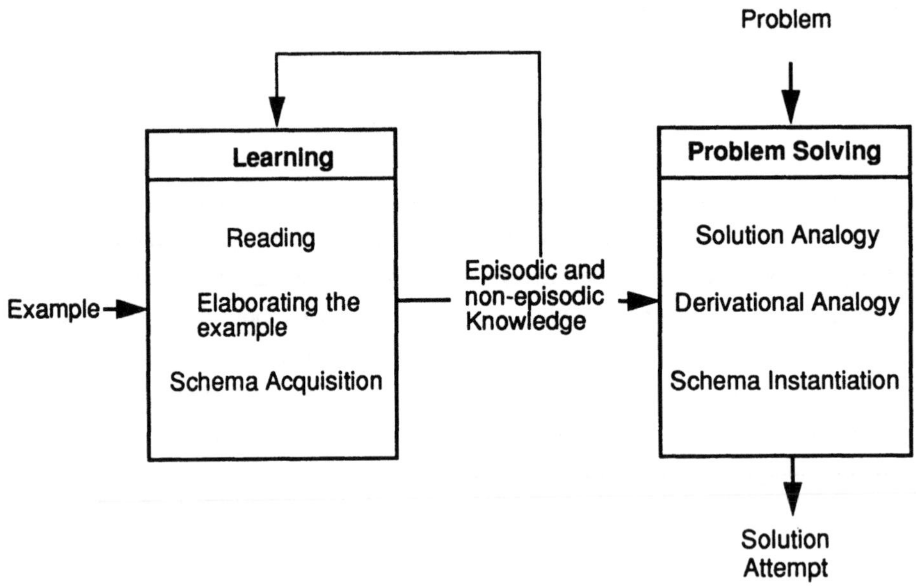

Figure 2. Relation between different forms of learning and problem solving.

Knowledge structures to be acquired

When reading through a worked-out example solution, the student is confronted with a sequence of *statements*. A statement specifies either a *goal*, an *action*, i.e., the application of a problem solving operator, or a *state*. For instance, statement 5 in the worked-out example shown in Figure 1 is a goal specification, statement 15 shows the application of a (trigonometry) operator, and statement 9 describes a state. Goal specifications provide the student with information that can help her to grasp the rationale of specific actions taken by the one who solved the example problem[4]. In worked-out examples, actions occur typically much more frequently than goal specifications. The intention of the learner is to process the example in a way that will help her later to solve problems on her own.

The learning phase can result in different forms of knowledge structures. For instance, the knowledge acquired can be episodic: a representation of the specific example. Or the student can acquire more abstract forms of knowledge, for instance, in form of generalized solution

4. In this specific example, the only goal statement that occurs specifies just the overall problem solving goal and is therefore of limited help to understand why the specific actions were taken.

plans. Since we are concerned with the early phases of learning, it can be assumed that the information acquired from first examples is more of the episodic kind. For the episodic representation we can further distinguish whether the example in stored in terms of its surface features only, i.e., close to the form as presented in the textbook, or whether the example is stored in an elaborated form, enriched with inferences made by the student. Content and form of example representation will influence how the example can be used for problem solving and further learning (see Figure 2). For instance, generalized solution schemata can be directly instantiated to solve a new problem. The usefulness of episodic example knowledge for problem solving depends on the information stored with the example. If the example is stored only in its surface form, then it can only be accessed if the surface features of a problem are very similar to the example features; and adaptation of the example solution to a new problem will be based on relatively simple, syntactical transformations. If information is added to the example as a result of elaborating, chances increase that the example can be accessed using more structural properties of the domain and that adaptation includes transformations anchored in the meaning rather than only the syntax of steps taken in the example. Akin to Carbonell [4], I call the first kind of analogy *transfer solution* and the second kind *derivational analogy*.

Together with non-episodic background knowledge, the elaborated example can without further generalizing be directly put to use for problem solving by derivational analogy [4, 5]. Derivational analogy is a form of problem solving where search is reduced by using solutions of similar past problems to guide the problem solver in constructing the solution to the new problem. Problem solutions are stored in an episodic knowledge structure modelled after Schank's concept of a *dynamic memory* [32]. In order for derivational analogy to work, steps in the original solution have to be augmented with justifications that essentially capture information about why a specific step has been taken and under which circumstances it seems reasonable to perform the same solution step again. A solution is represented as a sequence of problem solving operators and their variable bindings.

In the derivational analogy paradigm as introduced by Carbonell, justifications are generated based on problem solving attempts performed by the learner herself. In our case of learning from examples, the learner relies on examples to acquire the necessary knowledge. However, as we have seen, normal examples do by far not contain enough of the required information so that they could be used for derivational analogy. To gain this information from the example it is necessary to understand why the specific example solution works. The question then becomes: What are the inference processes that generate these elaborations? It is also necessary to consider exactly what information is required to perform analogical problem solving of various forms. Further, these two questions about the content of elaborations and the process of elaborating must be answered taking into account psychological constraints. Different from research in artificial intelligence, for cognitive modeling we must not only demonstrate that the required inferences can be generated by an algorithm, but that the inferences can be computed within the limitations of humans' cognitive capabilities. From such a psychological perspective, it is for instance improbable that humans store the huge amount of information with a solution episode as it is assumed in Carbonell's (AI-) model of derivational analogy.

Imposing constraints on a computational mechanism for a complex task and calculating the effects of these constraints on performance cannot be done without the help of computer simulation studies. In the next sections, I describe first the general components of a cognitive computational model for learning from worked-out examples and delineate then how to implement example elaboration strategies in this architecture.

Model components

The general design of the model is based on insights into the process of text comprehension and understanding [3] as well as on research concerning the dynamic character of human memory and its influence on problem solving and understanding [20, 32]. Figure 3 shows the main components of the model. It comprises modules for example comprehension and problem solving. The problem solving component demonstrates the competence of the system. It attempts to solve new problems building on knowledge mainly acquired by learning from examples. Both the problem solving module and the example understanding module are hybrid systems, combining a case-based approach that uses episodic knowledge about example solutions ("cases") with general but weak inference methods.

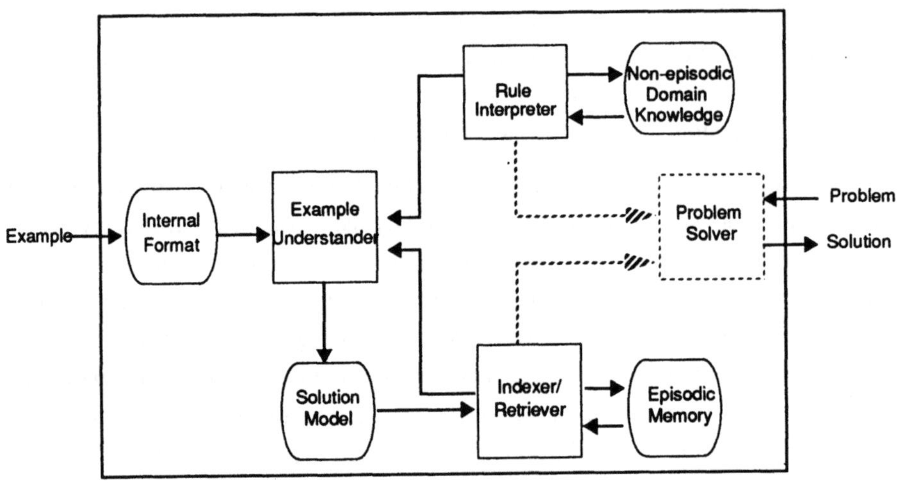

Figure 3. Model Components

Most important for the current discussion is the example understander. This component takes a worked-out example as input (in a propositional format) and generates a model of the example solution that is then stored into Episodic Memory. To capture the essential differences between successful and less successful learning from examples, the process of example elaboration is treated as a plan recognition task where the understanding system has to encode a given example solution in terms of problem solving goals, operators, and relations to domain concepts. To construct the example solution model, the system calls up episodic knowledge about former examples and non-episodic domain knowledge. The general control decision is to rely on episodic knowledge before attempting to use more abstract domain knowledge, this time for the example comprehension task.

Two learning-from-examples strategies are contrasted: an active, expectation-driven one that leads to a hierarchical representation of an example and allows for problem solving by derivational analogy (Carbonell, 1986), versus a passive example processing strategy leading to a flat representation that enables the problem solver to use an example by transformations based on syntactical similarities. The distinction between the two strategies as made it here is an idealization. In reality, the two strategies share more components than I assume here for purposes of exposition. It is further supposed that the passive learning strategy consists of a subset of the active strategy's components. Therefore, I describe the active strategy in some detail and suggest only shortly which components are missing in the passive strategy.

Active example elaboration strategy

Under this strategy, the system monitors its understanding of the example by checking for the steps in the example solution whether they cohere with its knowledge about problem solving in the domain. It tests its understanding in an active, expectation-driven manner: The system predicts the next example step and compares its prediction with the step actually appearing in the example. Problems in understanding are identified by false or missing predictions. Having identified an understanding problem the system tries to explain the offending part of the example, thus resorting to a cognitively more demanding mode of example processing. This includes attempting to derive the action(s) observed in the example from its episodic and general knowledge about the domain as well as to come up with alternate solution plans. In summary, this strategy reflects many aspects of the performance of successful subjects from the self-explanation study and incorporates general principles of comprehension [3].

This strategy is currently being implemented in a system called AXE, the Active eXample Elaborator, using KEE and CommonLisp. AXE "reads" an example statement by statement and attempts to formulate expectations about the content of the next statement. A statement comprises propositions describing either (a) a goal selection, or (b) an operator selection (law, equation, inference rule), or (c) a problem solving state. Thus, AXE tries to predict (a) which goal will be worked on next, (b) which operator will be applied, or (c) what effects a goal or operator selection will have on the solution. These predictions - or expectations - are generated as follows.

Step 1 - Case-based expectation formation: After having read the first lines of an example where the problem is described in terms of objects and their relations, values given, and values desired, AXE performs a look-up in its Episodic Memory to find out whether it has encountered a similar example before. Episodic Memory has the structure of a MOP- based memory [32] that indexes problem statements and solution steps as they occur in worked-out examples. Similarity at this stage is defined in terms of overlap in features of the problem statement. If AXE finds a similar example, it will be used to derive expectations for the new one. AXE retrieves the first operator from the old solution and uses it to predict the next state of the current example. Operators in Episodic Memory are indexed in terms of the goal(s) they serve, their necessary preconditions, effects, and further justifications. Justifications capture information not expressed in local goals and preconditions about the circumstances under which an operator can reasonably be applied (e.g., which other operators must be available, what the goal tree must look like) and in which circumstances the operator should not be applied. When transferring an operator from an old solution to the current example, AXE checks whether the operator can be applied (are the necessary conditions satisfied?) and whether it should be applied (are the justifications satisfied?). If these conditions are not met, the specific operator will not be applied. In this case, AXE will attempt to suggest at least the class of operators it expects to see in the next example statement. For instance, it may only be able to predict that the next solution step will deal with "rotating the reference frame" without being able to predict the details of that rotation.

Step 2 - Rule-based expectation formation: If no similar example is found, AXE attempts to formulate a prediction by using its non-episodic domain knowledge in a forward- reasoning mode. Non-episodic knowledge is represented in form of rules. Again, AXE can at times predict the class of operators instead of a specific operator since its knowledge about domain operators is organized in a rule hierarchy. After Step 1 or 2, AXE does either have an expectation about the next state of the example solution or didn't derive an expectation. Having no expectation is considered a failure and the system continues with Step 5, trying to repair this failure. If an expectation has been derived, it continues with Step 3.

Step 3 - Comparison: In the next step, the expectations about the example solution state are compared with the actual situation as described in the example statement. This comparison leads to an evaluation of the expectation. It may be (a) specific, (b) too general, or (c) insufficient, depending on whether a specific state can be predicted (a), or whether only a general state of affairs can be predicted (b), or whether the expectation covers only part of the effects resulting from the application of the operator (c). It is further determined whether the expectation is correct or wrong.

Step 4 - Continuation: In the case of a specific and correct hypothesis, AXE goes on and tries to formulate expectations for the next example state. Before that, the correctly predicted operator instantiation is stored into Episodic Memory in terms of its differences to the operator instantiation retrieved.

Step 5 - Case-based repair: In all the other cases (no expectation, wrong, insufficient or too general expectation) AXE tries to learn from the expectation failure. The failure-driven repair is performed first in a case-based manner, relying on episodic knowledge before attempting to use more general, but search-intensive repair strategies. AXE looks into Episodic Memory to see whether it has made a similar expectation failure before and has encountered a similar example state before. This information is used to update the example solution model. The now successfully explained solution step is stored into Episodic Memory and the system goes on predicting the next example statement.

Step 6 - Rule-based repair: If an expectation failure cannot be explained using Episodic Memory content, the system attempts to reconstruct the plan that might underlie the actions taken in the example part. This mode of example processing is particularly important during initial learning, since in this stage the learner possesses only a fragmented episodic knowledge structure and therefore cannot recognize an example as the "same old story". During plan reconstruction, the system resorts to its non-episodic domain knowledge and tries to find an operator or a sequence of operators that could have generated the problem solving state as displayed in the current example statement. If AXE finds a sequence that connects the statement in the example with a currently active problem solving goal, it stores the sequence into Episodic Memory. In effect, the system can from now on substitute matching for search when encountering similar example parts.

Step 7 - Copying the example solution step: If the plan reconstruction attempts fail, the system will store the specific representation of the example statement (an operator and its bindings) under the recent goal it was working on. Although AXE does not understand why this step was taken, it can at least copy the solution step when encountering a similar situation in a problem solving context.

An additional step - counterfactual reasoning: Whenever AXE succeeds either in predicting a part of the example solution (Step 1 or 2) or in recovering from an expectation failure (Step 5 or 6), it engages in a kind of counterfactual reasoning. After having established goal-action links by solution step recognition or reconstruction, starting from the goal it is asked: What other means exist to realize this goal? And why were they not chosen in the example solution? To come up with answers to these two questions, alternate solution paths are generated by working forward form the respective goal and applying domain rules. Doing that, AXE looks for problems with the alternate solution plans, problems that can be classified into categories such as *failed constraint*, *excessive costs*, *failed result*, or *bad side effect* [10]. The first step of the alternate sequence with the explanation of why it failed is stored in Episodic Memory. By learning from counterfactual reasoning, the system acquires decision heuristics, thus enriching the necessary conditions of operators with further justifications. Evidence for reasoning similar to this counterfactual inference component was found in the most successful student of the self-explanation study.

Passive comprehension strategy

A word on the passive strategy that would reproduce aspects of poor learners' performance. It can be described in terms of differences to the active elaboration strategy. Missing monitoring of understanding is explained by assuming that the learner does not form expectations about the next parts of the example solution, but that she first reads the statement and then tries to explain: the understanding process is data-driven. Further, the kinds of explanations derived for a solution step are more local ones: Instead of trying to relate the observed solution step to higher-level goals and to constraints stemming from principles of the domain, the passive learner only tries to relate the step to the last step observed by subgoal-, effect- and enable- links. Finally, the passive strategy does not include any attempts to engage in counterfactual reasoning.

This active/passive distinction captures the end-points of what more realistically is a continuum: "Real" students will hardly work so actively or so passively. I take these extremes as a starting point and will develop models that are psychologically more plausible by moving along the dimension spawned by these two points.

Conclusions

I have outlined a strategy to learn from worked-out examples where the learner engages in an active process of elaborating to gain the knowledge required for analogical problem solving. It should be noted that "active" refers more to the quality than the quantity of example processing. It is the engagement in trying to find the deep structure of an example rather than sticking with the surface features that characterizes the active learner in this sense. That this is not only characteristic for processing of examples but for text understanding in general is supported empirically by Ferguson-Hessler & De Jong [14], for example. They observed that successful and less successful problem solvers show characteristic differences during the learning phase which in their case consisted of studying a text about mechanics (without examples). Whereas good as well as not so good problem solvers showed about the same amount of activities during reading, those students who became good problem solvers attempted significantly more often to grasp the deep structure of the text. They for example imposed structure on the text that was not directly present and explicated procedural information that was given only indirectly in the text.

Based on the observations from the elaboration study [7] and from studies on learning and transfer, it seems save to say that only a minority of students will engage in such an active and cognitively demanding example study behavior. Instead, students mostly skim over examples and treat them as text material rather than as problems. It is one of the main convictions behind my approach that students can be motivated to become more active example elaborators if it can be demonstrated to them convincingly that elaborating examples is of great value both for understanding the domain and for problem solving. Motivation is not enough, students also should be taught *how* to learn from examples. It seems to me that this can best be accomplished by using computer-based tutorial programs which incorporate computational models of example elaboration skills. This research can be seen as a continuation of the work by Brown et al. [3, 11] who expressed the notion of strategies for understanding in the language of artificial intelligence, thereby making more explicit what precisely understanding means. However, this work did not result in implemented computational models which incorporated the strategies identified. One reason for this may be that Brown, Collins and co-workers strived for a domain- independent characterization of strategies, whereas computer implementations usually require to consider a fair amount of domain-specific details. By restricting modeling attempts to a specific sort of text - examples - in a specific and formal domain - mechanics - the chances for implementations increase. And those are the prerequisite for more advanced forms of tutoring students by computers.

A second feature besides being more domain-specific that distinguishes the research reported here from earlier attempts on how people understand formal text is the explicit consideration of the knowledge *access* issue. The knowledge representation format predominant in research on text understanding - the schema - gives an account of the structure of knowledge but says comparatively little on how knowledge is accessed during listening or reading and how it is acquired in the first place. With the notion of dynamic memory I use a knowledge representation format that builds on schema theory, but extends it by treating learning and access/ retrieval issues more explicitly. In particular, Schank's [32] notion of a MOP-based memory provides us with concepts to express the differences between episodic and abstract knowledge naturally and to describe how episodic information becomes integrated with abstract knowledge. This is particular important for early learning phases and provides the means to explain novice-expert shifts in knowledge organization [12].

The theory of dynamic memory does not impose constraints on the kind and amount of information to be stored in an episodic memory. As mentioned before, artificial intelligence models incorporating an episodic memory component [4] store huge amounts of data in it, capitalizing on its organization that is optimized for economical storage and quick retrieval. I try to limit the kind and amount of information stored with a case (an example) by means of the elaboration strategy: The system goes beyond the literal representation of the example to the degree required to relate problem solving decisions to goals. We have to await experiences with the performance of AXE to see whether this criterion is strong enough to focus it on the necessary and only the necessary information. I want further experiment with constraints that are located within the general architecture rather than the specific strategy employed; for instance, by introducing a focus of attention, or working memory.

Note that the active learning strategy outlined above results in the production of the kind of information other researchers would like to see integrated in the design of textbook examples: ". . . the critical factors [for the positive effects of examples on learning] are not whether worked examples are used, but rather whether the techniques used appropriately direct attention and impose a relatively light cognitive load" [37, p.4]. According to Ward and Sweller, attention must be directed toward problem solving states and connected moves. This is often better done by worked examples than by problem solving where attention is directed toward the difference between states, at least for novices using means-ends analysis. Exactly this focus on states and their connected moves is accomplished with the active elaboration strategy, in particular when the counter-factual reasoning strategy is employed. The strategy, however, does not reduce cognitive load. Instead, it is cognitively quite demanding.

What will the next steps be? In order to adapt instruction to the individual student, we need further knowledge about "non optimal" example study behavior. My intention is to explain these insufficient strategies in terms of deviations to the optimal strategy and to the optimal declarative knowledge base as they have been outlined in this chapter. But the first step will to demonstrate by means of implemented models that the proposed strategy works as expected.

References

1. Ashley, K.D., & Rissland, E.L. (1986). Compare and contrast, a test of expertise. *Proceedings AAAI.*
2. Bransford, J.R., Stein, B.S., Vye, N.J., Franks, J.J., Auble, P.M., Mezynski, K.J., & Perfetto, G.A. (1982). Differences in approaches to learning: An overview. *Journal of Experimental Psychology: General, 111*, 390-398.
3. Brown, J.S., Collins, A., & Harris, G. (1978). Artificial intelligence and learning strategies. In H. O'Neill (Ed.), *Learning strategies* (pp. 107-139). New York: Academic Press.
4. Carbonell, J.G. (1986). Derivational analogy: A theory of reconstructive problem solving and expertise acquisition. In R.S. Michalski, J.G. Carbonell & T.M. Mitchell (Eds.), *Machine Learning. An artificial intelligence approach Vol. 2* (pp. 371-392). Los Altos, CA: Kaufmann.
5. Carbonell, J., & Veloso, M. (1988). Integrating derivational analogy into a general problem solving architecture. In *Proceedings of the Case-Based Reasoning Workshop* (pp. 104- 124).
6. Chi, M.T.H., & Bassok, M. (1989). Learning from examples via self-explanations. In L. Resnick (Ed.), *Knowing, learning, and instruction. Essays in honor of Robert Glaser* (pp. 251- 282). Hillsdale, NJ: Erlbaum.

7. Chi, M.T.H., Bassok, M., Lewis, M., Reimann, P., & Glaser, R. (1989). Self-explanations: How students study and use examples in learning to solve problems. *Cognitive Science, 13*, 145-182.
8. Chi, M.T.H., Feltovich, P., & Glaser, R. (1984). Categorization and representation of physics problems by experts and novices. *Cognitive Science, 5*, 121-152.
9. Chi, M.T.H., & Vanlehn, K. A. (in press). The content of self-explanations. *Journal of the Learning Sciences.*
10. Collins, G.C. (1987). *Plan creation: Using strategies as blueprints.* PhD thesis, Yale University.
11. Collins, A., Brown, J.S., & Larkin, K.M. (1980). Inference in text understanding. In R.C. Spriro, B.C. Bruce & W.F. Brewer (Eds.), *Theoretical issues in reading comprehension* (pp. 385-410). Hillsdale, NJ: Lawrence Erlbaum.
12. Elio, R., & Scharf, P. B. (1990). Modeling novice-to-expert shifts in problem-solving strategy and knowledge organization. *Cognitive Science, 14*, 579-639.
13. Eylon, B., & Helfman, J. (1982). *Analogical and deductive problem-solving in physics.* Paper presented at the AERA meeting, New York.
14. Ferguson-Hessler, M.G.M., & De Jong, T. (1990). Studying physics texts: Differences in study processes between good and poor performers. *Cognition and Instruction, 7*, 41-54.
15. Gentner, D. (1983). Structure-mapping, a theoretical framework for analogy. *Cognitive Science, 7*, 155-170.
16. Greeno, J.G., Riley, M.S., & Gelman, R. (1984). Conceptual competence and children's counting. *Cognitive Psychology, 16*, 94-143.
17. Halliday, D., & Resnick, R. (1985). *Fundamentals of Physics.* New York: John Wiley & Sons.
18. Klein, G.A., & Calderwood, R. (1988). How do people use analogues to make decisions? In *Proceedings of the Case-Based Reasoning Workshop* (pp. 209-223). San Mateo, CA: Morgan Kaufmann.
19. Kolodner, J.L. (1984). *Retrieval and organizational structures in conceptual memory: A computer model.* Hillsdale, NJ: Erlbaum.
20. Kolodner, J. L., & Simpson, R.L. (1989). The MEDIATOR: Analysis of an early case-based problem solver. *Cognitive Science, 13*, 507-549.
21. Larkin, J. H., McDermott, J., Simon, D.P., & Simon, H.A. (1980). Models of competence in solving physics problems. *Cognitive Science, 4*, 317-345.
22. LeFevre, J., & Dixon, P. (1986). Do written instructions need examples? *Cognition and Instruction, 3*, 1-30.
23. Pirolli, P.L., & Anderson, J.R. (1985). The role of learning from examples in the acquisition of recursive programming skills. *Canadian Journal of Psychology, 39*, 240-272.
24. Reed, S.K., Dempster, A., & Ettinger, M. (1985). Usefulness of analogous solutions for solving algebra word problems. *Journal of Experimental Psychology: Learning, Memory & Cognition, 11*, 106-125.
25. Reif, F. (1985). Acquiring an effective understanding of scientific concepts. In L.H.T. West & A.L. Pines (Eds.), *Cognitive structure and conceptual change* (pp. 133-151). New York: Academic Press.
26. Reimann, P, & Chi, M. T. H. (1989). Human expertise in complex problem solving. In K. Gilhooly (Ed.), *Human and machine problem solving* (pp. 161-192). London: Plenum Press.
27. Ross, B.H. (1984). Remindings and their effects in learning a cognitive skill. *Cognitive Psychology, 16*, 371-416.
28. Ross, B.H. (1987). This is like that: The use of earlier problems and the separation of similarity effects. *Journal of Experimental Psychology: Learning, Memory, and Cognition, 13*, 629-639.
29. Ross, B.H. (1989a). Distinguishing types of superficial similarities: Different effects on the access and use of earlier problems. *Journal of Experimental Psychology: Learning, Memory, and Cognition, 15*, 456-468.
30. Ross, B.H. (1989b). Some psychological results on case-based reasoning. In *Proceedings Case-Based Reasoning Workshop* (pp.144-147). San Mateo, CA: Kaufmann.
31. Ross, B.H., & Thompson, P.T. (in press). Generalizing from the use of earlier examples in problem solving. *Journal of Experimental Psychology: Learning, Memory, and Cognition.*
32. Schank, R.C. (1982). *Dynamic memory: A theory of reminding and learning in computers and people.* New York: Cambridge University Press.
33. Sweller, J. (1988). Cognitive load during problem solving effects on learning. *Cognitive Science, 12*, 257-285.
34. Sweller, J.S., Chandler, P., Tierney, P., & Cooper, M. (1990). Cognitive load as a factor in the structuring of technical material. *Journal of Experimental Psychology: General, 119*, 176-192.

35. Sweller, J. & Cooper, G.A. (1985). The use of worked examples as a substitute for problem solving in learning algebra. *Cognition and Instruction*, *2*, 59-89.
36. Thagard, P., Holyoak, K., & Nelson, G. D. (1990). Analog retrieval by constraint satisfaction. *Artificial Intelligence*, *46*, 259-310.
37. Ward, M, & Sweller, J. (1990). Structuring effective worked examples. *Cognition and Instruction*, *7*, 1-39.

Fostering Conceptual Change: The Role of Computer-Based Environments

Stella Vosniadou

School of Education, Aristotelian University of Thessaloniki, GR-54006 Thessaloniki, Greece and Center for the Study of Reading, University of Illinois at Urbana-Champaign, 51 Gerty Drive, Champaign, IL 6182 USA

Abstract: Findings from a project which investigates the knowledge acquisition process in thc domain of astronomy are summarized and their implication for the construction of computer-based learning environments are drawn. It is argued that the process of knowledge acquisition involves the reinterpretation of certain entrenched beliefs which constrain students' understanding of scientific concepts. It is concluded that the situations in which learning takes place are the situations which make students aware of their entrenched beliefs and provide them with information which facilitates their eventual re-interpretation. Computer-based learning environments have a distinct advantage over traditional methods for achieving these objectives.

Keywords: conceptual change, learning, computer-based learning environments

Some of the contributors to this volume have noted the importance of analyzing the situations in which successful learning takes place as a means of understanding how to construct better computer-based learning environments. For example, Pea [this volume] proposes that we need to examine the situations under which successful learning takes place, use the results of this analysis to re-examine our epistemological assumptions about learning, and craft new learning environments based on these revised epistemological assumptions. I fully agree with this general recommendation and particularly with the need to create the social environments that reinforce students' motivation to learn. Nevertheless, I believe that our attempts to build learning environments that promote problem-solving and conceptual change will be incomplete if they are based only on an examination of learning situations without taking into account the learner and the learning process.

In this chapter I will review some recent findings from a project on the acquisition of knowledge in the domain of astronomy which provide information about the process of learning science. I will then draw the implications of these findings for the construction of computer based learning environments which promote problem solving and conceptual change.

More specifically, I will argue that the process of knowledge acquisition does not involve only the enrichment of existing conceptual structures but also their restructuring. That in the process of restructuring existing knowledge students must reinterpret certain entrenched beliefs. And, finally, that the situations in which successful learning take place are the situations which make students aware of their entrenched beliefs and provide information which facilitates their reinterpretation.

Conceptual Change: Enrichment vs. Restructuring

Our models of learning have traditionally been enrichment or accretion models. In these models learning is conceptualized as the addition of new knowledge to an existing knowledge base with some resulting differentiation or hierarchical integration of the existing knowledge structures as and when needed. Recent developments in research in cognitive science, science education and developmental psychology have suggested that the accretion view of learning may not be the only one possible. Sometimes the acquisition of new knowledge requires that the existing structures become radically reorganized. This is often the case in the learning of science.

A number of research findings have accumulated over the last years which clearly show that science naive individuals come to the science learning task having formed a naive understanding of the world which provides them with explanations of physical phenomena based on their everyday experience. It is often the case that these naive explanations are very different from the currently culturally accepted scientific explanations of the same phenomena [3, 4, 7, 13]. When there is a large discrepancy between naive and scientific explanations the knowledge acquisition process cannot be explained in terms of the enrichment to existing knowledge structures because the to-be-acquired structures are so radically different from the existing ones.

For example, in the case of astronomy, the research my colleagues and I have conducted has shown that children start with a naive model of the cosmos according to which the earth is flat and stationary and located in the middle of the universe [e.g., 8, 10]. The sun moves down and hides behind the mountains during the night. The moon has its own light and the stars are small objects found in the sky only at night. This cosmology is very different from the culturally accepted Copernican model of the solar system, according to which the earth is a sphere which revolves around the sun, the day/night cycle is caused because the earth rotates around its axis, and the stars are huge suns.

These two systems of knowledge are different in their individual concepts. The concept of a spherical earth which revolves around the sun is very different from the concept of a flat earth rooted on the ground. They are also different in their structure. A geocentric universe has a different structure from a heliocentric solar system. Finally, the naive cosmology differs from the scientific one in the range of phenomena it explains. The scientific system uses constructs such as the rotation and revolution of solar objects, their light, and shadow to provide explanations of seemingly different observable phenomena, such as the day/night cycle, the seasons, the phases of the moon and the eclipses, which belong to different categories of explanation in the naive system.

In order to acquire a model of the solar system compatible with the culturally accepted one children must restructure their knowledge base in a rather radical way. The research my colleagues and I have conducted attempts to understand how this conceptual change takes place. In the next section I will present a summary of some of our findings.

The Role of Synthetic Models

In a detailed investigation of the development of the concept of the earth [8, 9, 11], we have shown that many children (about 80% of the children in our sample) use in a consistent fashion a small number of mental models about the earth which are different from the culturally accepted model of a spherical earth. These mental models of the earth are shown in Figure 1. As can be seen, some children believe that the earth is flat like a rectangle or a disc. Others think that there are two earths a flat one on which people live and a round one which is up in the sky (dual earth mental model). Some children believe that the earth is like a hollow sphere with people living on flat ground inside it, or that it is shaped like a flattened sphere with people living only on the flat parts. We have called these mental models "synthetic" because they appear to be attempts from the part of the children to reconcile the information that the earth is spherical with the belief that the ground is flat.

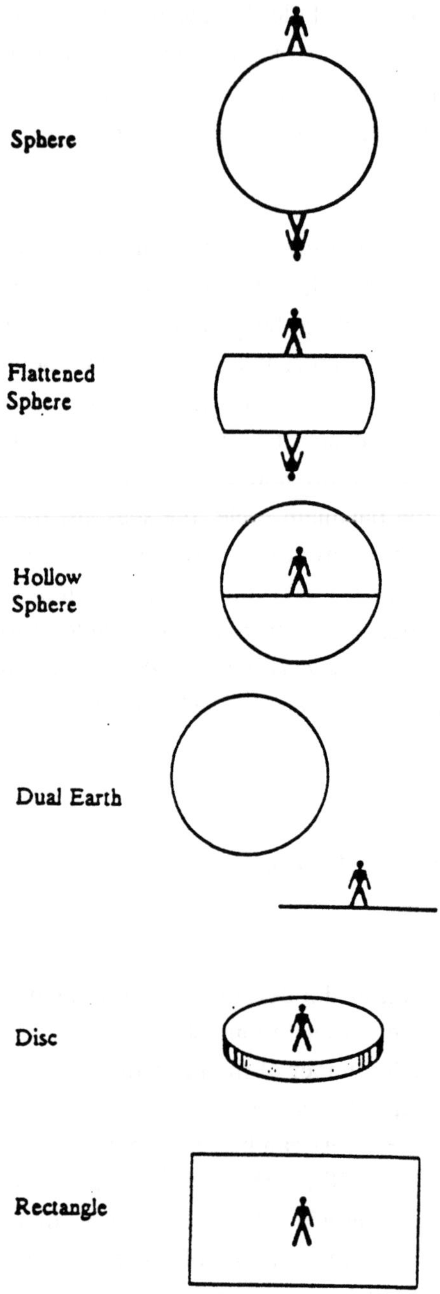

Figure 1. Mental Models of the Earth.

Similar finding shave been obtained in our studies of children's explanations of the day/night cycle [12], where, again, a small number of mental models of the day/night cycle were used consistently by the majority of children in our sample. Some of these mental models are shown in Figure 2.

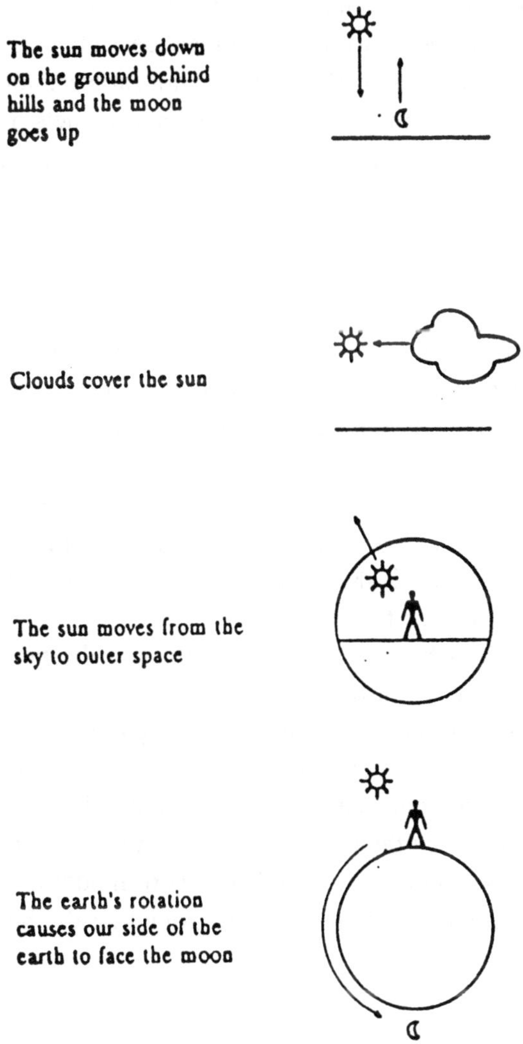

Figure 2. Explanation of the Day/Night Cycle.

Some children believe that the sun moves down and hides behind the mountains at night and the moon comes up. Others think that the sun does not move but clouds come and cover it up. Some children who believe that the sky covers the earth like a dome think that the sun moves out of the sky to outer space during the night and then comes back in. Finally, some of the children who know that earth rotates around its axis contribute the day/night cycle to the presence or absence of the moon. These children believe that night happens because the rotation of the earth causes our side of the earth to face the moon (which is supposed to be located in some part of the sky where it is always night).

Why do children come up with such strange ideas about the shape of the earth or about the causes of the day/night cycle which are not (at least not directly) reinforced by the adult culture? The formation of synthetic mental models or misconceptions raises important questions about the nature of naive knowledge and the process of conceptual change.

In a previous paper [8, 11], I argued that children's misconceptions regarding the shape of the earth represent attempts to reconcile the culturally accepted view that the earth is a sphere with two contradictory entrenched beliefs: (a) the belief that if something appears to be flat is flat (and therefore that the ground is flat), and (b) the belief that if something is not supported it will fall (and therefore that both the earth and the people on the earth need to be supported by something).

As shown in Table 1, the first belief (that the ground is flat) constrains all of children's mental models of the earth except, of course, the spherical earth model. In order to understand that the earth is a sphere children must reinterpret their belief that things that appear to be flat are always flat. They need to understand that it is possible for big round things to appear to be flat.

The second belief (the belief that the earth and the people and objects on the earth need to be supported) constraints children's naive mental models. The rectangle or disc earth is conceptualized to be supported by ground or water. The people and objects on the earth are also supported by the ground or water underneath them. During the acquisition process children seem to suspend their belief that the earth needs to be supported before they change their belief that the people and objects on the earth need to be supported. For example, in the hollow sphere model the earth is thought to be suspended in space but the people live on flat ground inside the sphere.

Table 1
Entrenched Beliefs Constrain Students' Mental Models of the Earth

Entrenched Belief		Mental Models of the Earth
"If something appears to be flat it is flat. Therefore, the ground is flat."	→	Rectangle Earth Disc Earth Dual Earth Hollow Sphere Flattened Sphere
"If something is not supported if will fall. Therefore, the earth needs to be supported."	→	Rectangle Earth Disc Earth (part of) Dual Earth
"Therefore, the objects and people on the earth need to be supported."	→	Hollow Sphere Flattened Sphere Sphere with people living only on its top part

Conceptual Restructuring Requires the Reinterpretation of Entrenched Beliefs

It appears that entrenched beliefs act as constraints on the kinds of possible mental models of the earth children can form. Understanding the culturally accepted model of a spherical earth requires the reinterpretation of these beliefs and their replacement with new explanations. It is important to note that when a belief is reinterpreted the observations that gave rise to it in the first place do not change. The earth still appears to be flat and things on the earth still fall down when not supported. What changes is the interpretation of these beliefs. I have claimed that this change in interpretation is a kind of theory change which is analogous in some respects to the kinds of theory change we observe in the history of science.

The view of learning which I am trying to promote requires an active, constructive system capable of using observation to draw inferences and construct hypotheses, not necessarily in a self-conscious type of way. In fact, I think that while children engage in theory construction types of process from very early, they are unaware that they are constructing theories. They seem to think that their beliefs represent the true state of affairs about the world.

Second, it is important to consider the kinds of constraints that operate on this learning process. Originally, constraints built into our biological make up constrain the kinds of hypotheses infants make about the nature of objects and people. Later infants' first ontological and epistemological beliefs constraint the kinds of mental models they construct of phenomena like the shape of the earth, the day/night cycle, the phases of the moon, etc. At a later stage, children's existing mental models themselves may act as constrains on subsequent mental models.

Our studies of knowledge acquisition in astronomy have shown that there is an interdependency among the various concepts that comprise the domain of astronomy, such that students' mental models of the earth may constrain their mental models of the sun, or their explanations of the day/night cycle, etc.

For example, there is substantial evidence supporting the contention that students' mental model of the earth constrains their understanding the scientific explanation of the day/night cycle. Students interpret the information that the earth rotates around its axis differently depending on the kind of mental model of the earth that they have formed. Children with a disc model of the earth interpret the information that the earth rotates to mean that the earth turns in circles, as in Figure 3A. However, when the rotation of the earth is conceptualized in this way it does not have any explanatory power with respect to the day/night cycle. As a result these children continue to believe that the alternation of day and night is caused because the sun goes down behind the mountains and the moon comes up into the sky.

Children with a hollow sphere model also interpret the rotation of the earth to mean that the earth turns sideways. These children often create the synthetic model of the day/night cycle presented in Figure 3B. According to this model, the hollow sphere consists of two hemispheres: the lower hemisphere which represents the earth on which people live and the upper hemisphere which represents the sky covering the earth like a dome. The sun and moon are located at two opposite sides within the upper hemisphere. As the earth turns people move from the day side to the night side.

Finally, children with a spherical earth model usually interpret the information that the earth rotates around its axis to mean that the earth rotates in an up/down way, as shown in Figure 3C. These children form the synthetic model of the day/night cycle previously mentioned according to which the sun and the moon are stationary at opposite sides above and below the earth. As the earth turns, our side of the earth moves from facing the sun (day time) to facing the moon (night time).

These results show that there is an interdependency among the various concepts that comprise the domain of astronomy. Information is interpreted in the context of the mental models students already have. When these mental models differ from those required for the accurate interpretation of the new information, then the new information will very likely be misrepresented to fit the existing model.

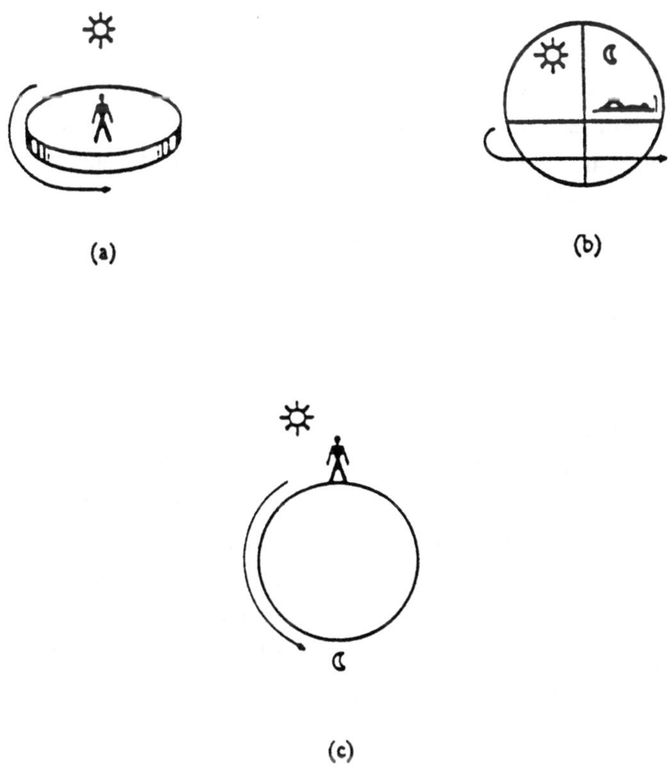

Figure 3. Explanations of the day/night cycle in terms of the earth's axis rotation for (a) a disc earth mental model, (b) a hollow-sphere mental model, and (c) a sphere mental model.

The view of knowledge acquisition I have been promoting depends critically on the reinterpretation of certain entrenched beliefs. This is certainly not the only view of the knowledge acquisition process. In a recent paper diSessa [3] argues that naive knowledge consists of fragmented knowledge pieces which he calls phenomenological principles (p-prims). These p-prims appear to be similar in many respects to what I call entrenched beliefs. diSessa does not view the process of conceptual change as the reinterpretation of p-prims. In his system, conceptual change is accounted for in terms of the enhancement or reduction in the priority of some p-prims or in terms of changes in their function. Change in function means that p-prims are no longer self-explanatory; they refer to much more complex knowledge structures, physics laws, etc.

While the changes in the function or salience of p-prims that diSessa describes may certainly be true, I think it is wrong to underestimate the inhibiting power that entrenched beliefs can have on the acquisition process. Our investigation of the knowledge acquisition process in astronomy as well as more recent studies into the knowledge acquisition process in mechanics and in the area of heat, confirm the view that students mental models are constrained by their entrenched beliefs and that the understanding of scientific concepts requires the reinterpretation of these beliefs. Furthermore, diSessa's [3] theoretical views cannot explain the presence of synthetic models or misconceptions during the knowledge acquisition process.

It has been argued that misconceptions do not represent well defined and consistent theories (as some researchers such as McCloskey [6] have argued), but that they are fragmented, self-contradictory and unstable pieces of knowledge. In the studies my colleagues and I have conducted we have been able to identify a number of well-defined misconceptions about the shape of the earth, the day/night cycle, the nature of heat and of heat transfer, and the concept of force which are used in a consistent way by individuals across a variety of problems. Nevertheless, I would not like to claim that misconceptions are well formed theories which necessarily exist as permanent structures in the knowledge base. In fact, I have maintained that individuals often construct synthetic models on the spot at the time when we ask them to answer our questions. From my point of view misconceptions are interesting because they reflect students' attempts to reconcile the scientific information with their entrenched beliefs and thus attest to the resilience of entrenched beliefs. What is persistent and difficult to change are the entrenched beliefs that give rise to misconceptions and not the misconceptions themselves.

Implications for the Design of Computer-Based Learning Environments

The findings discussed in the previous section have certain implications for the design of computer-based learning environments. For example, they suggest that in designing a learning environment we could profit a great deal from knowledge of the interdependencies among the concepts that comprise a given domain because these interdependencies determine to a large extent the order of acquisition of these concepts. Second, they indicate that we should pay particular attention to those modules of knowledge that appear to be the fundamental ingredients of naive knowledge, namely students' entrenched beliefs. Students' entrenched beliefs constrain the kinds of mental models they can form and lie at the root of their synthetic models and misconceptions. Given the highly intricate web of interrelations that exist among the various concepts that comprise a given domain it is possible that the presence of a small number of entrenched beliefs can cause misconceptions which affect students' interpretations of a wide variety of problems, as was discussed in the previous section. In designing learning environments it is important to create the circumstances that will help students (a) become aware of their beliefs, (b) understand the need to question them, and (c) replace them with a different explanatory structure.

In traditional instruction we do not usually take into consideration students mental models and entrenched beliefs. As a result, when instructed in a new domain students are often presented with factual statements which are counter-intuitive from their point of view. For example, the information that the earth is round like a ball or that the earth spins around its axis is counter-intuitive to a student who has never been exposed to this information before and whose knowledge is based on his or her everyday experience of a stationary, flat earth. When presented with such statements students seem to have two courses of action open to them. One is to simply add the new facts to their existing conceptual structures. This will result in the creation of internally inconsistent representations. The other is to distort the new facts to make them consistent with the existing structures. In this case the result will be a misconception.

In an experiment I recently conducted I examined the understanding of the day/night cycle by third grade children before and immediately after they read a two-page text from a leading science series which explained the day night cycle by stating that the earth rotates around its axis. Out of the 23 children investigated, 6 provided axis rotation explanations of the day/night cycle prior to reading the text and 8 after they read it. In other words, only two children changed from a naive or synthetic model of the day/night cycle to a scientific one. Most children were influenced by the text and changed their original models but the changes either produced internally inconsistent representations or resulted in some well defined misconception.

In order to understand counter intuitive information students must be given more than a simple statement of facts. They must be either given a new explanatory structure or be put in a situation to discover it. This new explanatory structure must contain an explanation of the counter-intuitive information. In other words, it must explain to the student how it is possible for the earth to be round when it appears to be flat, or how it is possible for the earth to move when we do not feel this movement. It must also show why the new explanation is a better one than the one the student has already constructed on the basis of everyday experience.

Computer-based learning environments have a distinct advantage over traditional methods to provide this type of instruction. One thing that we can do very well with computers is provide students with additional evidence that helps them extend their observations beyond what is available to them through everyday experience. Using computers we can model highly abstract and unobservable processes. We can show what happens inside a body or inside a cell. We can model the solar system and show how the planets move as well as their relative size and distance. We can go inside the earth or look at the earth from the sun or the moon.

Such simulations help students understand the limitations of their phenomenal experience and realize the need to replace their beliefs with alternative explanations which are more adequate with respect to the existing empirical evidence. In order to achieve this, however, it is important to make very clear the connections between the microworlds and the real world. Presentations of microworlds or simulations which are presented as isolated instances with unclear relations to everyday experience encourage students to create disconnected modules of knowledge for the two situations (what happens in real world is different from what happens in computers) and do not help students restructure their knowledge.

Another way in which computer based learning environments can help foster conceptual change is by creating metaconceptual awareness. Although students are active in constructing hypotheses that attempt to explain their phenomenal experience they seem to be largely unaware of the theoretical nature of their beliefs. They think that their beliefs represent the true state of affairs about the world. This can be inferred from the fact that when students are presented with scientific ideas which are contrary to their beliefs they do not seem to question these beliefs. Rather, they distort the scientific ideas to fit their existing conceptual structures, a process which leads to the formation of synthetic models or misconceptions. It is important to provide students with instructional experiences that will make them aware of the hypothetical nature of their beliefs and of the possibility that there may be alternative explanations than the ones they have constructed.

Again, computer-based learning environments have an advantage over traditional methods to do this. In computer-based learning environments we can provide children with tools that allow them to create their own representations of situations. In this way they can make their representations external, a necessary first step before being able to examine them and change them. A number of existing computer-based learning environments allow students to express and modify their representations. For example, in CSILE [Scardemalia & Bereiter, this volume], students are provided with icons which help them label their notes according to "thinking types." These activities help students make their representations public, allow them to examine them objectively, draw their implications, and receive feedback from others.

Conclusions

I have argued that the process of conceptual change often requires the re-interpretation of entrenched beliefs which students have constructed on the basis of their everyday experience. Computer-based learning environments can help foster conceptual change because they can provide students with tools that allow them to express their representations of situations and thus become aware of their beliefs. In addition, microworlds and simulations have a distinct advantage over traditional methods of instruction to extend students' observations beyond what is presented to them in their phenomenal experience. When students understand that there are empirical phenomena which their beliefs cannot explain they are more likely to question these beliefs and replace them with explanatory frameworks which are closer to ones currently accepted by the scientifically literate adults in our society.

References

1. Carey, S. (1985). *Conceptual change in childhood.* Cambridge, MA: MIT Press.
2. diSessa, A. (1982). Unlearning aristotelian physics: A study of knowledge based learning. *Cognitive Science, 6,* 37-75.
3. diSessa (in press). Toward an epistemology of physics. *Cognition and Instruction.*

4. Driver, R., & Easley, J. (1978). Pupils and paradigms: A review of literature related to concept development in adolescent science students. *Studies in Science Education, 5,* 61-84.
5. Kuhn, T. S. (1970). *The structure of scientific revolution.* Chicago: University of Chicago Press.
6. McCloskey, M. (1983). Naive theories of motion. In D. Gentner & A. L. Stevens (Eds.), *Mental models* (pp. 199-324). Hillsdale, NJ: Erlbaum.
7. Novak, J. D. (1977). An alternative to Piagetian psychology for science and mathematics education. *Science Education, 63,* 83-93.
8. Vosniadou, S. (1989). On the nature of children's naive knowledge. *Proceedings of the 11th Annual Conference of the Cognitive Science Society,* Ann Arbor, Michigan.
9. Vosniadou, S., Archodidou, A., & Kalogiannidou, A. (in preparation). *Mental models of the earth in children from Greece: A cross-cultural study.*
10. Vosniadou, S., & Brewer, W. F. (1987). Theories of knowledge restructuring in development. *Review of Educational Research, 57*(1), 51-67.
11. Vosniadou, S., & Brewer, W. F. (submitted). Mental models of the earth: A study of conceptual change in childhood. *Cognitive Psychology.*
12. Vosniadou, S., & Brewer, W. F. (in preparation). *Elementary school children's explanations of the day/night cycle.*
13. White, B. Y. (1983). Sources of difficulty in understanding Newtonian dynamics. *Cognitive Science, 7,* 41-65.

Computers in a Community of Learners

Joseph C. Campione, Ann L. Brown, and Michael Jay[1]

School of Education, University of California, Berkeley, California 94720 USA

Abstract. In this paper we describe the place of computers in a classroom designed to introduce students to an integrated curriculum in which they read and write in order to acquire and understand scientific content. We outline the ways in which students and teachers in different classroom contexts used the computers, and present some of the early results that have been obtained.

Keywords: learning, metacognition, technology

A major concern in contemporary American education is with students' ability to engage in higher order thinking. Trends gleaned from the National Assessment of Educational Progress (NAEP) suggest that performance on basic skills has improved over the past decade or so, with notably less improvement in higher order thinking. Although the best graduates of U.S. schools are as capable of independent learning or of solving complex problems as any in the world, the dropoff from this most successful group is quite steep. Indeed, it might be concluded that those who leave school highly qualified and empowered to guide their own learning do so despite the educational system, whereas those who are in need of more powerful instruction are unlikely to receive it.

Our own work has been aimed at developing instructional programs designed to improve higher order reasoning skills, with particular attention to the problems of

[1]Preparation of this manuscript and the research reported were supported by grants from the James S. McDonnell Foundation and the Andrew W. Mellon Foundation to the first two authors. Ann Brown also received support from the Evelyn Lois Corey Fund.

underachieving students. Although we have worked with students from first through ninth grade, here we will concentrate primarily on fifth and sixth grade learners. The programs we have devised stem from an analysis of the reasons educational practice has not been as successful as one would like. In this analysis, traditional education features: (1) a *didactic teaching style*, with the teacher teaching and the students functioning as passive recipients of incoming information; (2) a *basic literacies curriculum* (reading, writing, and arithmetic) aimed at instilling basic before higher level skills; (3) a *content curriculum* (history, geography, science) that emphasizes breadth over depth, with a resulting focus on factual information rather than understanding; and (4) a *fragmented curriculum*, in which the enabling literacies are taught as discrete strands separable not only from substantive content courses, but even from each other.

Our attempts to overcome these problems have progressed over the last decade in terms of both the ambitiousness of the programs and the number of these features we have attempted to modify. We began by focusing attention on teaching style and higher order literacy skills, an effort that lead to the design of a successful reading comprehension program, reciprocal teaching [4, 8]. This was followed by an extension of that program wherein students read coherent scientific *content*, thus emphasizing how to learn from reading [3]. More recently we attempted to build an *integrated curriculum* in which students receive practice and instruction on reading, writing, and doing library and experimental research as tools for understanding both scientific content and scientific inquiry [2].

In the course of extending the context of our programs, we saw a number of ways in which computers could be integrated into the learning environment as additional tools for learning; and it is this feature of the research that we emphasize here. Before doing so, however, we provide a brief overview of the overall program to illustrate how we have attempted to address the four limitations of traditional practice we reviewed earlier.

Course content

In typical science classes, students learn a broad array of factual information, but get to spend little time pursuing any topic in depth. As a consequence, they are seldom required, even encouraged, to explain what they are learning, or to understand the mechanisms underlying the phenomena to which they are exposed. They are afforded few opportunities to apply what they have learned previously to what they are encountering presently. In short, they are given neither instruction nor practice in critically evaluating what they are learning; thinking skills are simply not targets of instruction. With this in mind, we decided to examine the effect of exposing grade school children to science in, to borrow Bruner's term, "an intellectually honest way," that is, without oversimplifying the material until it is intellectually vacuous. One cannot think critically in a vacuum, and coherent scientific inquiry provides a fertile ground for promoting literacy skills in general.

Because we wanted students to develop skills of plausible reasoning about probable cause, and because we thought it desirable that we choose a topic that would arouse and sustain their interest, we selected the basic biological concept of interdependence in nature as the basis of the curriculum. To implement the program, we devised three main units, each with five subunits, to support collaborative learning activities: (1) *Changing populations* (extinct, endangered, artificial, assisted, and urbanized); (2) *Food webs* (producing, consuming, recycling, distributing, and energy exchange); and (3) *Survival* (reproductive strategies, defense mechanisms, protection from the elements, ecological niches, and communities).

Roles of students and teachers

Although the topic and themes were provided by us, the students were responsible for doing their own research within those guidelines. Thus, the roles of both the students and teachers were changed as they formed a community of learners jointly responsible for creating their own knowledge about biological systems. The students were involved in designing their own curriculum. They worked in cooperative groups based on the Jigsaw method [1]. Students formed five *research groups*, each assigned responsibility for one of five subtopics of each content area unit. They prepared illustrated books on their research areas. Next the students formed *learning groups* where one member of each group was an expert on one fifth of the material. This expert then used the student-prepared books, along with other materials, to guide seminars on that topic. All members of each learning group were thus experts on part of the content.

An essential part of the classroom is establishing a collaborative and cooperative atmosphere. Students are required to collaborate in their research groups, and in their learning groups when they share their expertise. For example, it is impossible to understand the notion of food webs without understanding all five subunits and the relations among them, so it is essential that each of the research groups provide their piece of the puzzle. Furthermore, in the course of doing research, students are bound to encounter information that would be helpful to other groups, and we encourage them to communicate those findings, electronically or in person. Students are encouraged to provide feedback to each other, including both constructive criticism and suggestions about additional information sources.

As a consequence of developing an atmosphere of student research activity, the teacher's role changed dramatically. Students had considerable responsibility for charting their own learning, and for setting their own agenda, so what role was the teacher to assume? We were aware that unguided discovery can be dangerous; however, we also knew that didactic teaching leads to passive learners. We know that challenging students' assumptions, providing them with counterexamples to their own rules, etc. are good teaching ploys. But how intrusive should the teacher be, when should she guide, when should she teach? In short, how can the teacher foster self-direction and at the same time furnish guidance? This territory, midway between "pure" discovery learning and direct teaching, is largely uncharted. It is this territory our teachers are exploring, continually asking themselves if they should provide guidance or whether this is an opportunity for students to take control of their own learning. Toward this end, our teachers maintain a detailed log of their activities, reflecting each day on the sessions, noting things they saw as positive and areas in which improvements appear to be needed.

The teachers' direct instructional activity was limited to a series of benchmark lessons that bring the whole class together to concentrate on the deeper consistencies and themes, and to challenge developing misconceptions (e.g., children who believe that plants eat soil do experiments with hydroponic gardening). The teachers' indirect instructional activities included modeling the role of scientific inquiry through thought and real experiments. We also believe that teaching (and learning) is more effective on a "need to know" basis. In our environment, teachers certainly do a great deal of teaching, but the ideal is that they do this opportunistically, responding to a student's felt need. For example, when introducing students to the computer environment, we teach them only what they need to get started.

After that, niceties of the system are taught only when students point out that they need some new function, or note that it would be helpful if they could accomplish some other task using the computer. Students thus motivated learn rapidly and are encouraged to pass their new expertise on to others. The trick is to arrange the classroom context in such a way that students regularly recognize the need for additional resources.

An integrated curriculum

One major goal of these efforts was to achieve an integrated curriculum, one in which students read and write in order to learn, and one in which they are constantly asked to evaluate and monitor their state of learning. Students ask their own questions, continually upgrading the level of those questions. They comment on the work of their group members and of the rest of the class, and communicate what they are learning to their peers (see also [9]). The students read to gain expertise in a portion of the curriculum and write to prepare reports that are then used to teach that expertise to others. When they do engage in teaching, students are forced to consider the state of both their own knowledge, and of their written products: when either is found wanting, additional research and revision of their written reports are cued naturally. The consequent cycle is that students read in order to answer research questions, write to summarize and clarify their thoughts, and revise their texts to improve communication.

The Role of Technology

It became clear at a relatively early stage of this program that the addition of state of the art technology would support these learning activities. In the remainder of this paper, we describe the gradual inclusion of technology into the program. We will concentrate on the initial and revised systems we introduced, the different classroom contexts with which we have worked, and on observations about the problems and successes we have encountered. Although "technology" here refers primarily to computers, we also include the use of video materials to enlarge students' access to research ideas. Specifically, students have recourse to a library of videotapes, CD-ROM drives, and videodiscs.

Our starting assumption is that the availability of a supportive computer environment would have a fundamental effect on the learning and teaching that goes on in classrooms [9, 10], although we recognize that there is opposition to this position [5, 6]. Currently, computers are used in grade school primarily for drill and practice or to teach students to program. But the problem with these activities is that most people do *not* use computers to program or for drill and practice -- they use personal computers as just that, personal tools to aid learning. They use word processing and desk top publishing (including graphics), and perhaps spread sheets They set up and access their knowledge files. They use electronic mail. They use computers, not for an hour a week following a set timetable, but when needed. Most claim to be at a loss without them. We believe that this is how grade school children should initially come to view computers -- as invaluable tools for their own sustained learning: building up a portfolio, maintaining and revising their personal and collaborative files, using graphics tools, and networking. Students should be engaged in planning and revising their learning goals, and monitoring and reflecting on their own progress, as they construct a personal and shared knowledge base. Although these were our goals from the outset, we began with a fairly simple computer environment, adding over time more extensive software and visual materials.

The first experiment

Our first classroom application involved three 5th-6th grade classrooms, each with about 30 students, with whom we worked for nine (three per class) one hour periods per week. We used a simple system involving 6 Macintosh IIs and a large capacity fileserver, used primarily for storing data. Software available in these classrooms was initially limited to Apple's HyperCard application. While some users proved able to successfully manipulate HyperCard's simpler programming capabilities, such as creating buttons to link presentation elements together, the only features that were regularly exploited were the word processing and graphics modes. When navigational buttons were used to link materials, moreover, they were used in ways that did not concentrate on the content of the materials; for example, students would use labels like "Jonathan's Card" that would take the users to a card that particular student had created, but such links were not motivated by thematic or structural connections. As Scardamalia and Bereiter [9] found, students' initial searches in the CSILE environment were by author, not topic, first to locate their own work, and then to locate that of a friend or knowledge expert Although HyperCard was useful in integrating word processing and graphics tools within the same application, it did not provide the students with the publishing and editing capabilities they required. Furthermore, for the novice user, HyperCard does not provide database capabilities necessary to handle the large amounts of data generated through group research.

Although the virtues of HyperCard were not readily transparent to the majority of the children, the file folder system of the regular Macintosh interface was iconically powerful. In order to find their notes on, for example, the crested rat, students had to know that this animal was an example of animal defense mechanisms, under which topic they needed to enter the file on mimicry, know that it was necessary to refer to visual mimicry, and only then would they find the animal in question, one that mimics a skunk to defend itself. Forced to organize information into files within files within files, the children regularly practiced the use of hierarchical organization. This practice was reflected in their writing, which progressed from no discernible organizational structure to quite sophisticated use of hierarchies. In the extinction outline shown in Figure 1, the students created a five level

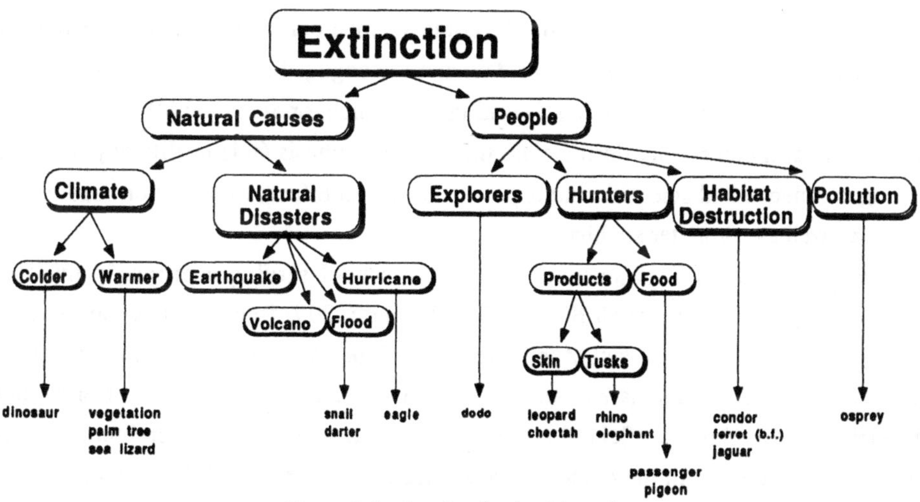

Figure 1. Students' extinction hierarchy.

hierarchy. In order to find the snail darter, a child needs to open first the Natural Causes file, then the Natural Disasters file, then the Flood file contained within it, to find her text on snail darters. These activities, practiced over a long period of time, reinforced this type of organizational structure, and it transformed the organization of their writing samples.

Also lacking in this first environment were networked storage devices. Though the six computers shared a printer, there were no other shared resources. Groups had to be assigned to a single computer or were forced to transport files on error prone and easily misplaced floppy disks. This was not intended, but was dictated by slow delivery of networking software. Nevertheless, the result was reduced utility of the computers, as it became more difficult for students to move information from one machine to another, to share group work, or to access personal files from whichever work space that happened to be available. From this early work, it was clear that the computer environment needed to be substantially improved in order to provide the flexibility and ease of use demanded by group research, writing and publication.

The second experiment

In experiment 2, we are currently involved in two sites. Although the main activities (research, writing, teaching, collaborating, etc.) are the same at the two schools, the sites differ in terms of the clientele being served and the specific setting in which the computers are placed. School A is a private school serving a multi-ethnic but generally middle class population, a significant portion of whom had prior access to computers. Our program here takes the form of a separate 6th-8th grade science class. Students come into the classroom for one 50 minute period per day, and are engaged in both library and "hands on" research. With some exceptions (working over lunch, or after school), students have access to our computers only during class time. The emphasis (science/literacy) is on the learning of science and scientific inquiry, with collateral effects expected on reading and writing, as well as computer sophistication.

In contrast, School B is an inner city public school serving a primarily minority population (approximately 60% Black, 15% Hispanic, 20% immigrant Asian, and 5% Caucasian). Seventy-five percent of the students are labeled "at risk" or are mainstreamed. Virtually none of the students had any significant contact with computers prior to their introduction in this class. At School B the emphasis (literacy/science) is on literacy skills, with those activities set in the context of a science curriculum. At School B, only token hands on science research is taking place. However, the computers are more deeply integrated into the class activities. The sixth graders at School B remain in this classroom for the entire day, and are allowed access to the computers for significant portions of each. In this case, rather than being a portion of a stand alone science class, the computers provide a resource that can be accessed and used for different purposes throughout the school day.

The modified computer environment

Although several extremely powerful computer environments have been developed [see particularly CSILE, 9], we chose (with one exception) to work with commercially produced and stable software available to any school, and capable of running on "low-end," relatively inexpensive hardware [for technical details, see 7]. The general setup consists of one high capacity fileserver connected to a set of individual workstations. We have installed these networks in the two schools. School A features a more elaborate system, but in School B, we attempted to replicate as much of that system as possible with a much less expensive configuration.

In School A we have a total of 13 Macintosh computers networked through a file server.[2] The local area network (LAN) has 9 computers in a main research room, three in the resource room, and one in the regular science/lecture room. Most of the machines are placed on mobile carts that can be shuttled between the available spaces as needed. Also available for student use at School A are two Apple CD-ROM drives (currently used for an on-line encyclopedia) and an interactive video station. Those on the LAN can print documents on a Laser printer.

The environment at School B was designed to be simpler, extremely cost efficient, and more quickly implemented. There are six Macintosh Plus computers and a Laser printer. The school's principal also maintains a Macintosh in her office as part of the LAN. Video equipment and materials are added on an "as needed" basis. This configuration, and the associated software, were designed to: (1) simplify student access to research materials, including books, magazines, videotapes, and videodiscs; (2) support writing, illustrating and revising texts; (3) allow for data storage and management; and (4) enable easy communication within and beyond the classroom.

Three pieces of software are included to facilitate student access to research materials, and to allow them to trade information. One of these, Browser, is the only piece of software we have developed ourselves.

---

[2] Hardware. The network installed at the School A site consists of nine Macintosh II's and four Macintosh SE's networked through a Macintosh IIci fileserver. The Macintosh machines contain at least 20 megabyte internal hard drives (the 20 megabyte drive being more than adequate for local storage needs) and a minimum of 2 megabytes of RAM. Machines connected to the scanner and videodisc player have 5 megabytes RAM. The fileserver uses two hard drives: an internal 80 megabyte drive and an external 650 megabyte drive. Students use the external drive while the internal drive remains dedicated to administrative needs. The fileserver back-ups are created using a 1.3 gigabyte DAT back-up unit. Archival "mirror image" back-ups are made at regular intervals. Also available for student use are an Apple flatbed scanner, two Apple CD-ROM drives, and an interactive video station consisting of a Pioneer 4200 videodisc player and a Macintosh II. Those on the LAN can print documents on an Apple LaserWriter II NT printer.

- *AppleShare* allows students to access files and documents on the classroom network (LAN) and on remote servers used to establish a wide area network (WAN). This increases the amount of information that students can access; they are not limited to the information stored on one computer. It also allows greater flexibility: by having the information stored in one place they can use any computer and not have to worry about storing information on a floppy disk or working at the same computer every day.
- *Browser*, illustrated in Figure 2 (below), is a piece of software written in HyperCard 2.0 that serves several functions. First, it allows filing of documents by topics (e.g., Animal Defense Mechanisms) and themes (e.g., Camouflage, Mimicry). Given a topic and theme, Browser generates a list of titles, and indicates the type (text, magazine, videotape, or videodisc) of each. If a specific selection is stored on the fileserver, the student can call it up directly. It also includes space for a summary of each selection. Browser also supports some interesting student activities. For example, when students choose to publish some of their own work, they must generate key words in the form of a general topic and subtheme. In addition, for each selection, students are required to provide a summary before they can publish their work. Finally, we provide example summaries for some of the basic entries, and students generate and discuss their own summaries after using specific materials. Over time, the library becomes progressively more annotated by the students themselves, with their entries providing us with important data about what they see as significant.
- *GOfer* lets students search local and networked hard drives for documents that contain words or phrases of interest to them. This enables a student to find documents relevant to a topic in which they are interested.

The next three applications support student writing and facilitate storage and management of data. Again all are commercially available.

- *Microsoft Works*, an integrated application that includes a word processor, database, and spreadsheet. This is the primary word processor that the students use to create reports. It allows database, spreadsheet and graphical information to be included along with the text of any report. Students use the database and spreadsheet components to analyze data collected from hands-on experiments.

- *MacDraw* is a drawing package that can be used when precision is required, and is most useful in producing charts or schematics.
- *MacPaint* is a graphics package that allows the children to create new pictures and/or customize artwork that has already been created (either scanned in or pre-packaged clip-art).

The final application is provided to support the electronic network.

- *QuickMail* is a user-friendly program that allows students to send information electronically. In addition to supporting communications between members of the classroom (LAN) and the extended research community (WAN, including the University community), the system can be customized by designing special forms that facilitate specific interactions such as requests to publish, comments, queries from junior scientists to scientists and vice versa, etc.

("Ospreys" is the selected Topic. Theme(s) associated with a selected Topic appear in the Theme window.)

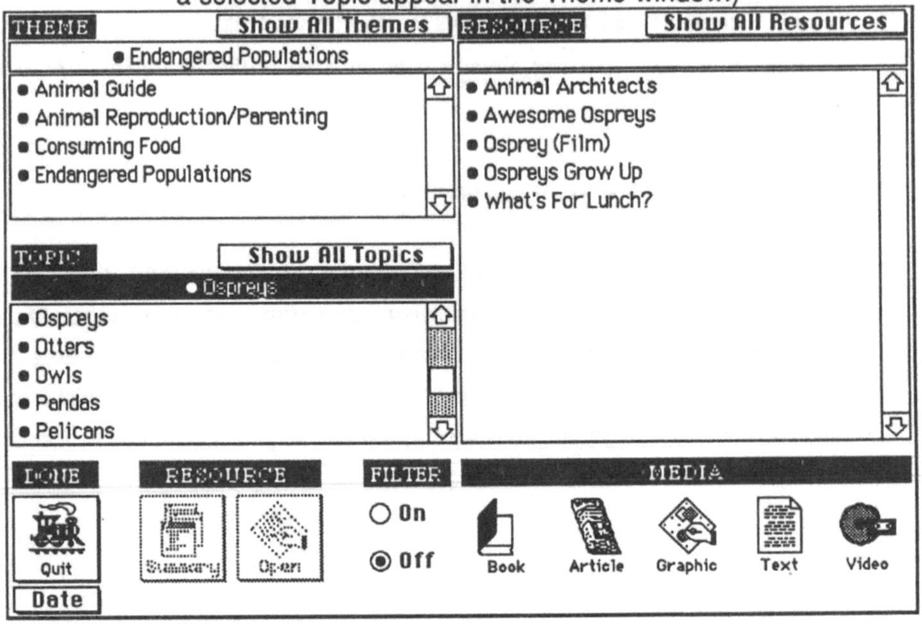

(Clicking on Summary button brings user to this screen.)

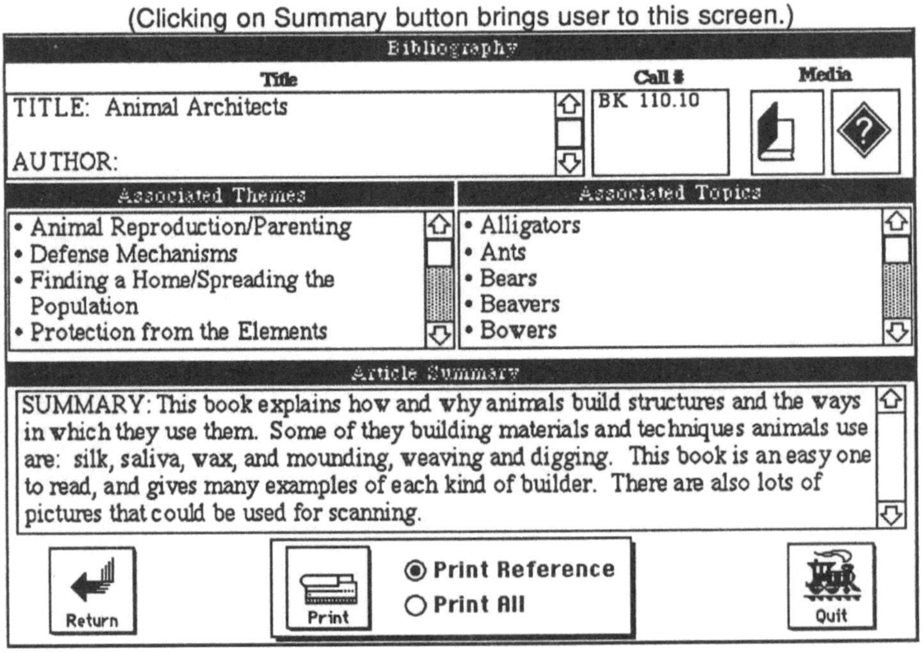

Figure 2. Browser.

Outcomes from the Collaborative Classroom

In this section, we sketch some of the chief results we have obtained thus far, culling information from the two experiments we have described. As the focus in this paper is on the computer component of the classroom, we begin with some data on computer use, including the introduction of the computers into the classrooms and the ways in which they came to be used by the students and teachers. The picture is different in interesting ways in varying classrooms. In some cases we present data, in others we simply outline our approach to evaluating student capabilities.

Computer data

User Software Support. Students were not expected to have any computer experience and training on computer use was provided early in the academic year. Initial student training consisted of short, hands-on lessons intended to orient the students within each of the applications available. Each lesson presented tasks for a pair of students to complete, usually within a forty minute time slot. The tasks were designed to tour the learner through the capabilities and characteristics of the given software. To facilitate this process, we designed a series of child-friendly manuals leading the students through the basic elements of each application; these manuals remained available for students to refer to whenever they wanted (the manuals were in fact quite effective, a main indicator being that new staff joining the project prefer to use them to become familiar with the envionment than any "supporting documentation" provided by the software developers). After completing this initial series of lessons, small groups composed and published a short document. that required the students to use and integrate the applications covered in the training. In accomplishing this task, the students were given help as needed, and we were interested in the kinds of help required for them to solve this problem (see section on computer sophistication)

Subsequent support initially included the daily presence of a computer expert. The majority of the computer support aides' class time was spent assisting students with software questions. Students had the most difficulty with the networking software, taking some time to become comfortable with data transfer and access control. The location of a given file often posed problems, with confusion centering on differences between the internal hard drives and the fileserver volume. As the classroom teachers became increasingly confident of their own abilities and more comfortable with the applications and network, they (and the more advanced students) began to fulfill the role of computer expert, and outside technological support was reduced to two or three class periods weekly.

Early in the project, classroom teachers were offered additional assistance outside of class. The introduction to the environment was comprehensive and dealt largely with the networking and e-mail software rather than the individual productivity applications. The teachers also had difficulty at the outset developing a conceptual map of the locations and functions of the network capabilities. Once this was understood, however, ease of use and transfer to novel management tasks increased rapidly.

Software Usage. In School A, student use of software focused initially on simple word processing. The word processor was treated very simply during the beginning of the first unit, most typically used to record research notes and begin construction of the documents desired. Students commonly worked individually or in pairs and generated text-only documents which they stored in their personal folders (lacking group access privileges) on the server volume. This contrasts sharply with Experiment 1, where the students mastered the graphics applications first, indeed to such an extent that we had to "ration" graphics use in order to encourage text production.

QuickMail was initially used for personal messages and casual conversation using the QuickConference function. The Browser tool was also used to gain access to a small amount of library information, but then neglected until later research was begun.

Eventually, use of the different applications became more integrated and purposeful. Networking capabilities became more fully exploited as the demands of group work forced a move from private to shared information. Appleshare access and management skills learned by students while working both individually and in pairs were observed to transfer to the group setting where work progressed to the compilation and editing of final documents.

As the end of a unit approached, documents were moved to group accessible locations (folders), and the concept of cutting and pasting text between documents became paramount. Some students became responsible for the merging of disparate work, editing as they cut and pasted portions together. The structure of the document often came under question, with paragraph structure and order revision emerging. Outlines generated earlier in the unit and until now neglected, came under closer scrutiny and revision. One or two group members took care of integrating graphics, cleaning up illustrations and attaching captions. Browser too came into full use at this stage, as group discussions uncovered "holes" in information and it became necessary to sort through themes and topics again to fill out the completed group reports.

QuickMail also became more than a social pipeline. As the groups spread out through the different rooms and became more focused on the task at hand, their use of QuickMail for constructive purposes increased as well. As deadlines approached, learners unwilling to leave their computers would notify group members in another room of finished work available on the server for inclusion in final documents.

An example of these expansions of group communication and information sharing occurred in a group nearing their publication deadline. The ethnographic field notes record the incident as follows:

> J. is talking to S. and E. J. and E. decide to go through the checklist. S. is concerned with how everyone's changes will get transferred into one document. A. tells S. that they're all supposed to QuickMail their changes to J, who will put them all into one document. Then A. QuickMails J.: "The corn report is done and the results final report [stored on the server volume] is now titled 'Results final new.'"

We also found that use of the networking capabilities were exploited more fully only as the demands of group work forced transfer from private to shared information. Overall, more sophisticated patterns of use developed with the pressure of production of the final document and the intensive communication demanded by that production.

The uptake of the computer supports was even more rapid at School B, partially because the School B teacher provided an enthusiastic model right from the start and had the luxury of incorporating the computer environment into the entire day. The School B teacher visibly models use of computers on a daily basis. She spends a minimum of one hour a day using the computers in the classroom, communicating through QuickMail or doing miscellaneous writing or planning tasks (ranging from organizing a kickball game to preparing homework assignments). As she put it, "They see me using (the computer) all the time."

The School B teacher's attitude toward both her own, and her students', use of computers is extraordinarily positive, and there is a strong sense in the School B classroom of the teacher enthusiastically *joining* with the students in the use of the computers. The teacher overtly encourages and supports student use of computers. In the classroom, she explicitly talks about the computer as a tool that can make many different activities easier. She explained:

> I really want the kids to see that they're . . . using the computer like they would use a pencil. Only it's more high tech, so it does some things nicely that they would otherwise have to do using a more laborious process. ... It's simply ways to make what you already are going to do easier . . . so you can get about the business of learning what you want to learn. . . . And I really don't want them to think that I couldn't do this unless I had a computer, but because I have a computer, I can do that much more.

The school B teacher also watches for opportunities to scaffold computer use. For example, in one case she noticed a student who sat at the computer for half an hour not getting anything written. With five minutes left before his time was up, she stepped in and asked him what he wanted to say. While she typed for him, he dictated a "really beautiful paragraph" which he could then print and edit, thus enabling intelligent computer use by a student who could not deal with the laborious task of typing in the initial material. As the teacher put it:

> with H I just decided that it was more important for him to have the text on the screen than it was for him to suffer through the hunting and pecking, . . . so I think you need to stand back and watch, is this a good point to step in? So that the kid gets what the final picture is going to be like.

Research issues in computer classrooms

Data collection in these classrooms is an enormous task. Project staff members who are in the classrooms keep daily journals of observations, as do the classroom teachers. There is also an ethnographer who takes detailed notes of classroom interactions (student/adult, student/student, student/machine). Contents of the video and audiotapes are annotated, appropriate sections transcribed, and entered into relational databases. Journal entries and ethnographic notes are also entered into the database. This material is then accessible to staff over the AppleShare network. Searches may be done according to site, observer, type of media, date, coded content areas, or combinations of the above.

From observational data files we can document that 73% of the School A students and 80% of the School B students are enthusiastic users of the computer tools available to them. Electronic mail use is widespread in School B but took longer to establish at School A. A primary difference between the two sites is that the School B teacher modeled the use of QuickMail as soon as it was available to her, and used QuickMail daily in their presence for communication with both the students and university personnel. Students readily began communicating with one another and the University staff, due no doubt in good part to the modeling and encouragement. Indeed, it has become necessary to "ration" QuickMail use to five minutes at the beginning of their official computer time, the remaining 25 minutes being devoted to composition. If assignments in other classes are completed rapidly, additional access to e-mail is permitted. As a result the students use e-mail as a routine part of classroom life.

We can illustrate the uses students have made through the following examples. The first set have to do with content issues. A student in School B addressed this query to a staff member, AG.

> Dear AG:
> I have been looking for this question all over and i can't seem to find it can you answer it why don't primates eat meat i hope you can answer my question and how come there are not predators.

AG passed the message (by QuickMail) on to SK, the science teacher at School A, who replied:

> Well, most primates WILL eat meat if they can catch it, even chimps. It's just that mostly, it is easier for them to eat vegetation and sometimes insects like termites. Their teeth are adapted for crushing plant stuff so, as they can get by on the easier diet of plants and fruit and bugs, they do.
>
> It's a very interesting question, N. I will also ask my friend GG who is a zoologist who has been studying Simangs (a kind of gibbon and a primate) in the wilds of the Borneo jungle. I'll send you another qm when I have a better answer. What made you think of your question?
>
> S (the teacher at School A)

The next example follows a student inquiry concerning the possibility that bamboo is becoming extinct, with consequences for the Pandas who eat it.

> [student from School B]: Is bamboo becoming extinct?
>
> [staff member RA, a non-biologist]: Bamboo is not becoming extinct, but the places where it usually grows are often being cut down and cleared to use as farms and housing. China, where the Pandas that depend on bamboo for food live, has BILLIONS of people. There are many more Chinese than Americans or people from any other country and they all need food and a place to live.

> This puts pressure on the land -- the Chinese have to decide how to balance saving a habitat for pandas and getting food and housing for their people. It's a hard problem to solve!
>
> [graduate student MJ, a biologist]: It is important for this question to understand that bamboo is a kind of plant, not a specific plant. Just like there are a lot of different types of roses, there are many different types of bamboo. These kinds of bamboo are different from one another in many ways: they live in different climates (wetter versus dryer, colder versus warmer, and a lot of other ways). At one time, they all were the same type of bamboo, but some of these bamboo may have grown better in moist soil than others. These bamboo survived better closer to streams and rivers where some of their "relatives" couldn't live so well. Their ability to live close to streams was passed on to their offspring, some of which also were better at living near water. After this group had many generations of offspring, they were probably able to live where the original bamboo couldn't and eventually couldn't have offspring with the original bamboo. This is how new species develop.
>
> So here is the answer to your question (sort of):
>
> - If the climate changes so that (for example) it becomes dryer,what will happen to the new type of wet bamboo?
> - What will happen to the original bamboo?
> - Do you think that all the bamboo types will become extinct?
> - Is there a chance that the wet bamboo type will survive?
> - If humans change the environment by planting new crops will this change how well different types of bamboo will survive?
>
> Hidden in these questions is the answer to your question.
>
> [science teacher at School A, SK]: I don't think so - so many people have it growing in their gardens. But it may be becoming extinct in the wild as people take up its habitat for agriculture. Giant pandas are very heavily protected in China and they live in bamboo forests, as you already know. So I don't think it is becoming extinct in the wild either. I did not have time to look it up for you though, so I could be wrong.

The following exchange took place after the students in School B had read about the unintended effects (on the food chain) of introductng DDT into Borneo to kill mosquitoes causing malaria.

> [student at School B]: I was wondering would a little D.D.T. last longer in the environment that more D.D.T?
>
> [science teacher at School A, SK]: What made you think of the question? I think that MORE DDT will last longer in the environment than if you start out with less. After DDT passes through several animals and goes back into the soil, some of it will

> be broken down by the digestion processes in the animals or by bacterial action. The more there is to start with, the longer it will be until it all gets broken down.
>
> [staff member RA, a non-biologist]: The rate at which individual molecules of DDT break down is independent of the amount that is found in any one area. Exposure to sunlight and harsh soil conditions (high or low pH, etc.) can speed breakdown; however, it remains reasonably unreactive. Concentration can on the other hand have impact on how long a lethal dose of DDT remains in the environment. So the point here is not that the rate of breakdown changes, but rather than exposure to lethal amounts of DDT are affected.

QuickMail has also been used to deal with procedural issues. The following response was sent to a student who had asked about a specialized QuickMail form and how she might be able to use it. (There are a variety of customized QuickMail forms we have developed for different kinds of notes.).

> [staff member SP, a computer expert]: I got that QuickMail form from someone else, so I'm not sure how it got drawn. You can install it on the machine you are working on (if you haven't deleted it yet) by first opening up that letter, then pulling down on the "QuickMail" menu and dragging down to "Install Form . . ." Give it a name, like "Calvin," and then click OK. It should appear under the form menu when you click on the "new" button.
>
> I forgot to mention to you all that you really should send your comments and questions about whether or not fire and the sun are alive to D (a graduate student and biologist). She would love to hear them, and I'm sure that she would have better answers for you than I do!
>
> She has a QuickMail address (provided), so pleas take a few minutes to write her a QuickMail note!
>
> It was a great reciprocal teaching session!

As these examples show, the students in both schools use QuickMail with some regularity and for a variety of purposes. For example, 40% of School B students spontaneously adopted QuickMail as their method of journal writing between student and teacher, 83% have communicated with the University staff about one query or another, and 13% have been in contact with biology experts at Berkeley. Students at School A spontaneously initiated using QuickConference to keep in touch with one another across the three rooms in which group members work. During a School B class discussion about students' views of a project-related activity, the idea was raised of informing the university staff of students' opinions. One of the children spontaneously remarked, " We could use QuickMail for that!" this began mass communication between the school and university, primarily involving students' opinions about the project.

It should be noted that this procedure is just getting under way, and we have been impressed with how readily the students incorporate it into their daily routine once the teacher begins using and modeling different kinds of uses. We are also increasing the network students will have access to by "signing on" contacts at the Lawrence Hall of Science, several museums in the area, and a number of graduate students representing other sciences. We are working toward establishing a larger community of learners/scientists, establishing conditions for continuous dialogue (rather than discrete question/answer episodes), and fostering increased student/student interchange.

Computer Sophistication. To evaluate computer sophistication, students are given a pretest, a guided test after their initial instruction, and a final posttest. In the pretest we seek information about the experiences students have had with computers, and elicit their knowledge of potential uses in different contexts. The posttest includes a parallel form of this instrument. After students have been given their initial instruction on the computers, they are tested on their ability to make combined use of the various software. In this, dynamic assessment, phase, we provide help as necessary to support their activities; and the amount and type of help they require is used to evaluate their capabilities immediately after their introduction to the software packages.

The students are given a general introduction to the task:

> I am going to ask you to make a story on the computer look like something you would see in a book. For example, if we look at this book, there is a lot of writing and mixed in with the writing are pictures. Sometimes there is a sentence under the picture to tell you what it is about. In this activity you will choose a story on the computer. The story you choose will have no pictures. Your job is to add pictures to the story you have chosen. Then I want you to add some words under the picture to describe it. You should try your best to do as much of this for yourself as you can, but I can give you help if you need it.

They then proceed to work on the task in phases, with each phase designed to assess their understanding of various pieces of computer management and use. Hints are given, proceeding from general to specific, to help them as they proceed. For example, the first subtask they work on is Story Selection, where they have to find, choose, and read a story. The hints proceed as follows: (1) Show me how you might find a story document; (2) The story document is inside one of these [indicating icons on screen]: (3) That was a good idea, but not quite right; (4) Double-clicking opens folders and documents; (4) Now open the "Stories" folder; (5) Double-clicking on an icon will open it; and (6) Let me show you ... now you try. Note that the hints become progressively more specific, eventually involving an actual demonstration of how to accomplish the set task, followed by having the student repeat the actions.

The next series of task/assessments taps students' ability to read a large document, involving scrolling in various ways. They then go on to locate and open documents that include pictures they might want to use to illustrate their chosen story. They proceed to select and copy the picture they identify, and paste it into their document. Their next task is to add a subtitle texplaining the picture's content. Finally, they make adjustments to the illustration, for example centering the picture and text.

In each phase, we provide graduated help as illustrated for the Story Selection task. With this procedure, students are scored, not purely in terms of unaided use of the various software packages, but also in terms of how much help they need to produce and distribute their report. Posttests consist of similar assisted, on-line tests of student capabilities. We have piloted and refined the procedure with several groups of students and are currently engaged in scoring the most recent version (administered in School B). We are interested in the nature of the help required across the various packages, as an indication of the relative ease of learning them, as well as the extent to which we can identify, shortly after instruction, the students who are likely to experience problems with the environment.

Outcomes other than computer use

Although the computer assessment is in an infant stage, we do have collateral information on the development of basic literacy skills and on the extent to which students come to understand the scientific content..

Reading. One immediate outcome of the collaborative classroom is the development of a community of learners acquiring and sharing a common knowledge base. The nature of the reading/learning discussions and the writing samples all reflect higher levels of reasoning skills than were apparent in a comprehension-based reading group covering the same material. Over time we saw increasing incidence of comprehension-extending activities in the student teaching dialogues. For example, the conditions of spontaneous use of deep analogy changed dramatically over the course of the program. First, impasse-driven analogy occurs only in the face of breakdowns in comprehension. This is followed by the use of analogy to help resolve annoying inconsistencies. Finally, spontaneous analogy is used in the absence of obvious inconsistencies or misunderstandings, as learners continually monitor, revise, and deepen their understanding of complex causal mechanisms. The same pattern was found for causal explanations, which increased in explanatory precision and coherence over time. Argumentation formats developed, permitting comparisons of different points of view and defensible interpretations. And finally, the nature and importance of spontaneous prediction evolved, with students going beyond predictions of simple outcomes to considering possible worlds and engaging in thought experiments about them.

The results of reading comprehension outcome measures, where students read and answered questions on passages *unrelated to the curriculum*, are shown in Figure 3. The fifth and sixth grade students were assigned to easy (third grade) or hard (sixth grade) passages depending on their standardized reading scores and teacher recommendationsStudents in the collaborative classroom (research) outperformed a comparison group reading the same passages via reciprocal teaching (RTC), a comprehension based reading program [4, 8], even though the RTC groups were given at least twice as much practice in collaborative reading procedures. A read-only control (ROC) failed to show improvement. A partial control group (PCont), consisting of sixth graders who were treated the same as the experimental group for the first semester and exposed to the computer materials, etc. in the second, also failed to achieve the same gains, even though they began as by far the most proficient readers.

Writing. Students' writing performance also improved dramatically, as they came to produce structures of increasing complexity. As the outset, they produced a perfect linear structure, with 22 examples on 22 different cards; but throughout the year, they progressed to hierarchical structures of several levels. Specific examples were now relegated to the fifth or sixth level where they served as supporting details for the higher level organization. A representative example (produced midway through the year) was shown previously in Figure 1. The hierarchy is impressive, but the specific examples confuse extinct and endangered species. Further revision would not be inappropriate! Overall, comparing the microgenetic change of our best students over a year, we estimate that they progressed from producing reports characteristic of poor grade school levels to generating organizational structures more typical of young adults.

Knowledge. The students also showed substantial gains on short answer questions.

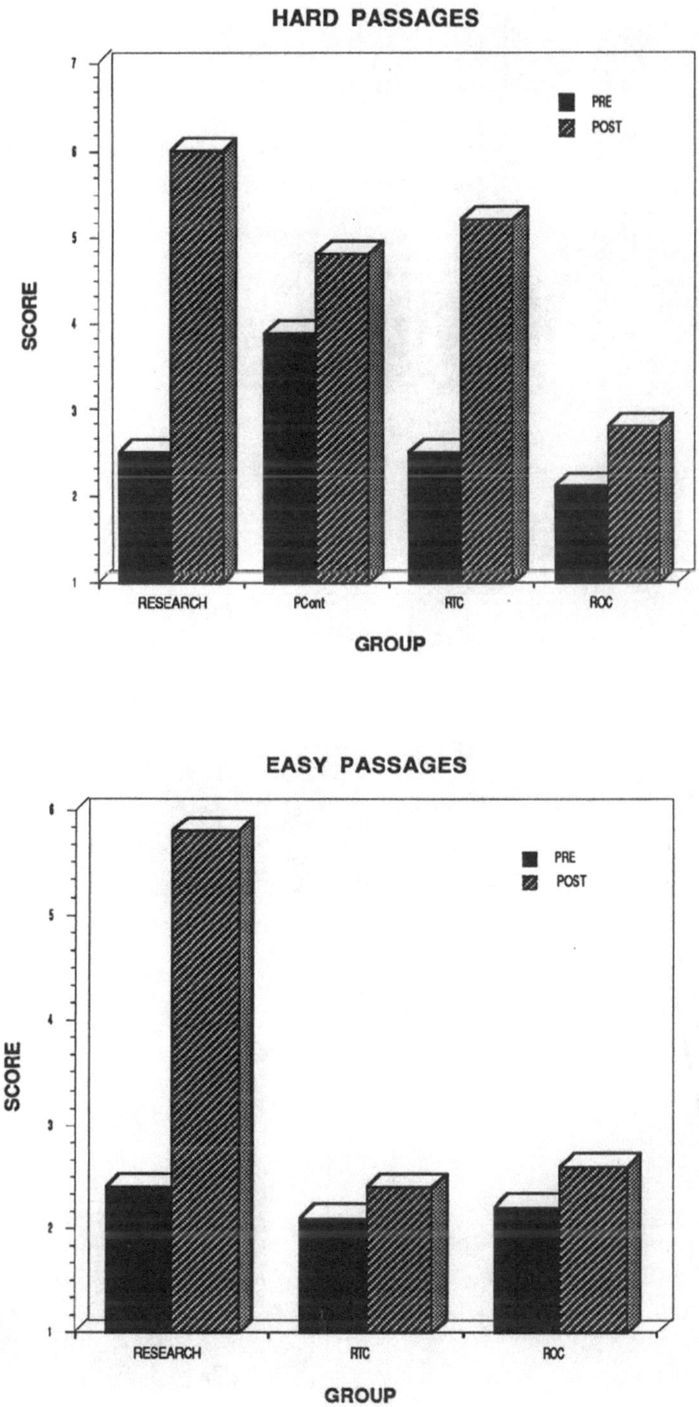

Figure 3. Reading Comprehension.

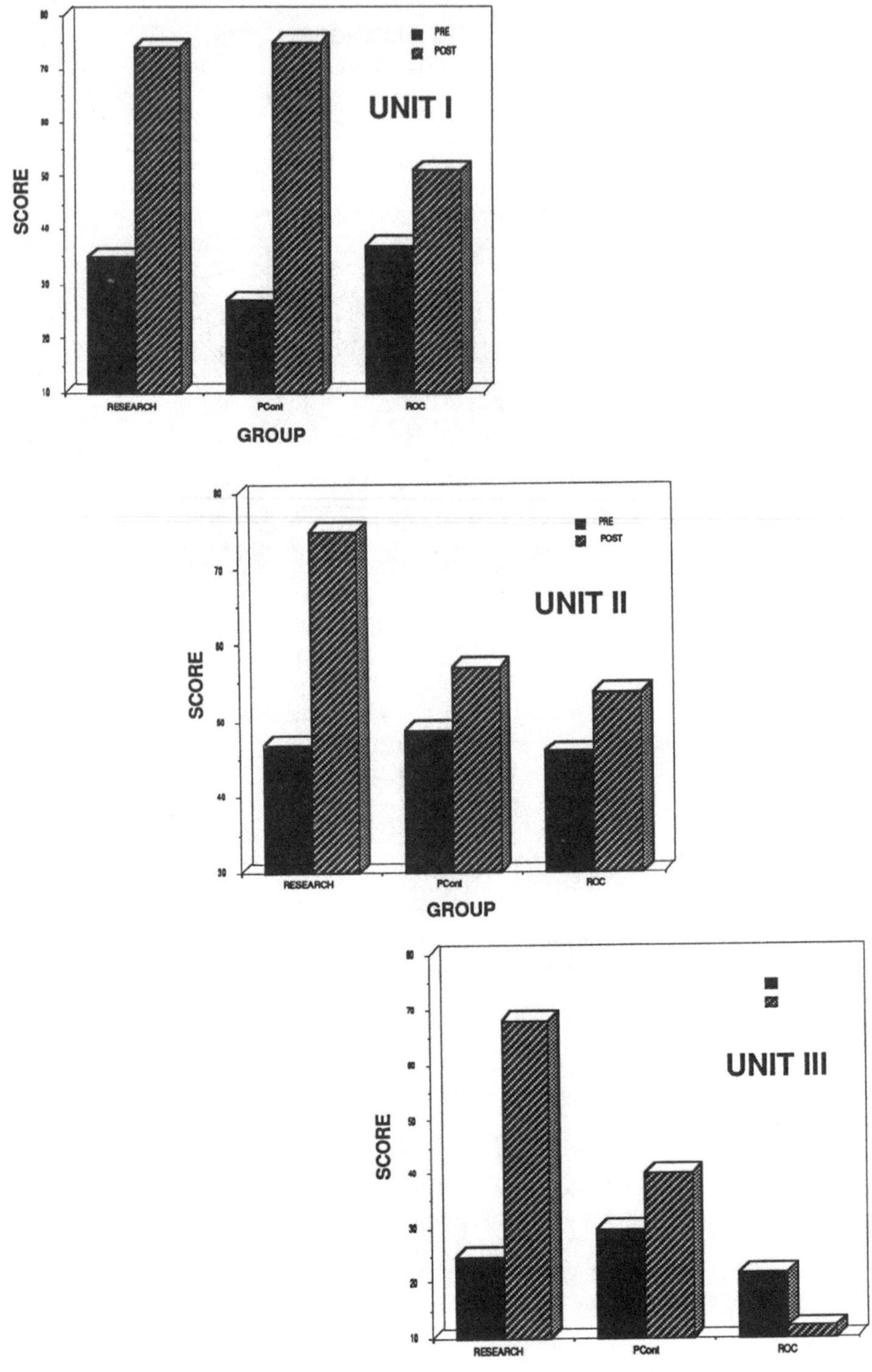

Figure 4. Content knowledge measures.

Pre- and posttest measures were given for Units 1-3. The partial control group (PCont) was treated identically to the research groups in Unit 1, but thereafter the teacher of that classroom (a science teacher) used his own approach to orchestrate Units 2 and 3. During this period, the PCont students had equal access to the materials and computers, spent the me amount of time on the subject, and pursued the same topics albeit under their own teacher's direction. In Unit 1, where the partial control and research groups were treated identically, both groups showed substantial gain, where the read only control did not. In the remaining two units, the research group outperformed the partial control groups even though the partial controls

In tests of application and transfer, the research children also outperformed control groups in their ability to apply biological principles to novel tasks, such as designing an animal to fit a habitat. In general, in terms of both in-depth analyses of on-line processes (dialogues, planning, designing instruction) and outcome measures of basic literacy, the collaborative research classroom was defensible. We are currently extending our assessment battery in this domain by developing an extended clinical interview in which students are probed in depth about their understanding of the major concepts and principles.

Conclusion

Over the course of the two experiments, involving five classrooms in three different schools, we have become confident that the classroom environment accomplishes significant portions of what we aimed for originally -- we would class it as an *existence proof* showing that an integrated curriculum involving basic literacies, computer use, and acquisition of content knowledge can be successfully implemented.

It is clear that students, even those with no prior computer experience, can easily be taught to use the basic elements of a variety of applications, and to use them in an integrated way. With some modeling and encouragement, they also take quickly to using the computers as a tool for communication, and with this, they make it easy to extend the community of learners beyond the school walls. Our students are quite excited about "using" outside consultants to get answers to questions that interest them and that cannot be answered using school resources. It also enables changes in teacher behavior. As one of our teachers pointed out to us, it supports activities like having children generate complex questions. Even if she cannot answer them (she is after all not a science teacher), she also has resources she can call upon. Rather than simply being in a potentially uncomfortable situation where she does not know how to answer a series of student questions, she can call upon her consultants to provide relevant information. This in turn results in her modeling research/learning activities -- seeking help, getting additional readings, etc. -- that we want our students incorporating into their own repertoires.

Finally, beyond computer use, and even beyond the fact that our students come away from the class having learned a great deal about the course content, we have strong evidence that the experience also leads to important improvement in basic reading and writing skills. Although not being "taught" to read or write, they engage in a lot of reading, and they write and revise a series of reports clearly designed to communicate what they are learning to their peers. We see this as an extremely important outcome, as one of the arguments against teaching a number of content courses (especially science) in the early school years is that it takes up time that should be devoted to reading and writing instruction. The results we have obtained thus far indicate that this concern may be quite unrealistic, and that in fact by situating reading and writing in the context of learning and communicating interesting content students may achieve greater gains than they would by being taught reading and writing as separate activities.

References

1. Aronson, E. (1978). *The jigsaw classroom.* Beverly Hills, CA: Sage.
2. Brown, A. L., & Campione, J. C. (1990). Communities of learning and thinking, or A context by any other name. In D. Kuhn (Ed.), *Contributions to Human Development, 21*, 108-125.
3. Brown, A. L., Campione, J. C., Reeve, R., Ferrara, R. A., & Palincsar, A. S. (in press). Interactive learning and individual understanding: The case of reading and mathematics. To appear in L. T. Landsmann (Ed.), *Culture, schooling and psychological development.* Hillsdale, NJ: Erlbaum.
4. Brown, A. L., & Palincsar, A. S. (1989). Guided, cooperative learning and individual knowledge acquisition. In L. B. Resnick (Ed.), *Knowing, learning, and instruction: Essays in honor of Robert Glaser* (pp. 393-451). Hillsdale, NJ: Erlbaum.
5. Cohen, D. (1989). Teaching practice: Plus ça change...In P. Jackson (Ed.), *Contributing to educational change: Perspectives on research and practice* (pp. 27-84). Berkeley, CA: McCutchan. (Also published by the National Center for Research on Teacher Education. Michigan: Michigan State University, 88-3, September 1988.)
6. Cuban, L. (1986). *Teachers and machines: The classroom of technology since 1920.* New York: Teachers College Press.
7. Jay, M., Paulukonis, S., Buhs, E., Brown, A. L., & Campione, J. C. (1991, April). *Scientific Literacy.* Paper presented at the Annual Meeting of the AERA, Chicago, IL.
8. Palincsar, A. S., & Brown, A. L. (1984). Reciprocal teaching of comprehension-fostering and monitoring activities. *Cognition and Instruction, 1*(2), 117-175.
9. Scardamalia, M., & Bereiter, C. (1991). Higher levels of agency for children in knowledge building: A challenge for the design of new knowledge media. *The Journal of the Learning Sciences, 1*, 37-68.
10. Schank, R. C., & Jona, M. Y. (1991). Empowering the student: New perspectives on the design of teaching systems. *The Journal of the Learning Sciences, 1*, 7-36.

Part II. Stimulating Higher-Order Thinking and Problem Solving

Introduction to Part II

Erik De Corte and Lieven Verschaffel
Center for Instructional Psychology and Technology (CIP&T), University of Leuven, Belgium

Transfer has been and still is a major issue in research on learning and instruction, as is shown by the substantial number of books and articles that have recently been published [see e.g., 1, 4, 5, 10, 11, 12]. In some respect this is not surprising as the issue is of theoretical and practical importance not only in the context of schooling but also in vocational and industrial training [9].

In the eighties the transfer problem has been raised expressly with respect to the acquisition of knowledge and skill in software environments, especially in computer programming [8]. Indeed, a major argument put forward to justify the teaching of programming in general education derives from the cognitive-effects hypothesis [6]. According to this hypothesis certain kinds of experience with computer programs, such as learning computer programming, would be particularly conducive to the development of important thinking and problem-solving skills, such as planning, debugging, and decomposing a problem into its constituent parts. Moreover, the acquired skills could be transferred to other learning and problem situations both within the same content domain and in other domains.

A series of investigations carried out in the first half of the past decade to test the cognitive-effects hypothesis reported rather disappointing outcomes [6]. But a number of more recent studies have yielded positive results with respect to the acquisition and transfer of thinking skills. In the latter investigations several shortcomings of the earlier research were taken more or less explicitly into account. A major feature was an effort to embed learning to program in rich and powerful teaching-learning environments [7].

A central focus of this section relates to the discovery, through empirical investigations, of instructional conditions, methods, and environments that can facilitate the acquisition and transfer of higher-order thinking and problem-solving skills through programming. This is the case for the chapters by Mayer, by De Corte, Verschaffel, and Schrooten, and by Clements and Nastasi. While they all report significant positive effects from programming instruction, their investigations attend to different aspects of the instructional treatment.

In a review of his own recent work, Mayer discusses three methods for promoting transfer of problem-solving skills which intend to help students overcome so-called obstacles in learning, namely

(1) to provide conceptual models of the computer, (2) to pretrain students in relevant prerequisite skills such as comprehending a list of instructions, and (3) to encourage them to develop problem-solving strategies of experts such as decomposing a problem into separate parts. The three methods were found to be successful in promoting what Mayer calls specific transfer of general skills, namely transfer from learning to program to solving other programming problems that were not explicitly taught.

While Mayer's studies clearly aimed at near transfer only, the goal of the research of De Corte et al. was more ambitious, aiming at near transfer as well as farther transfer to non-programming problem situations. They investigated the effectiveness of a Logo-based instructional environment for the acquisition and transfer of four higher-order thinking skills in sixth-grade children. They adapted the effective teaching strategies of cognitive apprenticeship as described by Collins, Brown and Newman [3]. De Corte et al. report that students gained substantial skill in solving Logo problems and also significant improvement on several far transfer tasks.

The studies of Mayer and of De Corte et al. are helpful in identifying aspects of instructional environments that can promote the acquisition of transferable higher-order thinking skills. Yet, they do not unravel the process characteristics of such environments that foster skill learning and development. The investigation of Clements and Nastasi represents an example of such a detailed analysis focusing on the role of social processes as mediators of cognitive effects for Logo computer programming. These researchers carried out detailed analyses of students' social behavior in Logo-based learning environments. They compared social interactions in Logo and other educational settings. They conclude that the enhancement of higher-order skills through Logo is mediated by the engagement in specific conflict-resolution strategies that are engendered by unique characteristics of Logo.

Programming environments are not the only type of computer-based learning environments from which the acquisition of generalizable higher-order thinking skills might be expected. Similar claims have been raised with respect to software designed to assist the processes of reading comprehension and text writing. In the first part of his chapter Salomon reviews three studies that have been carried out to address this question. Together, these three investigations support the argument that such programs can leave a generalizable cognitive residue in the form of improved higher-order thinking and self-monitoring skills which facilitate performance on later reading and writing tasks away from the computer programs. An important difference, though, between Salomon's work and the three studies mentioned before, is that Salomon's (meta-)cognitive guidance was built into the software. A major issue for future research concerns which parts of the instructional intervention should be integrated into the computer environment, and which parts should remain in the hands of the teacher.

Salomon also takes a rather explicit position on certain aspects of the research methodology concerning the cognitive effects of computer-based learning environments. He criticizes most of the related research (including his own) that employs controlled classroom experiments. In such investigations the computer is rather an add-on to the existing teaching-learning situation, and does not become part of the changed classroom environment. An example of such a more wholistic methodology - called the systemic approach - involving the modification of the entire classroom context and the organic integration of the computer, is briefly described. As Salomon states, the latter approach complements the analytic one.

This kind of more general reflection concerning the available research on transfer of computer-based learning experiences from a conceptual as well as from a methodological point of view is continued in Clark's chapter. He argues that the available research does not provide convincing evidence for the positive impact of computer instruction and computer programming on domain-general transfer. Indeed, a detailed analysis of the literature suggests that in those studies in which positive transfer was obtained, neither the computer nor the skills in programming constituted the critical components of the instructional treatments that produced these outcomes. Yet this does not lead him to a plea against further transfer studies. He rather argues that future research should be based on a new theoretical framework that focuses more on the cognitive processes involved in positive transfer. In this respect he stresses the important role of metacognitive activities such as selecting critical aspects of information and connecting knowledge. The importance of metacognitive processes for transfer is also emphasized by Campione and Brown [2].

Both Salomon's and Clark's point of view have clear methodological consequences for transfer research in the sense that it requires (1) attempts to integrate the use of computer tools into regular classroom activities or curricula and (2) the more direct study of the cognitive processes involved in domain-general transfer.

While in the former chapters the focus is on the development of higher-order thinking and problem-solving skills through computer-based learning environments, the last chapter by Vergnaud addresses the conceptual aspects of expertise and their relationships with thinking and problem-solving skills. He suggests how computer-based learning environments make available new possibilities of establishing better relations between these two aspects of skilled performance by giving examples from the domain of mathematics.

References

1. Butterfield, E. C., & Nelson, G. D. (1989). Theory and practice of teaching for transfer. *Educational Technology Research and Develpoment, 37*(3), 5-38.

2. Campione, J. C., & Brown, A. L. (1990). Guided learning and transfer: Implications for approaches to assessment. In N. Fredericksen, R. Glaser, A. Lesgold, & M. Shafto (Eds.), *Diagnostic monitoring of skill and knowledge acquisition* (pp. 141-172). Hillsdale, NJ: Erlbaum.
3. Collins, A., Brown, J. S., & Newman, S. E. (1989). Cognitive apprenticeship: Teaching the craft of reading, writing and matematics. In L. B. Resnick (Ed.), *Knowing, learning, and instruction. Essays in honor of Robert Glaser* (pp. 453-494). Hillsdale, NJ: Erlbaum.
4. Cormier, S. M., & Hagman, J. D. (Eds.) (1987). *Transfer of learning. Contemporary research and applications* (Educational Technology Series). San Diego: Academic Press.
5. De Corte, E. (Ed.) (1987). Acquisition and transfer of knowledge and cognitive skills. *International Journal of Educational Research, 11*, 601-712.
6. De Corte, E., & Verschaffel, L. (1986). Effects of computer experience on children's thinking skills. *Journal of Structural Learning, 9*, 161-174.
7. De Corte, E., Verschaffel, L., & Schrooten, H. (1990, March). *Cognitive effects of computer-oriented learning*. Paper presented at the Seventh International Conference on Technology and Education, Brussels, Belgium.
8. Mayer, R. E. (Ed.) (1988). *Teaching and learning computer programming. Multiple research perspectives*. Hillsdale, NJ: Erlbaum.
9. McKeachie, W. J. (1987). The new look in instructional psychology: Teaching strategies for learning and thinking. In E. De Corte, H. Lodewijks, R. Parmentier, & P. Span (Eds.), *Learning and instruction. European research in an international context* (Vol. 1, pp. 443-356). Oxford/Leuven: Pergamon Press/Leuven University Press.
10. Perkins, D. N., & Salomon, G. (1989). Are cognitive skills context-bound? *Educational Researcher, 18* (1), 16-25.
11. Prawat, R. S. (1989). Promoting access to knowledge, strategy, and disposition in students: A research synthesis. *Review of Educational Research, 59*, 1-41.
12. Singley, M. K., & Anderson, J. R. (1989). *The transfer of cognitive skills*. Cambridge, MA: Harvard University Press.

Teaching for Transfer of Problem-Solving Skills to Computer Programming

Richard E. Mayer

Department of Psychology, University of California, Santa Barbara, CA 93106 USA

Abstract: Computer programming is a problem-solving activity that requires transfer of training to programming problems that were not taught. Three effective methods for promoting transfer are to provide conceptual models of the computer during instruction, to pretrain the users in relevant prerequisite skills such as comprehending a list of directions, and to encourage users to develop the problem solving strategies of experts such as breaking a problem into separate parts. These instructional methods improved problem-solving performance substantially over conventional instruction. The results are consistent with Singley and Anderson's [27] theory of specific transfer of general skills.

Keywords: computer programming, problem solving, transfer.

Three Tries at Transfer

Computer programming tasks such as generating, comprehending, and modifying/debugging computer programs are problem solving activities [26]. The goal of most instruction in programming is problem solving transfer--that is, users are expected to be able to apply what they have learned to solving programming problems that they have not been taught [25, 27]. In practice, however, instruction generally emphasizes the syntax of a language, without providing the learner with opportunities to learn in ways that promote transfer [16, 18, 19, 22].

Transfer theorists [25] have distinguished between *high-road transfer*--effortful transfer than involves conscious attention--and *low-road transfer*--automatic transfer of highly practiced skills. Emphasizing low-road transfer, early work in the associationist tradition by Thorndike [28] supported the theoretical claim that transfer involved application of identical behaviors from the original task to the target task--that is, theories of *specific transfer*. Emphasizing high-road transfer, early work in the Gestalt tradition by Wertheimer [29] supported the theoretical claim that creative problem solving involved creation of novel behaviors through conscious restructuring of the problem--that is, theories of *general transfer*. During the ensuing years, most transfer studies have demonstrated strong limits on our ability to teach students to be creative problem solvers, resulting in an illusive search for teachable aspects of general transfer [17, 28]. A third approach, growing out of cognitive science, is to view high-road transfer as involving the *specific transfer of general skills* that are required in both the original and target tasks [27]. This is the theoretical approach taken in this chapter; we focus on teaching general problem-solving strategies, such as how to represent a problem or how to plan a solution, that are specifically relevant to writing and interpreting computer programs.

Consider the following example. A computer-naive student takes a two-hour mastery course in BASIC, learning 10 individual statements such as LET, INPUT, PRINT, GOTO, and IF-THEN. The student performs well on a retention test covering the material that was taught. For example, a retention question could be:

> Is the following a legal BASIC statement?
> If not, correct it.
> 10 LET X = 10

In contrast, the student performs poorly on a test of problem-solving transfer, such as having to interpret what a given program does. For example, a transfer question could be:

What does this program accomplish?

```
10    INPUT X
20    LET S = X*X
30    PRINT S
40    END
```

Good performance on retention and poor performance on transfer is a common outcome of conventional instruction in programming languages for novice users [12, 14, 16, 18].

How can we promote problem-solving transfer ability in new computer users? How can we create environments in which novice programmers can learn to become successful problem solvers? In short, what are the conditions for meaningful learning of computer programming? These are main questions addressed in this chapter. Although the history of research on teaching for transfer is disappointing [13, 15, 17], recent research in cognitive psychology allows us to analyze the issue into two sub-questions. First, what does a user need to know in order to be able to succeed on problem-solving transfer tasks? Second, which kinds of instruction promote the acquisition of this knowledge?

Addressing the first question, Mayer [20] has identified several differences in the knowledge possessed by expert and novice programmers.

Syntactic knowledge: Experts have stronger knowledge of language syntax, as measured by accurate recognition of correct versus incorrect code.

Semantic knowledge: Experts have a more integrated conceptual model of the computer system, as measured by verbal descriptions of what goes on inside the computer.

Schematic knowledge: Experts have more categories for types of routines, as measured by accurate sorting of routines by type.

Strategic knowledge: Experts have high-level planning strategies, as measured by the use of modularity in program design.

In this chapter, we focus on techniques for improving the semantic and strategic knowledge of novices.

Addressing the second question, Mayer [13, 15, 17] has identified several characteristics of successful programs for the teaching of thinking skills.

Component problem-solving skills: Successful programs tend to focus on teaching small component skills rather than a single general ability. Here, we focus on skills for understanding programming statements and for planning programs.

Specific problem-solving skills: Successful programs focus on skills that are specifically related to the to-be-tested problem-solving tasks (i.e., near transfer) rather than skills that are unrelated to the problems on the transfer test (i.e., far transfer). In the present studies, students learn skills outside of the domain of programming that are directly related to skills required within the domain of programming, such as how to follow or give directions in English.

Process of problem solving: Successful programs emphasize the process of problem solving rather than the final product. Although some have argued that a pure discovery approach is the best way to teach problem-solving processes, research strongly suggests that more guidance is needed [19]. In the present studies, students receive guided practice in how to interpret and produce programs.

In this chapter, we focus on guided discovery methods for helping students acquire directly relevant component skills for programming.

This chapter examines three attempts to help novices learn computer programming in ways that enable them to transfer what they have learned to new situations. In each case the learners are programming-naive (that is, they have never programmed a computer) and are mathematically unsophisticated (that is, they lack extensive knowledge of school mathematics). The first section focuses on the role of users' mental models of the computer: novices often lack appropriate conceptions of the internal operation of a computing system, so instruction in a conceptual model of the computer offers potential for promoting transfer. The second section focuses on the prerequisite cognitive skills that the user brings: novices often lack useful concepts of a procedure, so pretraining in prerequisite cognitive skills is a second possible way to promote transfer. The third section focuses on expert design strategies such as breaking a problem into parts: again, novices often lack design strategies, so direct instruction in strategies is a third possible road to transfer.

Promoting Transfer by Providing Conceptual Models

Obstacles to learning: Users' preconceptions

Learners enter the learning situation with preconceptions concerning the nature of programming. These preconceptions often conflict with concepts that actually underlie the to-be-learned language. For example, Fay & Mayer [7, 8, 24] identified two common preconceptions of children in Logo classrooms. First, 33% of the fourth-grade children had an egocentric view of turn commands. For example, the child assumes that the command RIGHT 90 means that the turtle will face the right side of the screen and that LEFT 90 means that it will face the left side of the screen. Instead, the appropriate conception is a turtle-centric view in which turns are made relative to the turtle's current orientation. For example, if the turtle is facing 90 degrees to the right, then RIGHT 90 means that it will face the bottom of the screen (i.e., 180 degrees) and LEFT 90 means it will face the top of the screen (i.e., 0 degrees). A second common misconception held by 57% of the fourth-grade students is to assume that turn commands mean to turn and move, similar to the idea that "turn right" often means "turn right and keep on going" in everyday navigational directions. For example, a child may assume that RIGHT 90 means "turn right and move 90 steps. " Instead, the correct conception is that turn commands tell the turtle to turn but not to move in a line so that RIGHT 90 simply means to rotate 90 degrees to the right.

Similarly, Bayman and Mayer [1, 2, 3] identified several damaging misconceptions displayed by beginning programmers who successfully completed a two-hour mastery course involving elementary BASIC programming. First, the majority of users did not possess the concept of addressable memory locations; for example, 43% said that LET A = B + 1 meant to write the equation "A = B + 1" into memory. Second, the majority of users did not possess the concept of how to enter data into memory; for example, 30% said that INPUT A meant to write "A" in the computer's memory. Third, a majority of users did not understand how to output data from the computer's memory; for example, 33% said that PRINT C means to print the letter "C" on the screen.

Mayer [13] has analyzed each BASIC into a list of *transactions*, with each transaction involving an *action* performed on an *object* in a *location* within the computer system. For example, the statement LET A = B + 1 consists of the following transactions:

Find the number in memory space B.
Put that number in the scratch pad.
Find the number indicated to the right of equals.
Put that number in the scratch pad.
Combine the numbers in the scratch pad.
Erase the number in memory space A.
Put the result in memory space A.
Move pointer to the next statement in the program list.

In elementary BASIC the most common actions are move, find, put, erase, decide, combine, and allow; the most common objects are numbers, program lines, and pointers; and the most common locations are memory address, input stack, output screen, program list, scratch pad, keyboard, and file holder. Research on users misconceptions reveals that users tend to lack a useful mental model of the locations in the computer [1, 2, 3].

Aids for teaching: Presentation of conceptual models

How can we help novices to acquire useful mental models of computer systems? Over the course of a long series of experiments conducted through our laboratory, we have found consistent evidence that providing a conceptual model prior to or during instruction tends to result in superior problem-solving transfer [9, 10, 11, 21, 22]. The models provide a semantic basis for interpreting what each programming statement means in terms of actions performed on objects in locations within the computing system. For example, Bayman and Mayer [3] report a study in which 95 low-ability, programming-naive college students spent approximately three hours learning about BASIC computer programming statements from one of two manuals: a standard manual that emphasized each statement's syntactic features or a conceptual manual that contained additional information emphasizing the underlying semantics from a manual. Both manuals presented the syntax of elementary BASIC statements such as PRINT, LET, GOTO, IF-THEN, IF-GOTO, INPUT, READ and DATA; however, only the conceptual manual presented the semantics of these statements. Figure 1 shows examples of sections from the two manuals.

What effect does conceptual model training have on learning and creative problem solving? First, the students given conceptual manuals performed 76% better than those given standard manuals on tests of semantic knowledge, such as stating the actions that occur inside the computer for statements such as LET A = B + 1. This result validates the idea that it is possible to provide direct instruction in underlying semantic knowledge; the conceptual manual included this information whereas the standard manual did not. Second, conceptual instruction resulted in 20% better performance than standard instruction on tests of programming problem solving transfer such as describing what is accomplished by the program:

```
10    INPUT A,B,C
20    LET D = A+B+C
30    LET D = D/3
40    PRINT D
50    END
```

A particularly interesting aspect of this study is that students who performed well tests of semantic knowledge also tended to perform well on tests of programming problem-solving transfer (r = .56). These results are consistent with the idea that semantic knowledge (i.e., conceptual models) can be directly taught and that increases in a user's semantic knowledge allow the user to perform better on tests of problem-solving transfer.

Excerpt From Standard Manual

In the memory of a computer there is a special area that is set aside to store values defined by the user. This area can be thought of as a group of memory spaces. Each memory space can contain only one value at a time. In most computer programming languages, including BASIC, the name given to a memory space is a variable. The value of a variable (or a memory space) changes each time it is assigned a new value by the user. A value is assigned to a variable by using the LET statement. For practical reasons, all the variable names used throughout this minicourse will be letters of the alphabet.

Now assume you type in:

```
LET A = 1
```

and press the RETURN key. The value, 1, will be stored in variable A.

Excerpt From Conceptual Manual

In the memory of a computer there is a special area that is set aside to store values defined by the user. This area can be thought of as a group of memory spaces. Each memory space can contain only one value at a time. In most computer programming languages, including BASIC, the name given to a memory space is a variable. The value of a variable (or a memory space) changes each time it is assigned a new value by the user. A value is assigned to a variable by using the LET statement. For practical reasons, all the variable names used throughout this minicourse will be letters of the alphabet.

Now assume you type in:

```
LET A = 1
```

and press the RETURN key.

The steps the computer carries out are:
1. FIND the number stored in memory space A.
2. ERASE the number stored in memory space A
3. WRITE the number 1 in memory space A.
4. WAIT FOR the next statement to be entered from the keyboard.

Before execution, the parts will look like this:

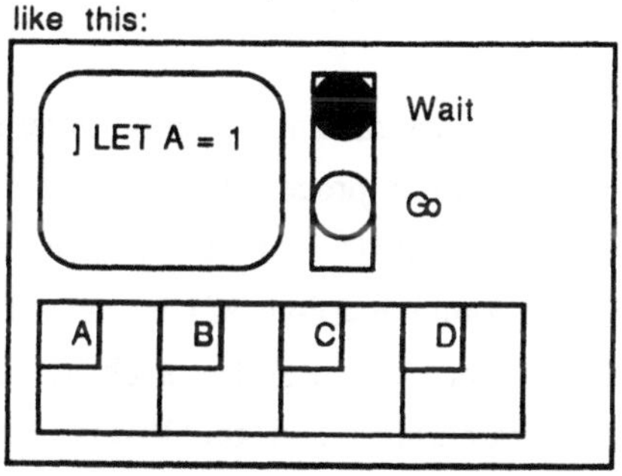

After execution, the parts will look like this:

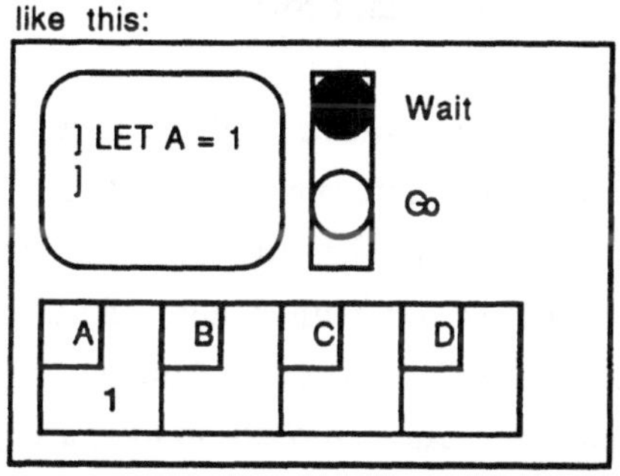

Figure 1: Conceptual Model Training.

Promoting Transfer by Pretraining Prerequisite Skills

Obstacles to learning: Users' prerequisite skills

Another major obstacle to learning a programming language is that the user may lack prerequisite cognitive skills, such as the concept of a program. Many programming-naive students enter the programming classroom without skills in how to generate or comprehend a procedure specification--that is, a list of actions that accomplishes some goal. For example, students average an error rate of approximately 33% on a procedure comprehension test containing items such as the following [4, 5]:

1. Put 5 in Box A.
2. Put 4 in Box B.
3. Add the number in Box A and the number in Box B, and put the result in Box C.
4. Add the number in Box A and the number in Box C, and put the result in Box A.
5. Write down the numbers from Box A, B, and C.

What is the output of this program?

a. 5, 4, 9
b. 14, 4, 9
c. 14, 9, 9
d. 9, 4, 9
e. none of the above

There is some evidence that skill in comprehension of natural language procedures is related to problem-solving performance in BASIC [23]. For example, procedure comprehension skill increases substantially (i.e., by 18%) as a student learns BASIC, and students who enter the programming classroom with higher levels of procedure comprehension skill are more successful in learning BASIC ($r = .44$). Apparently, lack of prerequisite skills, such as procedure comprehension, creates an obstacle to learning a programming language. If this is true, then pretraining in prerequisite skills should increase the ease of learning a programming language.

Aids for teaching: Pretraining in prerequisite skills

Dyck and Mayer [4, 5] have developed a computerized instructional procedure for helping 99 computer-naive college students acquire the cognitive skill of procedure comprehension--either with pretraining or no pretraining. In the pretrained treatment, the learner is seated in front of a computer terminal which presents a series of procedures stated in English; for each procedure the learner predicts the correct output by pressing a button and then receives feedback. After four sessions involving natural language procedures, the student is transferred to corresponding procedures stated in BASIC. In the no pretraining treatment all eight sessions involve BASIC procedures. Examples of the training in comprehending English and BASIC procedures are shown in Figure 2.

Example of Procedure Comprehension Problem in English

```
1. ASK THE OPERATOR FOR A NUMBER.
   PUT THAT NUMBER INTO BOX-B.
2. ASK THE OPERATOR FOR A NUMBER.
   PUT THAT NUMBER INTO BOX-R.
3. ADD THE NUMBERS FROM BOX-B AND BOX-R.
   PUT THE RESULT INTO BOX-K.
4. WRITE DOWN THE NUMBER FROM BOX-K.
5. THIS PROGRAM STOPS HERE

IF THE INPUT LIST IS 5 9 7 12
WHAT IS THE OUTPUT OF THIS PROGRAM?
(A) 5  (B) 9  (C) 5 9 9 9 9 9  (D) 14
```

```
1. ASK THE OPERATOR FOR A NUMBER.
   PUT THAT NUMBER INTO BOX-B.
2. ASK THE OPERATOR FOR A NUMBER.
   PUT THAT NUMBER INTO BOX-R.
3. ADD THE NUMBERS FROM BOX-B AND BOX-R.
   PUT THE RESULT INTO BOX-K.
4. WRITE DOWN THE NUMBER FROM BOX-K.
5. THIS PROGRAM STOPS HERE

IF THE INPUT LIST IS 5 9 7 12
THE OUTPUT OF THIS PROGRAM IS 14
```

Example of Procedure Comprehension Problem in BASIC

```
10 INPUT B
20 INPUT R
30 LET K = B + R
40 PRINT K
50 END

IF THE INPUT LIST IS 5 9 7 12
WHAT IS THE OUTPUT OF THIS PROGRAM?
(A) 5  (B) 9  (C) 5 9 9 9 9 9  (D) 14
```

```
10 INPUT B
20 INPUT R
30 LET K = B + R
40 PRINT K
50 END

IF THE INPUT LIST IS 5 9 7 12
THE OUTPUT OF THIS PROGRAM is 14
```

Figure 2: Procedure Comprehension Training.

What are the consequences of pretraining in natural language procedure comprehension? First, over the course of predicting the outputs of English procedures, students eliminated more than 75% of their errors. This result validates that the training was effective in helping students acquire procedure comprehension skill. Second, knowledge of procedure comprehension with English programs transferred strongly to comprehension of BASIC programs. The pretrained group made half as many errors in learning BASIC procedures as a comparison group that did not receive pretraining with English procedures. Pretrained learners also performed better on tests of programming problem-solving transfer even though they received half as much exposure to BASIC as a comparison group that did not receive pretraining.

Promoting Transfer by Modeling Expert Strategies

Obstacles to learning: Users' solution strategies

Novices and experts differ with respect to the strategies they use for solving programming problems [20]. Two particularly important expert strategies are *modularity*--breaking a task down into separate parts--and *reusability*--creating instructions that are repeated in a program. For example, suppose that the assignment is to generate a Logo program that will draw a house as shown in Figure 3. A modularity strategy would be to break the house-drawing task into smaller tasks such as drawing the frame, the door, and the roof. A reusability strategy would be to create a sequence of instructions to make a square that could be used both for the frame and for the door.

Although modularity and reusability strategies are commonly used by experts, novices who are in the early stages of learning Logo almost never use these strategies when they create a program; instead, novices produce a program as a single sequential unit without calls to subroutines [6].

Aids for teaching: Modeling of expert solution strategies

How can we help students to acquire expert solution strategies such as modularity and reusability? Fay [6] compared two methods for teaching computer-naive college students how to write Logo graphics programs--a strategy-trained group and a no-strategy-trained group. Twenty no-strategy-trained students read a 10-page Logo manual describing the graphics commands and participated in four sessions in which they wrote programs to generate experiment-provided pictures. Twenty strategy-trained students did the same but began by reading a six-page manual that described and exemplified the modularity and reusability strategies using natural language, were asked to apply the strategies in carrying out programming assignments, and received feedback concerning modularity and reusability on the first two assignments. For example, Figure 3 shows a portion of the instructional booklet that describes and exemplifies the modularity strategy.

Excerpt from Expert Strategy Manual

Modularization

One way to simplify a list of instructions is to break the list down into smaller, more manageable units. Each unit would consist of a shorter set of instructions which would be easier to follow.

First, you have to decide how to break a list into parts. With a drawing, one way to divide the list is by the parts of the drawing. In our house example, there are 3 different parts: the roof, the frame, and the door. Each part can be identified as a separate unit, as shown below.

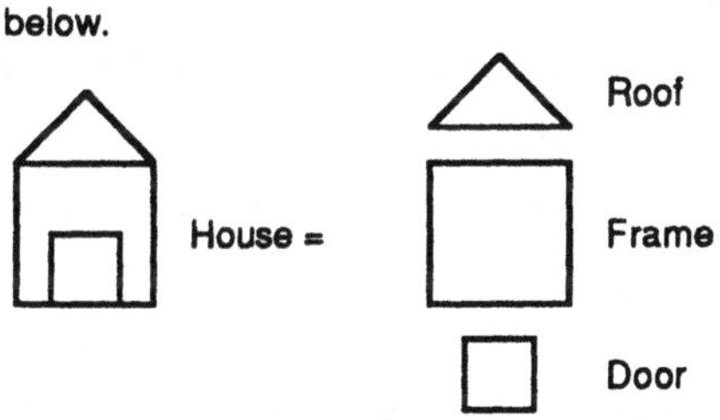

Second, take each part, one at a time, and write the instructions for that part.

To make the Frame
Move forward 1 inch
Turn right 90 degrees
Move forward 1 inch
Turn right 90 degrees
Move forward 1 inch
Turn right 90 degrees
Move forward 1 inch
Turn right 90 degrees
End of Frame

To make the Door
Move forward 1/2 inch
Turn right 90 degrees
Move forward 1/2 inch
Turn right 90 degrees
Move forward 1/2 inch
Turn right 90 degrees
Move forward 1/2 inch
Turn right 90 degrees
End of Door

To make the Roof
Move forward 1 inch
Turn right 120 degrees
Move forward 1 inch
Turn right 120 degrees
Move forward 1 inch
Turn right 120 degrees
End of Roof

Figure 3: Strategy Training.

Does modeling of expert strategies affect users' programming performance? First, when asked to write programs, students given strategy training created four times as many modules per program and reused instructions 19 times more often than students given conventional training. This result validates the effectiveness of the strategy training in helping students to acquire knowledge of modularity and reusability strategies. Second, on tests of problem-solving transfer such as interpreting what a given program would do, strategy-trained students performed 60% better than conventionally-trained students. This result is consistent with the idea that knowledge of expert strategies enables successful problem-solving transfer.

Conclusion

In conclusion, this report summarized three successful attempts to help novices learn computer programming in a way that promotes transfer. Novices tend to enter the programming classroom with unproductive preconceptions of what happens inside a computer as it follows a procedure, insufficient knowledge of how to comprehend a written procedure, and unsophisticated strategies for how to design a procedure. Conceptual model training promoted knowledge of the actions, objects, and locations inside a computer that are related to programming procedures; skill pretraining promoted knowledge of procedure comprehension; and strategy training promoted knowledge of expert design strategies such as modularity and reusability. In each case, training on the process of using specifically-relevant component skills resulted in improved problem-solving transfer.

The theme underlying this report is that instruction can affect a user's knowledge which, in turn, can affect problem solving transfer performance. In contrast to classic research which typically failed to demonstrate transfer, research on learning and teaching of problem-solving skills in programming offers several encouraging examples of successful transfer. The current results are consistent with Singley and Anderson's [27] theory of specific transfer of general skills: Positive high-road transfer occurs when students receive pretraining in English about high-level skills that are also specifically relevant to computer programming. Continued research is needed to focus on two links in the chain between programming instruction and problem-solving performance: Which aspects of conceptual and strategic knowledge enable improvements in problem-solving transfer and which instructional procedures are most effective in producing desired changes in user's conceptual and strategic knowledge?

References

1. Bayman, P. (1983). *Effects of instructional procedures on learning a first programming language.* Unpublished doctoral dissertation, University of California, Santa Barbara.
2. Bayman, P., & Mayer, R. E. (1983). A diagnosis of beginning programmers' misconceptions of BASIC programming statements. *Communications of the ACM, 26,* 677-679.
3. Bayman, P., & Mayer, R. E. (1988). Using conceptual models to teach BASIC computer programming. *Journal of Educational Psychology, 80,* 291-298.
4. Dyck, J. L. (1987). *Learning and comprehension of BASIC and natural language programming by novices.* Unpublished doctoral dissertation, University of California, Santa Barbara.
5. Dyck, J. L., & Mayer, R. E. (1989). Teaching for transfer of computer program comprehension skill. *Journal of Educational Psychology, 81,* 16-24.
6. Fay, A. L. (1990). *Computer programming instruction: The acquisition and transfer of design skills.* Unpublished doctoral dissertation, University of California, Santa Barbara.
7. Fay, A. L., & Mayer, R. E. (1987). Children's naive conceptions and confusions about logo graphics commands. *Journal of Educational Psychology, 79,* 254-268.
8. Fay, A. L., & Mayer, R. E. (1988). Learning logo: A cognitive analysis. In R. E. Mayer (Ed.), *Teaching and learning computer programming* (pp. 111-136). Hillsdale, NJ: Erlbaum.
9. Mayer, R. E. (1975). Different problem-solving competencies established in learning computer programming with and without meaningful models. *Journal of Educational Psychology, 67,* 725-734.
10. Mayer, R. E. (1976). Some conditions of meaningful learning for computer programming: Advance organizers and subject control of frame order. *Journal of Educational Psychology, 68,* 143-150.
11. Mayer, R. E. (1979). A psychology of learning BASIC. *Communications of the ACM, 22,* 589-593.
12. Mayer, R. E. (1981). The psychology of how novices learn computer programming. *Computing Surveys, 13,* 121-141.
13. Mayer, R. E. (1983). *Thinking, problem solving, cognition.* New York: Freeman.
14. Mayer, R. E. (1985). Learning in complex domains: A cognitive analysis of computer programming. In G. Bower (Ed.), *The psychology of learning and motivation, Volume 19* (pp. 89-130). Orlando, FL: Academic Press.
15. Mayer, R. E. (1987). *Educational psychology: A cognitive approach.* Boston: Little, Brown & Co.
16. Mayer, R. E. (1987). Cognitive aspects of learning and using a programming language. In J. M. Carroll (Ed.), *Interfacing thought: Cognitive aspects of human-computer interaction* (pp. 61-79). Cambridge, MA: MIT Press.
17. Mayer, R. E. (1987). The elusive search for teachable aspects of problem solving. In J. A. Glover & R. Ronning (Eds.), *Historical foundations of educational psychology* (pp. 327-347). New York: Plenum.
18. Mayer, R. E. (Ed.). (1988). *Teaching and learning computer programming.* Hillsdale, NJ: Erlbaum.
19. Mayer, R. E. (1988). Introduction to research on teaching and learning computer programming. In R. E. Mayer (Ed.), *Teaching and learning computer programming* (pp. 1-12). Hillsdale, NJ: Erlbaum.
20. Mayer, R.E. (1988). From novice to expert. In M. Helander (Ed.), *Handbook of human-computer interaction* (pp. 569-580). Amsterdam: North-Holland.
21. Mayer, R. E. (1989). Models for understanding. *Review of Educational Research, 59,* 43-64.
22. Mayer, R. E., Bayman, P., & Dyck, J. L. (1987). Learning programming languages: Research and applications. In D. S. Berger, K. Pezdek & W. P. Banks (Eds.), *Applications of cognitive psychology* (pp. 33-45). Hillsdale, NJ: Erlbaum.
23. Mayer, R. E., Dyck, J. L., & Vilberg, W. (1986). Learning to program and learning to think: What's the connection? *Communications of the ACM, 29,* 605-610.
24. Mayer, R. E., & Fay, A. L. (1987). A chain of cognitive changes with learning to program in logo. *Journal of Educational Psychology, 79,* 269-279.
25. Salomon, G., & Perkins, D. (1989). Rocky roads to transfer: Rethinking mechanisms of neglected phenomenon. *Educational Psychologist, 24,* 113-142.
26. Shneiderman, B., & Mayer, R. E. (1979). Syntactic/semantic interactions in programmer behavior: A model and experimental results. *International Journal of Computer and Information Sciences, 8,* 219-238.
27. Singley, M. K., & Anderson, J. R. (1989). *The transfer of cognitive skill.* Cambridge, MA: Harvard University Press.
28. Thorndike, E. L., & Woodworth, R. S. (1901). The influence of improvement in one mental function upon the efficiency of other functions. *Psychological Review, 8,* 247-261.
29. Wertheimer, M. (1945). *Productive thinking.* New York: Harper & Row.

Note

Conceptual model training as exemplified in Figure 1 is based on a doctoral dissertation by Piraye Bayman [1]; procedure comprehension training as exemplified in Figure 2 is based on a doctoral dissertation by Jennifer Dyck [4]; strategy training as exemplified in Figure 3 is based on a doctoral dissertation by Anne Fay [6].

Cognitive Effects of Learning to Program in Logo: A One-Year Study With Sixth Graders

Erik De Corte, Lieven Verschaffel, and Hilde Schrooten

Center for Instructional Psychology and Technology (CIP&T), University of Leuven, Vesaliusstraat 2, B-3000 Leuven, Belgium

Abstract: In this contribution we report a study that aimed at the development of a powerful Logo environment for the acquisition and transfer of four thinking skills. The investigation was carried out in three sixth-grade classes according to the pretest-posttest design with control group. A 60-hour Logo course, involving the explicit training of a systematic programming strategy, was taught in two experimental classes (E1 and E2). In E2 additional explicit instruction for transfer was given. The results show that the Logo knowledge and programming skills were mastered fairly well, and that positive transfer effects were obtained for three out of the four thinking skills. Starting from our own findings on the one hand, and the results of other recent successful investigations concerning the cognitive effects of learning to program on the other, some crucial characteristics of powerful learning environments are identified and suggestions for future research are formulated.

Keywords: Logo programming, thinking skills, transfer

Theoretical Background

Acquiring thinking and problem-solving skills is nowadays a primary educational objective [26]. Recent research has revealed that learning to solve problems requires the integrated acquisition of three categories of abilities [7]: 1. flexible application of a well-organized domain-specific knowledge base, involving concepts, rules, principles, formulas, and algorithms; 2. heuristic methods, i.e., systematic search strategies for problem analysis and transformation; and 3. metacognitive skills, involving knowledge concerning one's own cognitive functioning on the one hand, and the self-monitoring and regulation of one's own cognitive processes on the other. Learning to solve problems also supposes that the acquired knowledge and skills can be applied in new learning and problem situations within a specific domain and even in other domains; in other words, one assumes that transfer will occur.

Since the early eighties it has often been argued that learning computer programming would be particularly conducive to the development and transfer of important thinking skills [22]. However, a series of investigations about the cognitive effects of programming - especially of programming in Logo - carried out over the first part of the past decade, did not report positive transfer results [9, 10, 31]. Nevertheless, the so-called cognitive effects hypothesis has not been abandoned, because the sobering outcomes of the previous investigations could be attributed to several weaknesses of this research [9, 10, 21]. First, the pupils in those studies did not acquire sufficient Logo knowledge and programming skills, due to the short duration of the hands-on experience (on the average 25 to 30 hours) on the one hand, and to the absence of systematic instruction on the other [e.g., 4, 16, 20, 24]. Second, an explicit and systematic orientation towards transfer was lacking, in the sense that the skills were not generalised by showing their applicability in different content domains [23, 25].

Several recent studies in which an attempt was made to overcome one or more of these shortcomings have produced positive transfer effects [3, 5, 15, 18, 19, 30]. In a one-year Logo teaching experiment we have also found encouraging results. We will present the hypothesis, the design and the results of this study.

Hypothesis

The major objective of this study was the development, implementation and evaluation of a powerful Logo learning environment aiming at the acquisition and transfer of general thinking skills. We focused on a subset of problem-solving skills that are considered to be sensitive to programming experience: two metacognitive skills (planning and debugging), and two heuristics (problem decomposition and construction of an external problem representation).

Taking into account the results of earlier work about the cognitive effects hypothesis on the one hand, and the recent cognitive-psychological literature on the other, we hypothesized that fulfilment of the following three conditions is crucial in order to attain transfer of cognitive skills:

1. the pupils have acquired sufficient domain-specific knowledge (i.e., the Logo language features and concepts);
2. they have achieved mastery of the heuristics and metacognitive skills within the Logo environment;
3. they have learned how to apply the skills taught in the programming environment in at least one other content domain.

As argued before, previous research has already convincingly shown that fulfilment of the first condition alone is not sufficient in order to obtain transfer of thinking skills. Therefore, the present study focused on the necessity of the transfer conditions 2 and 3. The general hypothesis was formulated as follows: When the first two conditions are fulfilled, transfer of thinking skills will occur; fulfilment of the third condition will enhance the transfer effect.

The project was carried out in two stages. During the school year 1986-87 an exploratory study was undertaken, aiming at the development, tryout and revision of a Logo learning environment, and at the construction of instruments for measuring pupils' knowledge and skills within this environment [see 12 for more detailed information]. To test the hypothesis mentioned above, a more systematic teaching experiment was undertaken during the school year 1987-88.

Method

Subjects

Three sixth-grade classes with 24 pupils each participated in the study (n=72). The pupils (12 years old) had no prior experience with Logo. Two classes served as experimental groups (E1 and E2), while the third one was a control group (C). In the beginning of the school year an intelligence test and a school achievement test were administered in all three classes. No significant differences between the three groups were found.

Design

The experiment was carried out according to the pretest-posttest design with control group. In E1 and E2 fulfilment of the first two transfer conditions mentioned above was pursued through a 60-hour Logo turtle graphics course, involving the systematic instruction of the primitives of the Logo language on the one hand, and of a strategy to write programs on the other. This strategy consists of two main phases, namely a planning phase and an executing-and-testing phase, and constitutes the specification of the previously mentioned general thinking skills (planning, debugging, problem decomposition, and external representation).

In E2 the third transfer condition was also pursued. After the Logo course, the children were taught during a series of five lesson periods how to apply the learned skills in another context, namely in solving multi-step mathematical word problems. The control class was a non-treatment group.

Logo learning environment

A Logo course outlined in Table 1 was taught in the two experimental groups one afternoon each week during the whole school year (approximately 60 hours).

Members of the research team took care of the Logo instruction, in cooperation with the classroom teacher. The computer room, which was directly accessible form the regular classroom, was equipped with nine Philips computers. The course consisted of two major components, namely teaching of the Logo primitives and concepts (transfer condition 1) on the one hand, and training of the programming strategy (transfer conditon 2) on the other.

Teaching Logo primitives and concepts (transfer condition 1): The content of the Logo course was limited to the so-called turtle graphics. Children were only taught those Logo primitives and concepts considered to be necessary with a view to the acquisition of the intended programming strategy. The following topics were treated: making simple drawings using the basic Logo primitives (FD/BK, RT/LT, PU/PD), working with the REPEAT command, writing procedures and superprocedures.

Starting from a moderate constructivist conception of learning [7], we assumed that a powerful learning environment is characterized by a good balance between discovery learning on the one hand, and systematic instruction and guidance on the other.

Table 1
Overview of the Activities in the Logo Course

Unit	Activity
1	Acquaintance with the computer
2-4	Elementary primitives: FD/BK, RT/LT, PU/PD
5	REPEAT command (the list to be repeated only contains elementary primitives)
6-7	Procedures
8-9	Superprocedures
10	REPEAT command (the list to be repeated also contains procedures)
11	Demonstration of the entire programming strategy
12-14	Demonstration and practise of the separate components of the programming strategy
15-16	Practice of the programming strategy with simple assigned drawings
17-19	Practice of the programming strategy with complex assigned drawings
20-22	Practice of the programming strategy with self-generated drawings First administration of mastery tests
23-25	Practice of the programming strategy with assigned and self-generated drawings containing circles Remedial instruction / Exploratory projects Second administration of mastery tests

The Logo primitives and concepts were presented and taught in a well-considered sequence. The teacher only passed on to the next primitive or concept when the results of an intermediary test showed that the former one was sufficiently mastered.

The teaching of each new Logo primitive or concept was performed in a systematic way, either via demonstration at the computer or in a discussion format. During these demonstrations and discussions the children's typical errors and misconceptions which were observed in the previous exploratory study were taken into account. For example, to avoid the typical misconceptions and errors with respect to angles [12], the teacher inserted activities like playing turtle, estimating angles, and measuring angles using a protractor. The children were given ample opportunity to practise the newly learned primitive or concept in small groups of two to three pupils working on challenging tasks at the computer. Also game-like microworlds were presented, in which the children could exercise certain Logo notions in an intensive but attractive manner. The "direction game", for instance, aimed at practising the primitives LT and RT. The goal was to send a moving turtle to its nest, using LT and RT instructions. At the end of the practising phase a discussion was held about children's difficulties, findings and strategies.

Finally, the children had each time the occasion to practise the newly mastered Logo knowledge in more extensive self-chosen projects, offering also the opportunity for integration with previously learned concepts.

Teaching a strategy for writing Logo programs (transfer condition 2): The Logo programming strategy that was taught, consists of two main phases: a planning phase, and an integrated executing-and-testing phase.

In the planning phase, carried out independently from the computer, three steps are distinguished (see Figure 1):

1. making a drawing of the intended screen effect;
2. constructing a tree-like diagram in which the complex drawing is subdivided in building blocks that are easy to program; this diagram involves at the same time the sequence in which the different parts have to be drawn on the screen;
3. making separate drawings of the different building blocks, indicating for each part the lengths and angles as well as the start and the end position of the turtle.

Obviously planning as a metacognitive activity takes place in this phase, and elicits also the application of the two heuristics, namely problem decomposition and construction of an external representation.

Once the planning is completed, the integrated executing-and-testing phase at the machine can begin. This activity which involves debugging, is guided by two principles, namely top-down programming and immediate testing and debugging.

Top-down programming means that the children are taught to start with the most global procedure, called the "mother procedure", which consists of the names of the subsequent parts from the second level of the tree-like diagram, together with the names of the "connecting links" that move the turtle from the end position of a previous part to the start position of the next building block[1]. Subsequently each component of this procedure is specified until the lowest level of the tree-like diagram is reached.

The second principle consists of immediate testing and debugging of each newly defined procedure after defining it. By calling the "mother procedure" the result appears on the screen and can instantly be evaluated; furthermore the error message ("There is no procedure named...") indicates which procedure has to be written next.

To illustrate these principles, we present the interaction with the computer of a child who applies the strategy for drawing the car in Figure 1. In accordance with the principle of top-down programming the child first defines the mother procedure (CAR):

```
TO CAR
  WINDOW CL-WM MOTOR CL-MW WHEELS²
  END
```

Then, following the second principle, the child types the name of the mother procedure (CAR), resulting in the error message "THERE IS NO PROCEDURE NAMED WINDOW AT LEVEL 1 OF CAR". Next the WINDOW procedure is defined:

```
TO WINDOW
  REPEAT 2 (FD 15 RT 90 FD 40 RT 90)
  END
```

As a result of the subsequent testing (CAR), the turtle draws the window of the car. The error message at the bottom of the screen indicates that the connecting link between the window and the motor (CL-WM) is the next procedure to be written. The child types:

```
TO CL-WM
  PU RT 90 BK 10 PD
  END
```

When the CAR procedure is executed again, the turtle draws the window and then moves to the starting position for the motor.

This process is repeated until all building blocks and connecting links of the MOTOR procedure, and subsequently all parts of the remaining components of the mother procedure (CAR) are defined.

---

[1] In the present study it was agreed that the name for a connecting link between two building blocks consists of the following three parts: The characters "CL" (from Connecting Link), followed by a dash and the first characters of the former and the following block (e.g., connecting link between WINDOW and MOTOR: CL-WM).

[2] For the sake of clarity, the children were taught to start a new line for each procedure in a superprocedure. However, in this text all procedures are on one line because of the limited space available.

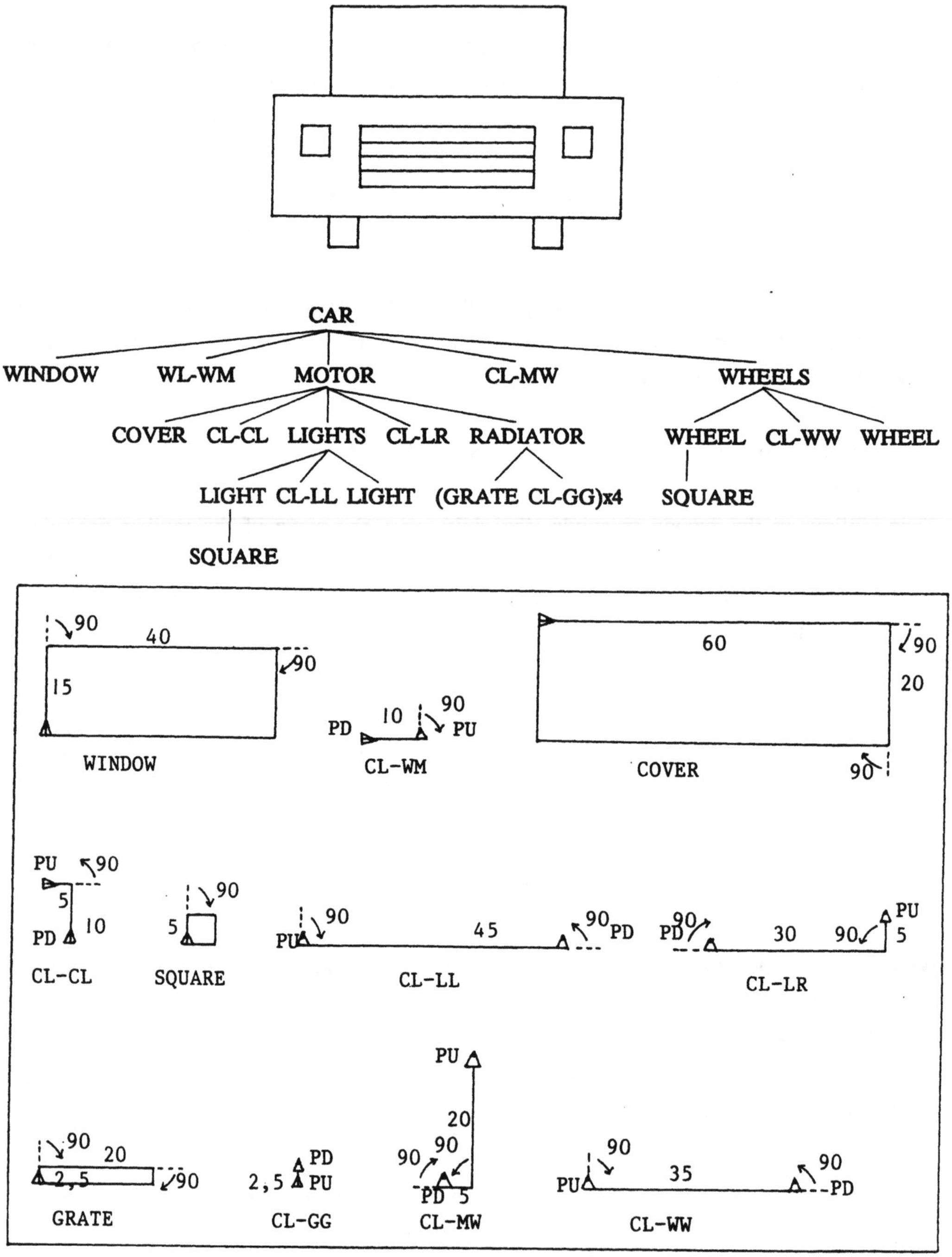

Figure 1. Plan for writing a Logo program for drawing a car.

The teaching of this programming strategy constituted the major component of the Logo course, and involved the application of teaching strategies that are considered in the recent literature to be effective for the instruction of general thinking skills: modelling, scaffolding, coaching, fading, articulation, reflection and exploration [6].

At first, the entire strategy was demonstrated by an "expert" (= the teacher). This allowed the children to build a conceptual model of the steps and processes required to carry out the task (*modelling*).

Then each component of the strategy was taught and practised separately. In this phase *scaffolding* was used: The teacher provided direct support to the pupils in carrying out those parts of the task that they could not yet perform independently. For instance, a special computer program was used to help the children in constructing a tree-diagram. This program guides the decomposition process during the planning phase through a series of simple questions (e.g., Is MOTOR simple enough? Which are the constituting parts of MOTOR?). Through the repeated use of this program for constructing diagrams for a diversity of drawings, internalization of the planning strategy was pursued.

Next, the children were given ample opportunity to practise the whole strategy, working on progressively more difficult problems in small groups of two to three pupils (*exploration*). In order to help them perform the strategy autonomously, the teacher provided several charts containing the different steps of the strategy and other useful information. In the beginning the pupils were guided intensively using hints, explanation, and feedback (*coaching*). In doing so the teacher again took into consideration the errors, misconceptions and difficulties observed in our exploratory study [12]. Later on teacher help and supervision decreased with increasing student competency (*fading*).

Throughout instruction and practice of the strategy, the pupils were stimulated to spell out the procedures and skills they used (*articulation*), and to compare them with those of other pupils and with the expert model (*reflection*). With a view to attaining transfer, explicit and intentional abstraction of the trained thinking skills was pursued.

Explicit instruction for transfer (transfer condition 3): In the second experimental group (E2) explicit instruction for transfer was provided in an additional course in word-problem solving. During a series of five lessons the pupils learned to apply the skills acquired in the Logo context (namely planning, debugging, decomposition, and external representation) in solving multi-step arithmetic word problems. They were taught a procedure involving the same two main phases as the programming strategy. The planning phase consists in subdividing the multi-step problem into subproblems that can be solved easily; this activity can also be represented in a tree-like diagram (see Figure 2). In the integrated executing-and-testing phase each subproblem is solved and the outcome is immediately tested.

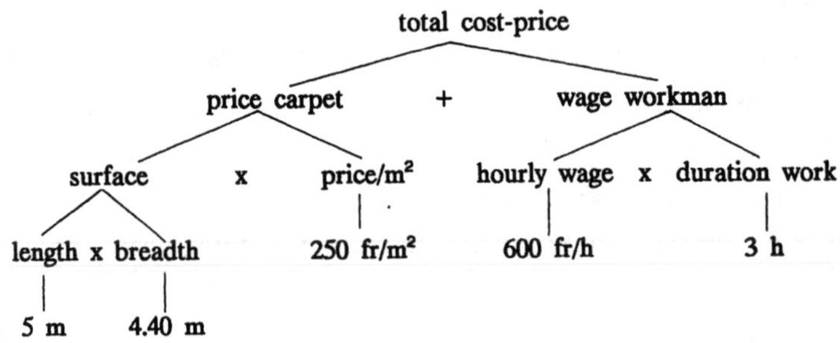

Figure 2. Plan for solving a multi-step word problem.

Mastery tests

Before testing the transfer effects, we first examined fulfilment of the transfer conditions. In both experimental classes the first two conditions were assessed one month before the end of the course, using one or more mastery tests for each condition. For each test an individual mastery criterion was put forward: 75% of the children had to obtain at least 2/3 of the maximum score on the test. When this criterion was not attained, we considered the related condition as not sufficiently fulfilled. In that case remedial instruction was provided for those children who did not meet the predetermined criterion, while the other children worked on exploratory projects. Afterwards, condition fulfilment was re-assessed. In E2 the tests with respect to word-problem solving (transfer condition 3) could be administered only once, because there was no time available for remedial instruction and retesting.

The test about the knowledge and understanding of the Logo primitives and concepts *(transfer condition 1)* consisted of a series of items in which the children either had to indicate which of four alternative Logo commands or procedures would produce a given screen effect, or to select the appropriate drawing for a particular Logo command or program. The incorrect alternatives were based on children's typical errors as observed in the exploratory study [12].

To assess mastery of the programming strategy *(transfer condition 2)*, three tests were administered. Two paper-and-pencil tests dealt with the construction of a tree-like diagram (= planning stage). In the first planning test each item consisted of a drawing with two tree-like diagrams, only one of which was correct. The children were asked to indicate the correct tree diagram for the drawing. The errors included in the tree-diagrams were again developed on the basis of children's solutions on tests administered in the exploratory study [12]. In the second planning test the pupils had to construct autonomously an appropriate tree-like diagram for a given drawing. The third test concerned the integrated executing-and-testing phase. Contrary to the two previous tests, this test was administered individually at the computer. The child was asked to write a Logo program for a relatively complex drawing starting from a given plan. To enable us to analyse children's debugging activities, four errors were deliberately included in this plan.

To assess mastery of the strategy for solving multi-step word problems *(transfer condition 3)* in E2, analogous tests were used. As in Logo, two tests related to the planning phase. In the first one the children were offered a series of word problems for each of which they had to indicate the best diagram out of two given ones. In the second test the children were asked to construct autonomously appropriate tree-like diagrams for five given word problems. The third test dealt with the integrated executing-and-testing phase; the task was to solve a number of problems starting from given diagrams.

Transfer tests

At the beginning and at the end of the school year the pupils of the two experimental classes and of the control group were given a series of transfer tasks in order to evaluate their ability to apply the thinking skills trained in E1 and E2 to other situations.

The "*mazes*" test, which was designed to measure planning skill, was administered individually at the machine. The computer subsequently offered three mazes, containing a hungry beatle and three salads (see Figure 3). The task consisted in sending the beatle by the shortest road to the three salads. Planning was required to find the best sequence.

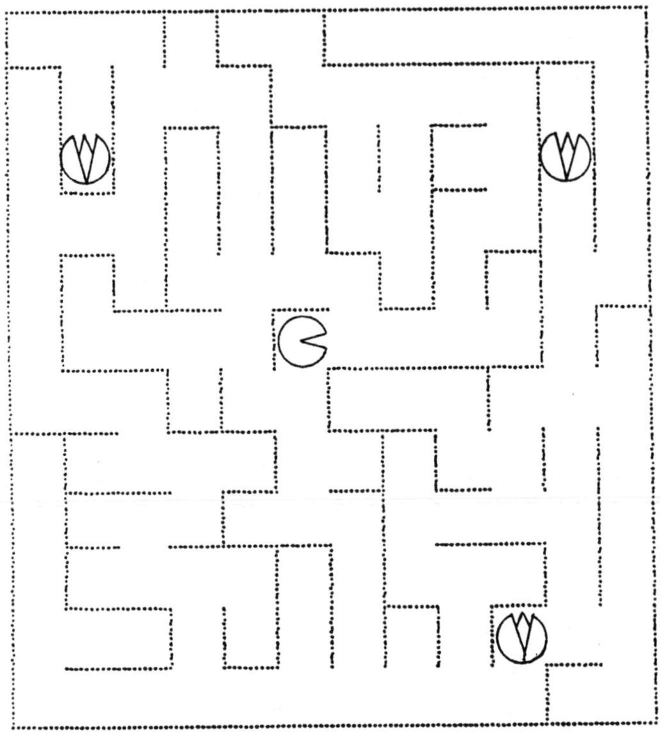

Figure 3. Example of an item from the mazes test.

In the "*blocks*" test the pupils had to describe complex figures in terms of their component parts, chosen form a series of elements of different complexity; the least possible elements had to be used. Solving this test efficiently requires application of the decomposition heuristic.

The "*error detection*" test, inspired by the work of Carver [2], measured both the planning skill and the ability to test and correct a series of instructions. It contained three stories, each one telling about person A giving person B directions. Person B followed the directions perfectly, but the result was not as intended because one of the instructions was wrong. The task was to find and to fix the bug in the directions, so that the anticipated result will be obtained (see Figure 4).

Mrs. Fisher was moving into a new apartment with the help of two movers. She asked them to arrange the furniture in the apartment and gave them a list of directions. The movers followed the instructions perfectly; however there was a bug in the directions; consequently the furniture was not arranged correctly. Picture A shows how Mrs. Fisher wanted her apartment to look like; picture B shows how the furniture was really arranged by the movers. Try to find the bug in Mrs. Fisher's directions and correct it.

Picture A

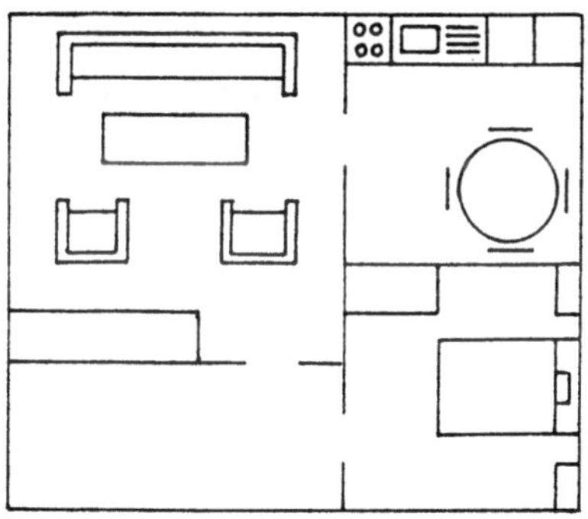

Picture B

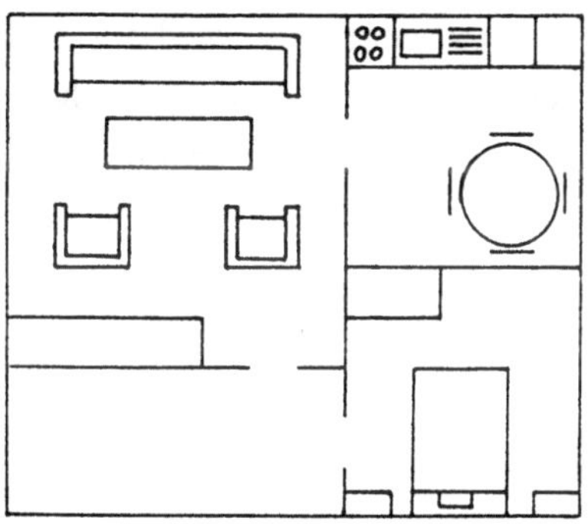

These are Mrs. Fisher's directions.

In the kitchen:

Put the electric cooker at the left of the dresser.

Put the automatic dishwasher at the other side of the dresser.

...

In the living room:

Put the sofa against the northern wall.

Put the two seats facing the sofa.

...

In the bedroom:

Put the head of the bed against the southern wall.

Put a pedestal on each side of the bed.

Put the wardrobe against the wall of the kitchen

Figure 4. Example of an item from the error detection test.

The "*weekdays*" test consisted of ten items such as "If tomorrow is Saturday, which day is the day before yesterday?" [1]. The most efficient strategy to solve such tasks involves the use of an external representation.

Children's mastery of the use of the latter heuristic was also tested in the "*boxes*" test, containing ten items, such as "There is one big box. Inside the big box there are two separate medium boxes. Inside one of the medium boxes there are three small boxes. Inside the other one there are four boxes. How many boxes are there altogether?" [2].

The major technique for data analysis was a MANOVA mixed design with one between (experimental treatment) and one within factor (pretest and posttest). Separate analyses were performed for E1 and C on the one hand, and E1 and E2 on the other. According to our initial hypothesis, E1 would make more progress from the pretests to the posttests than C; moreover E2 would outperform E1. Space restrictions do not allow us to describe in detail the scoring procedure for the tests and the statistical analysis techniques [see 13 for more detailed information].

Results and Interpretation

Mastery tests

Table 2 shows the number of pupils in E1 and E2 who reached the predetermined mastery criterion (2/3 of the maximum score) on the different tests by the end of the experiment.

The results on the Logo knowledge test *(transfer condition 1)* indicated that by the end of the experiment most children in both experimental classes (21 and 17 in E1 and E2 respectively) had acquired a sufficient level of knowledge and understanding of the Logo primitives and concepts.

The outcomes on the measures relating to the programming strategy *(transfer condition 2)* were also very positive. The results on the first planning test showed that in E1 and E2 19 and 23 children respectively were able to indicate the correct diagram for a series of drawings. The second planning test revealed that in E1 and E2 20 and 19 pupils respectively succeeded in constructing autonomously an appropriate tree-like diagram for a given drawing. On the test concerning the executing-and-testing stage of the programming strategy, namely writing a Logo program for a relatively complex drawing starting from a given plan, most of the children (20 in E1 and 22 in E2) were able to apply the instructed strategy efficiently: They used the top-down programming style fairly consistent; each time a new procedure was defined, it was tested by calling the so-called mother procedure; and when an incorrect screen effect was obtained, detection, analysis, and debugging of the programming error was performed in a systematic and straightforward way. The preceding results justify the conclusion that the children were able to apply the intended general thinking skills within the Logo environment. In summary, by the end of the Logo course, both the first and the second transfer condition were fulfilled.

Table 2
Number of Pupils in E1 and E2 who Attained the Mastery Criterion on the Tests for the Three Transfer Conditions

Transfer Conditions	Mastery Tests	Results (n=24)	
		E1	E2
Logo knowledge	Logo knowledge	21	17
Thinking skills in Logo	Planning 1	19	23
	Planning 2	20	19
	Executing/testing	20	22
Thinking skills in math	Planning 1	-	22
	Planning 2	-	10
	Executing/testing	-	22

In our opinion, these positive results can be attributed to the carefully constructed instructional environment, in which the shortcomings of the early studies were overcome. Two important characteristics of this learning environment were: 1) the time-intensive character, and 2) the good balance between exploratory learning activities on the one hand, and systematic instruction (e.g., the explicit teaching of a programming strategy) on the other. In evaluating these findings appropriately, one should certainly take into account that the results were obtained in a rather atypical situation (60 hours of Logo instruction, nine computers and several experienced teachers for a single class); this is certainly not representative of the current situation in Belgian elementary schools.

Table 2 also shows that the results on the tests with respect to the *third transfer condition*, namely transfer to other domains, were less positive. In E2 only 10 out of 24 children were able to apply autonomously the planning strategy in the domain of arithmetic word problems (second planning test). Although their performance on the other two tests was quite good, we consider the third condition as not sufficiently fulfilled, the argument being that the second planning test was undoubtedly the most critical measure of strategy mastery. Since the school year was at its end, there was unfortunately no time available for remedial instruction with respect to word problems.

A plausible explanation for this disappointing finding is the existence in children of inefficient strategies, attitudes and beliefs about word-problem solving, which they had acquired over the years during their normal math lessons, and which were difficult if not impossible to change substantially within the available amount of time [29].

Taking into account this latter finding, a conclusive test of the second part of our main hypothesis, namely that the fulfilment of the third condition enhances transfer, became impossible in the present study.

Transfer tests

The multivariate analyses showed that the Logo treatment was successful in obtaining transfer for three out of the four thinking skills that were taught, namely problem decomposition, construction of an external representation, and debugging. Indeed, Table 3 shows a significant difference between the first experimental group and the control group for these skills, and further analysis revealed that the E1 group made more progress from pretest to posttest on the corresponding measures than the C group.

The MANOVA on the measures for planning skill showed no significant differences between E1 and C. Both groups performed equally well on the tests requiring this skill. This is somewhat surprising, since the planning component was the core component of the trained programming strategy.

As we expected on the basis of the non-fulfilment of the third transfer condition, E2 did not make more progress than E1. Unexpectedly, we even found significant effects in the opposite direction. E1 outperformed E2 on the tests for problem decomposition, construction of an external representation, and debugging. Although systematic data are

Table 3
Comparison of E1 vs. C, and E1 vs. E2 on the Transfer Tests

Thinking Skills	Transfer Tests	Results	
		E1 vs. C	E1 vs. E2
Planning	Mazes, Error detection	.249	>.500
Decomposition	Blocks, Weekdays	<.001	.002
External representation	Weekdays, Boxes	<.001	<.001
Debugging	Error detection	.001	.016

lacking, there are several indications that the replacement in E2 of the Logo lessons by the course in word-problem solving resulted in a decrease in children's motivation and enthusiasm. This might explain why less transfer was found in E2.

In our opinion it would be premature to conclude from these findings that explicit training for transfer has no or even a negative influence on the development of general thinking skills. Indeed, the absence of the expected transfer effects, can be explained by the non-fulfilment of the third transfer condition. We tend to hold on the hypothesis that if one succeeds in realizing the explicit instruction for transfer properly, additional transfer effects will occur [17].

Discussion

The Logo learning environment developed and implemented in the present study produced positive transfer results for three out of four thinking skills that were taught. As mentioned before, other recent investigations also reported transfer effects [3, 5, 15, 17, 19, 30; for more detailed information, see 11]. These obtained successes are, in our opinion, to a large degree due to the fact that in those studies a thorough attempt was made to overcome one or more of the shortcomings of the earlier research mentioned in the first section of this contribution, namely the weaknesses of the Logo instruction and the lack of explicit teaching for transfer.

In all the recent successful studies Logo was embedded in a learning environment that differed substantially from the open learning situation in the earlier investigations. First, the Logo courses in those studies lasted much longer; the children had between 50 and 60 hours of Logo experience [e.g., 3, 5, 15], which is considerably more than in the older studies. Second, systematic and direct instruction of programming skills was provided in those environments. While the concrete design of this instruction differed from study to study, nevertheless recurrent ingredients can be indicated that correspond surprisingly well to the seven teaching strategies described by Collins et al. [6; see also 8] in their cognitive apprenticeship view of learning and teaching, namely modelling, scaffolding, coaching, fading, exploration, abstraction and reflection. Third, some researchers investigated whether the Logo knowledge and the programming skills were actually mastered at the end of the intervention [e.g., 3, 17]. When this was not the case, remedial instruction was provided. If one neglects to evaluate the learning outcomes before measuring transfer, failure to achieve transfer of a particular concept or skill becomes difficult to interprete [see also 18].

Most investigators now share the idea that in order to achieve transfer of thinking skills in learners, it is necessary to teach explicitly for transfer. This standpoint was also taken into account in the investigations in which transfer was observed. Using Salomon and Perkins' [27] terminology, two instructional strategies can be distinguished in these efforts to pursue transfer: mindful abstraction of the skills one wants to transfer, and decontextualisation of those skills by demonstrating their usefulness in other situations and/or other content domains. This distinction is echoed in Littlefield et al.'s [17] two instructional techniques, namely: framing, involving "the act of relating a specific set of behaviors to a broader framework of problem solving"; and bridging, meaning "the act of relating processes that occur within one context to similar processes occurring elsewhere" [17, p. 356] [see also 14].

In the study by Littlefield et al. [17] the first technique was realized in relating the notion of using subprocedures in Logo programs to the general context of analyzing problems and breaking them down into manageable components. The second strategy was applied in linking problem-solving strategies in Logo with familiar actions of Indiana Jones. Clements [5] used the "homunculi" device to abstract the learned skills; he then offered problems in a diversity of contexts to bring about decontextualisation. In our own work mindful abstraction was realized by explicitly labeling the thinking skills that were embedded in the trained programming strategy; decontextualisation was pursued in teaching the children how to apply the learned skills while solving multi-step word problems. Taking into account the available research, it seems reasonable to expect most perspective from studies in which both instructional strategies are used in teaching for transfer.

The previous discussion suggests that Logo in itself is not a vehicle for thinking, but that it can be a useful device for the acquisition and transfer of thinking skills, if embedded in a powerful learning environment [see also 3].

With respect to the design of such powerful learning environments the following important question raises: To what extent is it desirable to build the problem-solving instruction into the programming environment, and which part of it should remain in the hands of the teacher. In other words, would children profit from an "intelligent" Logo system that can perform a detailed analysis of their programming skills and strategies, and provide individualized coaching? Or should the major role of the computer consist of creating an environment that stimulates learners to use and practise their own knowledge and thinking skills, while the coaching is performed by the teacher [28]?

Recently we started the construction of a Logo-based computer environment involving built-in support to foster the acquisition of planning and debugging skills. Two levels of support are implemented. The first level consists of tools that guide and facilitate planning and debugging activities. When the planning tool is activated an empty "working page" appears, on which the child can easily and flexibly build up a plan for his program in the form of a tree-diagram. The debugging tool allows procedures to be carried out step-by-step, enabling the pupil to see where a possible error occurred. A computer coach constitutes the second level of support. This coach analyzes the pupil's activities, including the construction of the tree-like diagram, and - if necessary - gives comments and orienting help, based on a dynamic comparison of these activities with the expert programming strategy described earlier.

Another major issue warranting further research relates to what could be called Logo's "cognitive power". Since the positive cognitive effects observed in recent studies seem not to be due to Logo as such but rather to the entire learning environment, can we still consider Logo as a privileged domain for teaching thinking skills? Or can the investment of the same amount of time, effort and expertise in problem-solving instruction in more traditional content domains produce equal results?

Littlefield et al. tend to answer the latter question positively: "We think that perhaps the best way to view Logo is as one context that can be useful for facilitating the development of general problem-solving skills. It is a very good context for this purpose for reasons outlined previously (i.e., it is a problem-rich environment, a procedural language, and is accessible to young children), but it is one of an array of tools that can be utilized to foster the development of problem-solving ability." [17, p. 133]

Others, like Clements [5], adhere the standpoint that Logo has something to offer to the transfer process over and above other materials as it is particularly suited to the implicit evocation of thinking skills. Swan's [30] study, which was in part designed to answer this question, seems to offer some empirical evidence for Clements' argument: Children receiving problem-solving instruction in Logo outperformed others who had followed analogous instruction with concrete materials (cut-paper manipulatives). Swan considers this result as a confirmation of Papert's [22] original idea that learning to program in Logo truly offers unique qualities, because it supports what Papert calls "transitional objects to think with", that is, computer representations of abstract ideas that can be manipulated in seemingly concrete ways to help bridge the gap between concrete and formal thought.

It seems to us that further research is required with respect to both questions raised above. Indeed we still need a better understanding of the optimum features of powerful Logo-based learning environments, as well as of the possible unique contribution of Logo to the acquisition of thinking skills in such environments.

References

1. Bransford, J. D., & Stein, B. S. (1984). *The ideal problem solver*. New York: Freeman.
2. Carver, S. (1986). *Transfer of Logo debugging skill: Analysis, instruction and assessment*. Unpublished doctoral dissertation, Department of Psychology, Carnegie-Mellon University, Pittsburg, PA.
3. Carver, S. (1988). Learning and transfer of debugging skills: Applying task analysis to curriculum design and assessment. In R. E. Mayer (Ed.), *Teaching and learning computer programming. Multiple research perspectives* (pp. 259-297). Hillsdale, NJ: Erlbaum.
4. Carver, S., & Klahr, D. (1986). Assessing children's Logo debugging skills with a formal model. *Journal of Educational Psychology, 76*, 1051-1058.
5. Clements, D. H. (1990). Metacomponential development in a Logo programming environment. *Journal of Educational Psychology, 82*, 141-149.
6. Collins, A., Brown, J. S., & Newman, S. E. (1989). Cognitive apprenticeship: Teaching the craft of reading, writing and mathematics. In L. B. Resnick (Ed.), *Knowing, learning, and instruction. Essays in honor of Robert Glaser* (pp. 453-494). Hillsdale, NJ: Erlbaum.
7. De Corte, E. (1990a). Towards powerful learning environments for the acquisition of problem-solving skills. *European Journal of Psychology of Education, 5* (1), 5-19.

8. De Corte, E. (1990b). Acquiring and teaching cognitive skills: A state-of-the-art of theory and research. In P. J. D. Drenth, J. A. Sergeant & R. J. Takens (Eds.), *European perspectives in psychology. Volume 1* (pp. 237-263). London: Wiley.
9. De Corte, E., & Verschaffel, L. (1986). Effects of computer experience on children's thinking skills. *Journal of Structural Learning, 9*, 161-174.
10. De Corte, E., & Verschaffel, L. (1989). Logo, a vehicle for thinking. In B. Greer & G. Mulhern (Eds.), *New directions in mathematics education* (pp. 63-81). London: Routledge.
11. De Corte, E., Verschaffel, L., & Schrooten, H. (1990, March). *Cognitive effects of computer-oriented learning*. Paper presented at the Seventh International conference on Technology and Education, Brussels, Belgium.
12. De Corte, E., Verschaffel, L., Hoedemaekers, E., Schrooten, H., & Indemans, I. (1988, April). *Acquiring planning skills in Logo: An exploratory study with sixth graders*. Paper presented at the annual conference of the International Association for Computing in Education, New Orleans.
13. De Corte, E., Verschaffel, L., Schrooten, H., Indemans, R., & Hoedemaekers, E. (1990). *Logo als springplank voor het verwerven van denkvaardigheden bij zesdeklassers* [Logo as a spring-board for the development of thinking skills in sixth graders]. Leuven: Acco.
14. Delclos, V. R., & Burns, M. S. (1989, March). *Mediational elements in computer programming instruction*. Paper presented at the annual meeting of the American Educational Research Association, San Francisco.
15. Lehrer, R., Guckenberg, T., & Sancilio, L. (1988). Influences of Logo on children's intellectual development. In R. E. Mayer (Ed.), *Teaching and learning computer programming. Multiple research perspectives* (pp. 75-110). Hillsdale, NJ: Erlbaum.
16. Leron, U. (1985). Logo today: Vision and reality. *The Computer Teacher*, 26-32.
17. Littlefield, F., Delclos, V. R., Lever, S., Clayton, K. N., Bransford, J. D., & Franks, J. J. (1988). Learning Logo: Method of teaching, transfer of general skills, and attitudes toward school and computers. In R. E. Mayer (Ed.), *Teaching and learning computer programming. Multiple research perspectives* (pp. 111-135). Hillsdale, NJ: Erlbaum.
18. Littlefield, J., Delclos, V. R., Bransford, J. D., Clayton, K. N., & Franks, J. J. (1989). Some prerequisites for teaching thinking: Methodological issues in the study of Logo programming. *Cognition and Instruction, 6* (4), 331-366.
19. Mayer, R. E. (Ed.) (1988). *Teaching and learning computer programming. Multiple research perspectives*. Hillsdale, NJ: Erlbaum.
20. Milojkovic, J. D. (1983). *Children leraning computer programming: Cognitive and motivational consequences*. Unpublished doctoral dissertation, Stanford University, Department of Psychology.
21. Palumbo, D. B. (1990). Programming language/problem-solving research: A review of relevant issues. *Review of Educational Research, 60*, 65-89.
22. Papert, S. (1980). *Mindstorms. Children, computers, and powerful ideas*. New York: Basic Books.
23. Pea, R. D. (1985). *Transfer of thinking skills: Issues for software use and design*. Paper presented at the National Conference on "Computers and Complex Thinking" at the National Academy of Sciences, Washington, DC.
24. Pea, R. D., & Kurland, D. M. (1983). *On the conitive effects of learning computer programming* (Technical Report No. 9). New York: Bank Street College of Education, Center for Children and Technology.

25. Perkins, D. N., & Salomon, G. (1989). Are cognitive skills context-bound? *Educational Researcher, 18* (1), 16-25.

26. Resnick, L. B. (1987). Instruction and the cultivation of thinking. In E. De Corte, H. Lodewijks, R. Parmentier, & P. Span (Eds.), *Learning and instruction. European research in and international context. Volume 1* (pp. 415-442). Oxford/Leuven: Pergamon Press/Leuven University Press.

27. Salomon, G., & Perkins, D. N. (1987). Transfer of cognitive skills from programming: When and how? *Journal of Educational Computing Research, 3*, 149-169.

28. Scardamalia, M., Bereiter, C., McLean, R. S., Swallow, J., & Woodruff, E. (1989). Computer-supported intentional learning environments. *Journal of Educational Computing Research, 5*, 51-68.

29. Sowder, L. (1988). Children's solutions of story problems. *Journal of Mathematical Behavior, 7*, 227-238.

30. Swan, K. (1989). Logo programming and the teaching and learning of problem solving. *Journal of Artificial Intelligence in Education, 1* (1), 73-92.

31. Swan, K., & Black, J. B. (1987). *The cross-contextual transfer of problem-solving skills* (CCT Report 87-3). New York: Columbia University, Teachers College, Department of Communication, Computing and Technology.

The Role of Social Interaction in the Development of Higher-Order Thinking in Logo Environments

Douglas H. Clements[1], Bonnie K. Nastasi[2]

[1]State University of New York at Buffalo, Department of Learning and Instruction, 593 Baldy Hall, Buffalo, NY 14260, USA

[2]Dept. of Psychology, 435 DeGarmo Hall, Illinois State University, Normal, IL 61761, USA

Abstract: This paper reviews research examining the role of social processes as mediators of the effects of Logo computer programming on elementary-grade students' higher-order thinking. Specifically, we examined the extent to which engagement in conflicts or disagreements during problem solving with Logo accounted for cognitive gains. In three separate studies, students were observed working in pairs in Logo and control (computer-assisted instruction) environments and behaviors reflecting conflict and its resolution were identified. After completion of the computer sessions, students were administered problem-solving posttests. Overall, results suggest that Logo facilitates conflict resolution and development of higher-order thinking. Furthermore, the enhancement of cognitive skills may be mediated by engagement in specific conflict resolution strategies that are engendered by the Logo environment. Specifically, the critical social process fostered by Logo may be the interpersonal coordination of divergent ideas. This research advances our understanding of the specific features of Logo computer environments that foster changes in higher-order thinking and extends existing theory by suggesting that specific types of interpersonal conflict resolution facilitate cognitive growth.

Keywords: higher-order thinking, Logo, social-cognitive processes

Research on the efficacy of Logo computer programming environments to enhance students' higher–order thinking is plagued by contradictions and omissions. Even when substantive effects have been reported, the reason for these positive findings is often not clear. Clements [6; 8; 9] has explained these positive effects in terms of completeness of the educational environments. That is, multiple features were integrated to create an environment conducive to developing higher-order thinking processes; for example, these processes were discussed explicitly, students engaged in all phases of problem solving, and teachers used a variety of teaching methods. There are two shortcomings with this explanation. First, the extent to which children engage in specific social interactions has not been adequately elaborated. Second, the contribution of various aspects of the teaching-learning context such as these social interactions has not been determined. Correspondingly, this paper has two purposes. First, we will provide a theoretical rationale for examining social interactions within Logo environments. Our hypothesis is that specific interactions mediate the effect of Logo on higher-order thinking. Second, we will review research conducted to (a) measure the social interactions that occur in Logo environments and (b) determine their contribution to this cognitive growth.

Theoretical Perspective

Social interactions may play an important role in facilitating cognitive growth. Research suggests that students prefer to work cooperatively rather than individually at computers [13]. Furthermore, research on cooperative learning has consistently shown that students who solve problems in pairs or small groups perform better on subsequent cognitive measures than those who work alone [27]. Thus, collaboration may facilitate cognitive growth more than solitary endeavors. When individuals engage in a process that we call *reciprocal sense-making,* they attempt to extract meaning, generate ideas, or solve problems through a process of discourse. In such an interaction, Student A poses a solution to a problem, Student B elaborates on the idea of Student A, Student A then elaborates on the idea of student B, and so on. In this manner, the students create a solution that represents a combination or consensus of two viewpoints. This type of social coordination requires that students synthesize their actions with those of other students; otherwise, the students act independently and the social "interaction collapses into parallel and uncoordinated activities" [3, p. 216] and consequently is less likely to facilitate cognitive growth. This position is supported by research: Learning contexts in which students are placed in small groups and instructed to work as a team and to seek concurrence have resulted in higher learning and cognitive gains (e.g., on measures of problem solving) than those in which students are instructed to work alone or to compete with each other for rewards [23; 27].

Further, cooperative learning situations in which students were encouraged to disagree and challenge the thinking of their partners have been shown to result in even greater gains (i.e., on measures of achievement and perspective taking) than situations in which they were encouraged to seek concurrence and avoid disagreement [21; 22]. In a similar vein, Doise and Mugny [19] report that students who engaged in cognitive conflict (posed opposing viewpoints) in the process of discussing ideas made the greatest cognitive gains. Those who discussed their ideas without apparent conflict, however, still made greater cognitive gains than those who worked alone. Such results support social-cognitive theory that suggests that certain types of collaborative interactions may be more beneficial than others for facilitating cognitive growth; namely, those in which individuals must resolve discrepant viewpoints (interindividual "conflicts of centrations" [29]). That is, development occurs when a conflict between opposing egocentric viewpoints is resolved. According to this view, cognitive development requires that "individuals face others who contradict their own intuitively derived concepts and points of view, and thereby create cognitive conflicts whose resolutions result in the construction of higher forms of reasoning" [3, p. 203]. Within a Logo context, such behavior may be manifested as students pose different solutions to a programming problem, discuss the relative benefits of the various solutions, and come to consensus regarding a single solution.

In sum, theory and research suggest that although collaborative interactions are more beneficial than individualistic efforts, the structuring of controversy and students' consequent efforts to resolve these disagreements (i.e., "conflicts of ideas") may provide the optimal setting for enhancing learning and cognitive development. Although such enhancement of learning has been documented in Logo [2; 8; 9; 16; 17], the exact nature of the students' interactions in these and other educational settings has not been extensively examined (previous research has relied too heavily on anecdotal evidence [20]). In our research, we examined social processes that distinguished Logo from other computer environments and the role of these processes in facilitating the cognitive gains that were observed. We identified situations in which peer collaboration occurred. In addition, we analyzed each occurrence of interindividual conflict (e.g., partners pose discrepant solutions to a problem) and the specific strategies used to resolve these conflicts. Finally, we investigated the mediational role of these processes and strategies in facilitating the development of higher-order thinking.

Social Interactions in Computer Environments

Logo is a computer language specifically designed as a learning environment. Students construct programs that create graphics by directing the movements of a cybernetic "turtle," a triangular pointer, with commands such as FORWARD 50 (FD 50) AND RIGHT 90 (RT 90; i.e., rotate 90° clockwise). There are reasons to posit that certain Logo environments engender the types of social interactions that facilitate cognitive growth. Attempts at *sense-making* (extracting meaning, generating ideas) are most likely to occur when individuals work with self-selected projects and engage in every step of problem solving, from problem definition to strategy selection to solution monitoring [14]. Furthermore, *reciprocal sense-making* is more probable when the participants are encouraged to work cooperatively and share in the decision making (e.g., students work at computers in pairs and are encouraged to devise and implement mutually agreed upon Logo programs). Under these conditions, students are likely to negotiate their problem definition and solution strategies to proceed successfully. Participants in such a coherent problem-solving interaction each attempt to make sense of, or give meaning to, the other's actions, and continually modify their own goals in the process. If their goals or solution processes diverge too radically, their actions will interfere with each other. Therefore, students must continually validate their own problem-solving activity in view of their partners' activity, always striving toward mutual agreement. An example of such an interaction during a Logo session follows. Note that each student contributes ideas for problem solution and questions the other's thinking. The solutions seem to evolve out of the integrated contributions of the two. At times, one student even finishes the other's statement. We observed informally that partners over time seemed to develop implicit understandings and to share a common knowledge base that was unique to their interactions (such that a third party who had not observed them over time would have difficulty understanding the connections of their explicit ideas).

Student 1: F... (typing).
Student 2: ...D 20 (looking at the screen).
Student 1: Why 20?
Student 2: (shaking her hand in air) I mean 50. 50. (getting ready to type)
Student 1: (typing along with Student 2) 9. Now RT...I mean LT (abbreviations of right and left, respectively).
Student 2: (pointing to the screen) No, RT. (typing)
Student 1: RT, Yes, RT. RT 50.

In contrast, the beginning of the following interaction possesses the quality of reciprocal sense-making. As the interchange progresses, however, the students become engaged in two different conversations. The latter type of interaction is less likely to facilitate learning and problem solving.

Student 1: Let's try this.
Student 2: Oh yes, 38 minus 68. OK, it stops here....
Student 1: So, we go left 90, forward 35, then we get to that, right? We got to get to the middle. So we got to go forward 35 again....
Student 2: Oh, that looks pretty cool, lets keep it....
Student 1: It did something, really.
Student 2: Computers are too smart. People don't understand computers that much.
Student 1: What's 35 and 35?
Student 2: Come back next week for the conclusion of.... (talking to the camera as if on television)
Student 1: What's 35 plus 35?
Student 2: Computers alive!
Student 1: 70. Oh, I forgot to do something.
Student 2: I don't even understand what you are doing.
Student 1: Um, this has to be...there.

It has been demonstrated that peer interactions focused on learning and problem solving are more prevalent in Logo environments than in others such as traditional classroom tasks or computer-assisted instruction (CAI) drill [5; 7]. In addition, Logo environments encourage self-directed problem-solving (i.e., students solve problems that they themselves have posed, with the support but not direction from the teacher) and mutual "ownership" of the problem [5; 10; 24; 32]. Of course, social interaction may also be facilitated when students are paired within CAI environments [25]. In CAI environments, however, self-directed learning and mutual commitment to the problem are less probable. The computer structures the problem, the solution process, and the interaction. In addition, one student's response is not necessarily contingent upon, or even related to, the other's; thus, an egocentric point of view may be more easily maintained.

In our early research, we looked at the types of social interaction that occurred in different computer environments [11]. Forty-eight first and third graders worked in pairs for 14 weeks with either Logo programming or CAI (drill and practice). Using a naturalistic observational procedure, we examined the social processes in which these students engaged. In particular, we identified situations in which, as directed, students engaged in *cooperative work*; that is, worked in a collaborative manner, each contributing ideas toward the completion of a mutual goal. We distinguished these situations from those in which students worked individually or engaged in off-task behavior. For example, the following interchange would characterize a collaborative interaction that we labeled *cooperative work*.

Student 1: (looking at screen) OK. Go FD 50.
Student 2: Now let's go RT 90.
Student 1: Return! (pointing to the RETURN key)
Student 2: What else did you want to do?
Student 1: OK, I'm going to do something real cool. Real cool. Put FD 1000, 1000. Put another zero. Oh, oh (looking at screen and beginning to type). Now I'm going to do FD, FD, FD 6. What number should I put?
Student 2: Oh! You wanted RT. (typing and looking at screen.)
Student 1: That's right, RT 90.
Student 2: And *then* FD....

In addition, we identified situations in which *conflicts* occurred between the individuals. In this early work, we did not distinguish between *social conflicts* (e.g., possession of the keyboard) and *cognitive conflicts* (e.g., disagreements about problem solution). The following interaction provides two examples of conflict (these are labeled in brackets; note that throughout this paper such labels serve to highlight particular interactions; protocols are not fully labeled).

Student 1: Just do LT. (typing)
Student 2: OK, now.... (bringing her right arm around to type)
Student 1: Wait, wait. (reaches for keyboard as she blocks Student 2's arm) [*conflict*]
Student 2: Oh.... (in a laughing tone)
Student 1: Wait a minute, I don't know where.... (begins to type) RT 90.
Student 2: You're wrong. (looking at screen) I told you so! [*conflict*]

Finally, we determined whether the pair successfully resolved the conflict. In our early work, we coded successful *conflict resolution* whenever the students were able to resolve the disagreement without adult supervision. We used a broad definition of resolution and did not attempt to distinguish among methods. The conflicts characterized above would be considered to be successfully resolved when the students independent of teacher intervention were able to end the conflict and resume the task at hand.

Results of this research revealed that students from both the Logo and the drill and practice groups spent a substantial (and similar) percentage of their time working cooperatively (about 60-70%). They also exhibited similar amounts of conflict (about 1%). The Logo group, however, demonstrated a higher percentage of conflict *resolution* (with Logo students resolving almost all of their conflicts; CAI students almost none). In addition, the students working in Logo outperformed those working in the drill and practice group on a posttest measure of higher-order thinking in mathematics [2]. Examination of their problem-solving behaviors during treatment sessions indicated that Logo students exhibited significantly more behaviors we categorized as *rule determination* (making plans, suggesting steps to task solution, defining the nature of the problem, or establishing parameters of the problem situation). Students in the Logo group engaged in rule determination behaviors approximately 30% of the time, compared to approximately 5% for the students in the drill and practice group. Qualitative analysis revealed that this category was composed mainly of high-order thinking processes.

Additional observations provided confirmation of this qualitative analysis and verified that Logo students were using higher-order thinking processes to a greater degree than were control students. We defined higher-order thinking according to Sternberg's [31] theory of metacomponential functioning. Metacomponents are executive or control processes responsible for planning and evaluating all information processing. They include deciding on the nature of the problem (determining what the task requires; e.g., "we've got to draw a square with a triangle on top"), choosing and combining performance components relevant to the solution of the problem (choosing and sequencing the necessary steps to problem solution; e.g., "we'll make the turtle go up this way about 10, then RIGHT 90 and 10 down, then forward 5, and RT 90 again..."), selecting a representation ("Let's draw the figure on paper first and measure the angles"), monitoring solution processes (keeping track of progress, recognizing an error in thinking, etc.; e.g., "Let's think about what we're doing. Is it gonna work?"), responding to external feedback ("It went 'blub, blub, blub'. That means we were wrong"), and allocating resources ("That's enough time talking. We should draw."). Performance components, in contrast, operate during task execution. The verbalizations of students working in Logo more frequently reflected the use of certain metacomponents—defining the nature of the problem (mean percentage of verbalizations for Logo, 8.7; Control, .2), selecting performance components (Logo, 8.2; Control, 3.2), combining performance components (Logo, 2.5;

Control, .6), and monitoring of solution processes (Logo, 6.6; Control, 3.1). In contrast, students working in drill and practice more frequently made verbalizations that reflected responsiveness to feedback from the computer (also a metacomponent; Logo, 11.6; Control, 14.3) and actual task performance (performance components; e.g., typing in the commands without indication of preplanning; Logo, 17.0; Control, 23.4). These findings indicated that students working in Logo spent more time engaged in task conceptualization and resolution of conflicts than those in drill and practice, suggesting that the Logo environment facilitates peer interaction in the aid of higher-order thinking. The relative contribution of specific social interactions to cognitive gains, however, was not assessed in this study.

Social Processes as Mediators

Veridical measurement of the contributions of social interactions to cognitive change demanded three alterations to our initial research design. First, problem-solving, rather than drill, CAI had to be used with the control group. Otherwise, differences in the occurrence of behaviors such as conflict and conflict resolution could be attributed to the problem-solving nature of the Logo treatment relative to CAI drill. We hypothesized, however, that Logo and problem-solving CAI still possess fundamentally distinct qualities that would differentially evoke certain social behaviors (e.g., CAI could still proceed via turn taking). Second, the construct of conflict had to be elaborated. In this study, we differentiated between social and cognitive conflict. Third, connections between these interactions and outcome measures of higher-order thinking had to be established.

Thus, we conducted a study to determine (a) if students working in Logo and CAI problem-solving environments exhibited different amounts of behaviors indicative of social and cognitive conflict and conflict resolution and (b) if these behaviors were related to higher-level thinking [26]. We hypothesized that the Logo group would exhibit a greater frequency of cognitive (but not social) conflict and of resolutions of this conflict. In addition, we expected that engagement in cognitive (but not social) conflict and its resolution would be related to later performance on a measure of higher-order thinking.

We worked with 12 fourth graders and 28 sixth graders randomly assigned to one of two groups, Logo or CAI problem solving. The goal of both experimental treatments was to increase mathematical problem-solving ability. The teacher used whole-class instruction to introduce new topics. The remainder of the time, students worked in pairs at the computers while the teacher circulated about the room trying to help them discover answers to problems posed. The focus of Logo activities was on turtle graphics as an environment for geometric problem solving. The teacher introduced the concept of procedurality during the first sessions and students completed independent drawing projects. Students then constructed regular polygons and star polygons. The teacher proposed specific projects as challenges. Finally, the teacher introduced recursion and students explored various recursive programs. After students finished work on a particular project or challenge, the teacher encouraged them to pose and undertake problems of their own invention. The CAI group worked on problem-solving computer programs, most developed by Sunburst, Inc. (e.g., using two tanks of given capacities, try to fill, empty, and transfer liquid to produce a given amount; guide a frog through a pattern of lily pads; induce patterns in a series of numbers; use imaginary chemicals to design a monster).

Towards the end of 42 sessions, we videotaped students working in pairs. We coded the videotapes for the occurrence of several social behaviors: Cooperative work, social conflict, cognitive conflict, and resolution of conflict. *Social conflict* was defined as verbal or physical altercation related to the interpersonal aspects of collaboration. This category included behavior such as arguing about turn taking or control of keyboard, criticizing the partner, name calling, or hitting. *Cognitive conflict* was defined as a disagreement concerning task conceptualization or solution (the conflict of centrations discussed previously). For example, students suggest alternative steps to solving the same problem. Continuing the previous interaction, we see examples of both types of conflict.

Student 2: We wanted RT 180. You dumbo. [*social conflict*] You want FD 9.
Student 1: 9, no, 20, 20. [*cognitive conflict*]
Student 2: (typing 20) Hey if it's wrong, don't blame it on me.
Student 1: I *want* it that way.

Finally, we distinguished between attempts at conflict resolution and successful resolution. An attempt at resolution might be evident in one partner's plea to consider another point of view, or in his or her suggestion of a compromise position. Successful resolution might be characterized by one partner's acquiescence to the other's dominance or in both partners accepting a compromise position. For example, this pair neither attempted to resolve nor successfully resolved the social conflict ("You dumbo"). There was an attempt to resolve the last instance of conflict in Student 2's plea ("Hey if it's wrong..."), and, eventually, a successful resolution by way of Student 1's dominance ("I *want* it that way"). We did not attempt in this work to distinguish optimal (e.g., those reflecting consensus or synthesis) from nonoptimal (e.g., those reflecting dominance by one partner) resolution strategies. After the completion of the sessions, students were administered a test of higher-order thinking in mathematics [2].

Differences in total amount of social interaction were not expected. Consistent with the previous study [11], no significant differences were found in the percentage of time spent in cooperative work (approximately 75%). Similarly, the groups did not differ significantly in the amount of social conflict, though the CAI group's percentage was somewhat higher (Logo,

4.4% Control, 8%). As hypothesized, however, the Logo group demonstrated *cognitive* conflict more frequently than did the CAI group (Logo, 10.9%; Control, 7.1%). Furthermore, the previous research had indicated that resolution occurred more frequently in Logo than in CAI (drill) environments. The separation of cognitive and social conflict and resolution in the present study delineated these findings: Attempts at and successful resolutions of cognitive conflict were significantly more frequent in Logo than CAI (problem solving) environments (Logo, 12.7%, 6.8%; Control, 5.1%, 2.3%, respectively), but differences between these groups in the frequency of attempts at or successful resolutions of social conflict were nonsignificant (Logo, 2.0%, 1.2%; Control, 2.5%, 1.6%, respectively). Thus, the Logo group spent more time engaged in cognitive conflict as well as in attempts to resolve such conflict. In addition, they successfully resolved cognitive conflicts more frequently. The Logo group also improved more than the CAI group on the posttest of problem-solving [2]. Moreover, only those behaviors indicative of cognitively–oriented (as opposed to socially-oriented) interactions were significantly correlated with scores on the problem-solving test (r ranged from .31 to .46). In particular, the results suggested that the more frequently students engaged in attempts to resolve cognitive conflicts, and the more they were successful in this endeavor, the higher they scored on the problem-solving test.

Furthermore, this study extended previous research in testing and confirming the *mediational* role of cognitive conflict in facilitating problem-solving performance. We used a mediational model to determine which social behaviors accounted for group differences in problem-solving (posttest) performance [1]. According to this model, a social behavioral variable functions as a mediator if (a) the groups exhibit different amounts of the social behavior, (b) the groups perform differently on the posttest measure, (c) the social behavior is correlated with posttest performance, and (d) group differences in posttest performance are reduced when variations in the social behavior are controlled. Regression equations were used to test each component of this model. Results indicated that resolution attempts and successes, but not the frequency of conflicts, were effective mediators of group differences. The fourth criterion provides a metric of the strength of each mediator. When variations in the frequencies of resolution attempts were controlled, the variance in problem-solving scores accounted for by group placement was reduced from 14% to 6%. For successful resolution, the reduction was 14% to 5%. When the relative mediational influence of attempts and successes was compared (by entering the two social behaviors into a single regression equation), it was *successful resolution* more than attempts at resolution of cognitive conflict that accounted for the differential effects of Logo treatment on problem solving.

Also consistent with previous research, the Logo group engaged in a greater percentage of rule-making behavior (Logo, 15.5%; Control 9.7%). Furthermore, these rule-making behaviors were correlated with individual posttest performance for both groups (r = .46).

Students within either environment who engaged more frequently in problem definition and planning also scored higher on the posttest measure of problem solving.

These results may reflect the different environmental conditions. The greater need for collaborative decision making within the Logo environment may enhance specific problem-solving skills such as problem formulation and resolution. Logo may encourage students continually to negotiate their problem definition and solution strategies, while students in CAI may be more likely to allow the program to structure the interaction rather than to try to negotiate a solution with one's partner. This confirms our hypothesis that in CAI students may disagree and yet successfully continue because their interactions with the computer environment are inherently more structured.

In sum, this research supported the idea that the structuring of controversy during problem solving and students' subsequent efforts to resolve these disagreements may provide the optimal setting for enhancing learning and cognitive development. More specifically, success in resolving these interpersonal discrepancies may facilitate cognitive growth. Yet to be identified were specific strategies students use to resolve cognitive conflicts and the mediational role of these strategies in developing higher-order thinking.

Conflict and Conflict Resolution Strategies as Mediators

In our most recent study, we altered the control treatment to more closely approximate features of the Logo environment [26]. Instead of drill-and-practice or problem-solving software, we used computer-based instruction in writing (CBI-W), deemed an open-ended problem-solving environment more similar to Logo. Observations of students during CBI-W have revealed collaboration regarding technical aspects as well as content of writing [18]. CBI-W, however, still may allow students to collaborate yet maintain their own egocentric points of view. They could share ideas, react to each other's writing, and create a composition together that intermingles but does not integrate divergent ideas. Thus, students writing together could continue without resolving disagreements (e.g., including alternative and even conflicting ideas). In contrast, the nature of Logo is more likely to require that students reach consensus before proceeding with a problem solution, at least at the level of producing programming code to obtain a certain effect. That is, the precision of a computer programming language contrasts with the tolerance and ambiguity of natural language. Therefore, although social interactions may be facilitated by both environments, those focused on collaborative problem solving and conflict resolution may occur more frequently in Logo than in CBI-W environments.

In addition, we extended our study of conflict resolution by examining specific strategies for conflict resolution and the extent to which these strategies explained cognitive gains. That is, we differentiated conflict resolution strategies (for both types of conflict) that were socially based from those that were cognitively based. In *socially-based conflict resolution*, students determine the outcome based on *social dominance* (one partner imposes the solution and/or the other partner acquiesces; e.g., one partner takes over the keyboard by physical force, or one student ignores the other's suggestion for a problem solution and proceeds with his/her idea) or *social negotiation* (through mutual negotiation on a purely social basis; e.g., students agree to alternate control of the keyboard at 10-minute intervals, or students agree to take turns selecting a solution—"we used you idea last time, so we'll use my idea this time"). In *cognitively-based conflict resolution*, students determine the outcome based on the quality of the ideas offered within ongoing dialogue. For example, in resolving a social conflict the partners discuss the relative benefits of alternating keyboard control based on time intervals versus typing skill. The following protocol exemplifies *cognitively-based resolution* of a cognitive conflict. Students are creating a hexagon. They have already made two sides; Student 1 suggests turning the turtle 180° to move around the opposite way, but Student 2 disagrees.

Student 1: Right 180.
Student 2: No! *[cognitive conflict]* Because 180 would go (gesturing)...we want it to go, shh-shh-shh-shh-shh (with gestures indicating that the hexagon can be completed by continuing the same way), we only have to repeat that 5 times. [*attempt* at *cognitively-based resolution*, with an idea-based rationale accompanying the objection]
Student 1: Oh, OK. [*cognitively-based resolution* successful]
But *4* times. [*cognitive conflict*]
Student 2: No, 1, 2, 3, 4, *5*. 'Cause that one's already there. [*attempt* at *cognitively-based resolution*]
Student 1: 1, 2,... 1, 2, *3, 4, 5, 6.*
Student 2: *4.*
Student 1: Yes. [*cognitively-based resolution* successful]

For these students, the process of decision making involved logical consideration of the relative merits of different solutions for achieving goal completion. A consistent interactional pattern was for one student to disagree, then the other to defend his original idea. Towards the end of this interaction, this pattern changes, as Student 1 presents his rationale for a proposed solution even *before* his partner objects (as if in anticipation of a counterproposal).

Student 1: We have to put the left 125, because the turtle will be there, and it will slip around and go shh, shh, shh (gesturing). [an attempt at *cognitively-based resolution*] Try it over. Now right 125.
Student 2: Left. [*cognitive conflict*]
Student 1: Left 125. Yes. [*cognitively-based resolution*] Back down. Left 125. Repeat 6, to make sure it will do the whole thing. [*idea-based rationale* for the proposed solution, even before an objection can be raised] Oh, you forgot the bracket. Bracket, forward, space left, left, 60, bracket, end. Try it.... Ugh!
Student 2: It's no good.
Student 1: Look how close we got it though. It's off, we didn't have it at a steep enough angle, we have to make it 130. Oh, and it's so close. 130. Too much.
Student 2: Put it right 10.
Student 1: 120, like I said. Left 120, try it. We did it! Yes! Teacher, we got it!

Despite the *cognitively conflictual* nature of the interchange (they disagree about *ideas* related to problem solving), the partners are still collaborating. Such interactions were typically very productive and reflected a friendly affective tone. Students were disagreeing about ideas, not making ad hominem attacks against their partners. The focus was on goal attainment; who contributed what seemed to be unimportant.

Based on earlier findings, we hypothesized that students working in Logo would engage more frequently in cognitive (but not social) conflict and successful resolution of these conflicts. In addition, we expected that the Logo group would more frequently resolve conflicts on a cognitive (rather than a social) basis. Finally, we expected that conflict resolution, particularly cognitively-based resolution of cognitive conflicts, would account for treatment differences in posttreatment higher-order thinking.

We worked for 26 weeks with 48 third-grade students, randomly assigned to one of two conditions, Logo computer programming and control. Again, we videotaped all subjects while they worked in pairs on an independent project during the last few weeks. At the end of these sessions, after a delay of one week, we interviewed students individually to determine their use of higher-order thinking.

Measurement of higher-order thinking

Due to the unique nature of the posttest interview and its importance to the results of this study, a description is provided. We designed a 21-item dynamic interview instrument to measure metacomponential functioning [12]. Students are presented problems whose successful solution is assumed to depend on intensive use of one of the following metacomponents: (a) deciding on the nature of the problem, (b) choosing and combining performance components relevant to the solution of the problem, (c) selecting a representation, and (d) monitoring solution processes. They are encouraged to solve the problems without help; if unsuccessful, they are provided a series of five increasingly specific prompts. We made two basic assumptions concerning the prompts. First, if students are successful on an item emphasizing a certain metacomponent (without or without the aid of prompts), then they are using that metacomponent. Second, the number of prompts needed is inversely related to the degree of retrievability of that metacomponent.

An example item, designed to assess cognitive monitoring, follows. "When Albert was 6 years old, his sister was 3 times as old as he. Now he is 10 years old and he figures that his sister is 30 years old. How old do you think his sister will be when Albert is 12 years old?" The prompts are: (a) "Do you have to watch out for mistakes when you do this problem?" (b) "Is there something in the problem that could trick you if you weren't careful?" (c) "Is Albert right when he figures that when he is 10 years old his sister is 30 years old?" (d) "Will his sister always be 3 times as old as Albert? Is that a mistake? Should you multiply or add years?" (e) "Don't make a mistake. When Albert was 6, his sister was 18. 12 years older. What would his sister be 1 year later, when Albert was 7? One year later?"

This instrument has been validated in previous research [8]. Reliability was adequate (coefficient alpha, .84; interrater agreement, 84-99%), as were both construct and criterion-related validity (e.g., the instrument shared a moderate amount of variance with another measure of metacognitive processing administered in similar format and likewise with a measure of academic achievement).

Logo and writing computer environments

The computer environments created were altered and thus require additional description. The Logo tasks per se were similar to those used previously. In this study, however, teachers also introduced "homunculi," cartoon anthropomorphisms of the metacomponential processes, as follows. (a) The Problem Decider was a person thinking about what a problem means (via a "think cloud"). The Problem Decider often asked questions such as, "What am I trying to do?", "Am I doing what I really want to do?", "Have I done a similar problem before?", "How do the parts of the problem fit together?", and "What information do I have or need?" (b) The Representer was an artist, thumb raised, looking off into the distance. She was surrounded by a piece a paper with a graph or chart, another piece of paper with writing, a drawing, and a three-dimensional model. These served as metaphors for various ways to represent a problematic situation. Teachers introduced specific representations (e.g., drawing a diagram or picture) when appropriate. (c) The Strategy Planner was an intelligent-looking man with pencils and pens in his pocket, holding a notebook. Spaced over the remaining sessions, teacher introduced useful strategies in the Strategy Planner's repertoire, such as specific programming steps, decomposing a problem, and systematically guessing and testing. (d) The Debugger was an exterminator; a metaphor for cognitive monitoring (which is more omnipresent in problem-solving than is "debugging" proper; e.g., monitoring includes watching for errors in thinking, not just "fixing bugs"). To develop this more general cognitive monitoring, teachers frequently asked students: What exactly are you doing? Why are you doing it? How does it help you? Does this make sense?

Teachers introduced these homunculi as a part of the Logo-programming and problem-solving process, and used them to support four teaching methods: explication, modeling, scaffolding, and reflection. The goal of *explication* was to bring problem-solving processes to an explicit level of awareness. The teacher used the homunculi to describe processes one uses to solve many types of problems. In addition, teachers would *model* the use of the homunculi-based processes in solving actual problems. *Scaffolding* was employed during students' independent work with Logo; teachers would try to ascertain the process with which the student was having difficulty and would offer prompts and hints focusing on this particular process. When necessary, the teacher would model the use of the process directly. Finally, teachers used *reflection* to elicit discussion about students' use of homunculi in solving programming problems. They reviewed and encouraged use of the metacomponents as students worked on various projects emphasizing basic geometric figures, variables, regular polygons, seasonal interests, contests, list processing projects, and collaborative work on a mural.

Students in the control group received computer experience with different content and without the aid of the homunculi. The goal of this treatment was to develop creative problem solving and literacy. Students composed stories using Milliken's Writing Workshop, an integrated package of prewriting programs, a word processor, and editing programs. A composition process model, including the phases of prewriting (e.g., brainstorming), writing, revision (conferencing), and editing (e.g., checking spelling and grammar), served as a framework for instruction. Teachers introduced and discussed each phase with the entire group. Students then worked in pairs, with teachers supplying guidance via the previously-described four teaching methods. Students wrote and illustrated compositions, based on self-selected topics as well as "theme" topics such as seasonal interests and specific tasks (e.g., write a column for a group newsletter). Like students in the Logo treatment, those in the control condition were encouraged to solve these open-ended problems cooperatively.

Mediation of higher-order thinking

The Logo group outperformed the CBI-W group on the total score of the higher-order thinking (metacomponential) interview [8]. Consistent with previous research, the groups did not differ concerning the percentage of time engaged in cooperative interaction (approximately 50%). Also, the groups did not exhibit differing amounts of social conflict (Logo, 4.4%; Control, 5.0%), cognitive conflict (Logo, 1.3%; Control, 1.0%), attempts to resolve social conflict (Logo, 0.8%; Control, 0.1%), or attempts to resolve cognitive conflict (Logo, 1.9%; Control, 1.2%). Group differences were revealed, however, in resolution strategies for both types of conflicts. The Logo group more frequently used social negotiation to resolve social conflicts (Logo, .1%; Control, 0.0%). In resolving cognitive conflicts the Logo group more frequently used cognitively-based strategies (Logo, 0.9%; Control, 0.4%), whereas the CBI-W group more frequently used social negotiation (Logo, 0.0%; Control, 0.1%).

Finally, we again used a mediational model to determine which of these conflict resolution strategies accounted for group differences in higher-order thinking. Results indicated two mediators: (a) the use of social negotiation to resolve social conflicts and (b) the use of cognitively-based strategies to resolve cognitive conflicts. The variance in problem-solving scores accounted for by group placement was reduced from 10% to 5% for (a) and from 10% to 4% for (b). When the relative mediational influences of these two were compared, it was cognitively-based resolution of cognitive conflicts that accounted for the differential effects of Logo treatment on problem solving. Thus, results support the role of both social process variables as mediators, with stronger support for cognitively-based resolution of cognitive conflicts.

Discussion and Conclusions

A fundamental reason that certain Logo environments facilitate the development of higher-order thinking is their inclusion of social aspects of learning. Students work in pairs and teachers encourage them to solve problems cooperatively. Observations of students' social behavior indicate that (a) different types of educational environments engender different types of social interactions and (b) the enhancement of cognitive skills is mediated by engagement in specific conflict resolution strategies that are engendered by the Logo environment. In particular, both social and cognitive strategies that involve negotiation of perspectives are important. It is social negotiation, not social dominance, that mediates group differences in cognition. Similarly, the majority of cognitively-based resolutions (also mediators) reflect the use of *negotiation*; that is, partners engage in a process of trying to reach a solution that is mutually acceptable. At the least, this solution reflects a compromise position (e.g., a third position that incorporates aspects of both ideas but is not a true synthesis). At the optimal level, the solution reflects an integration of the two ideas into a qualitatively different third perspective (reflecting the notion of the dialectic combination of two opposing ideas into a coherent, higher-level idea). Furthermore, despite the importance of resolution of social conflicts, the resolution of cognitive conflicts (particularly resolution on a cognitive basis) is the stronger mediator between Logo environments and gains in higher-order thinking skills.

Why should the resolution of social conflict be less influential? Social negotiation requires the consideration of an alternative perspective, but does not necessarily involve consideration of the extent to which the quality of the alternative perspective has particular merit. Partners need only to reach a compromise concerning their social needs; whether one person's thinking is in some way superior to the other is not an explicit consideration. In contrast, cognitively-based resolution requires that individuals contemplate the relative efficacy of discrepant problem-solving strategies. Thus, they must consider the perspective of another person *and* confront in a direct and concrete manner the nature of the discrepancy and the relative merits of each proposed solution.

In sum, these studies suggest that experience in a Logo environment can facilitate the development of higher-order thinking by engendering a specific resolution strategy, the interpersonal coordination of divergent ideas. The process of coordination may result in a restructuring of the thinking processes on an interpersonal plane that leads to a restructuring on an intrapersonal plane [33]. Thus, partners might revise the manner in which they conceptualize and solve the immediate problem. In addition, they may devise new socially-constructed methods for problem solving that reflect a synthesis of individual methods. Finally, the thinking processes of the individuals may be modified as they internalize these new socially-constructed methods. As noted in the last protocol, one student showed evidence of altering his presentation of solution strategies, incorporating earlier justifications made in response to the partner's objections (i.e., presenting his rationale for a proposed solution as if in anticipation of a counterproposal).

Our studies have two implications for social-cognitive research and theory. First, they provide further evidence of the criticality of conflict resolution. Previous research has suggested that engagement in conflict will not necessarily lead to cognitive advances; indeed, merely disagreeing with others has been shown to impede advances in moral reasoning [15]. Students have to go beyond mere disagreement to benefit from the conflict of ideas [4; 15]. Our research strongly supports such findings; it is conflict resolution, not the occurrence of conflict itself, that accounts for growth in higher-order thinking. Second, our findings extend existing social-cognitive research by determining that certain resolution strategies—those that involve attempts to synthesize viewpoints—are of particular importance. We did not examine the extent to which the final solution reflected a true synthesis (combination of the ideas into a third, well-integrated coherent and qualitatively different idea). The partners needed only to achieve some compromise position that was reached by considering the quality of the divergent views. Future research could address such distinctions.

Research also needs to determine whether cognitive growth can be attributed to direct links between conflict resolution and higher-order cognitive activity as students solve problems, or whether it results in a more indirect way from the experience of taking another's perspective. In addition, we observed social interactions occurring spontaneously as students worked together. Researchers might directly intervene to encourage the use of processes such as cognitively-based conflict resolution, thus experimentally evaluating their efficacy.

Finally, findings such as those reviewed here do not imply that explicit interindividual conflict and its resolution are solely responsible for gains in higher-order thinking (these social processes typically account for only 10 to 15% of the variance in higher-order thinking scores). Perhaps the changes in cognitive ability result from the *intra*individual conflicts of centrations that occur even when conflict between partners is not observed. That is, a conflict of centrations may be engendered within one or both of the partners via the exchange of ideas, even without observable disagreement (i.e., reciprocal sense-making). Finally, cognitive disequilibrium may result from an individual's interactions with the physical environment regardless of his or her interactions with a partner. In these situations, the resolution of the discrepancy may occur internally (within the individual) rather than externally (between individuals). Such a notion is consistent with non-social components of Piagetian theory. For example, disequilibrium resulting from individual's reflections on actions made on the physical environment are oft reported [30]. It may be that some combination of the three postulated mechanisms of cognitive change— (1) interindividual conflict, (2) collaboration without observed conflict, (3) intraindividual conflict (possibly with a minimal level of one or more of these)—are necessary for optimal growth.

Of course, previously described cognitive aspects of educational environments also may be important [8]. For example, in these environments, higher-order thinking processes are made explicit, students engage in all phases of the problem-solving process, and teachers use a comprehensive set of pedagogical approaches. Higher-ordering thinking processes are articulated as explicitly and thoroughly as possible, as they arise in different contexts. Teachers ask students to verbalize their goals and solution procedures, as well as their use of these thinking processes, before attempting a solution on computer. The project approach to Logo engages students in all aspects of problem solving, including determining the nature of the problem, representing the problem in different modalities, selecting strategies, and monitoring. In addition, both modeling and scaffolding allow students to perceive the full task. Modeling provides students a schema for applying the homunculi-based processes. Scaffolding encourages successive approximation of the entire range of skills necessary for task completion. Future studies should investigate the contribution of these cognitive aspects of the pedagogical environment, as well as the interaction of the social and cognitive aspects.

What, then, is the role of Logo in all this? The language itself is undoubtedly neither a necessary nor sufficient condition. It did, however, play a major role in these experimental environments. Logo can serve as a tool that facilitates the role of the teacher as a mediator of higher-order thinking. In addition, Logo environments encourage collaborative problem solving and engender conflict and negotiation and resolution of that conflict. Perhaps most important, even compared to other problem-solving, open-ended computer environments, Logo environments engender a high-level type of conflict resolution involving coordination of divergent perspectives. Thus, evidence suggests that the use of Logo is not an arbitrary decision. Nevertheless, future research might ascertain whether other tools can serve a similar function. Some may also argue that the beneficial characteristics should be integrated into the total educational environment rather than emphasized within Logo sessions. Although such integration is advisable, infusion often leads to diffusion, severely limiting the focus and impact of the program. Therefore, a program that emphasizes higher-order thinking, such as the Logo environments described here, may make a significant instructional contribution. Indeed, Logo environments appear to support the development of motivated learners who seek to validate their ideas not only through their own reasoning, but also through meaningful communication with others [7]. Development of this type of student is consonant with recent positions of such organizations as the U.S. National Council of Teachers of Mathematics, which supports the cultivation of students who believe "that they have the power to do mathematics and that they have control over their own success or failure" and "explain, conjecture, and defend [their] ideas" [28, p. 29, 78].

References

1. Baron, R. M., & Kenny, D. A. (1986). The moderator-mediator variable distinction in social psychological research: Conceptual, strategic, and statistical considerations. *Journal of Personality and Social Psychology, 51*, 1173-1182.
2. Battista, M. T., & Clements, D. H. (1986). The effects of Logo and CAI problem–solving environments on problem–solving abilities and mathematics achievement. *Computers in Human Behavior, 2*, 183-193.
3. Bearison, D. J. (1982). New directions in studies of social interaction and cognitive growth. In F. C. Serafica (Ed.), *Social-cognitive development in context* (pp. 199-221). New York: Guilford Press.
4. Bearison, D. J., Magzamen, S., & Filardo, E. K. (1986). Socio-cognitive conflict and cognitive growth in young children. *Merrill-Palmer Quarterly, 32*, 51-72.
5. Carmichael, H. W., Burnett, J. D., Higginson, W. C., Moore, B. G., & Pollard, P. J. (1985). *Computers, children and classrooms: A multisite evaluation of the creative use of microcomputers by elementary school children.* Toronto, Ontario, Canada: Ministry of Education.
6. Clements, D. H. (1986). Effects of Logo and CAI environments on cognition and creativity. *Journal of Educational Psychology, 78*, 309-318.
7. Clements, D. H. (1986-91). Logo: Search and research [research review column]. *Logo Exchange.*
8. Clements, D. H. (1990). Metacomponential development in a Logo programming environment. *Journal of Educational Psychology, 82*, 141-149.
9. Clements, D. H., & Gullo, D. F. (1984). Effects of computer programming on young children's cognition. *Journal of Educational Psychology, 76*, 1051-1058.
10. Clements, D. H., & Nastasi, B. K. (1985). Effects of computer environments on social-emotional development: Logo and computer-assisted instruction. *Computers in the Schools, 2*(2-3), 11-31.
11. Clements, D. H., & Nastasi, B. K. (1988). Social and cognitive interactions in educational computer environments. *American Educational Research Journal, 25*, 87-106.
12. Clements, D. H., & Nastasi, B. K. (1990). Dynamic approach to measurement of children's metacomponential functioning. *Intelligence, 14*, 109-125.
13. Clements, D. H., & Nastasi, B. K. (in press). Computers and early childhood education. In T. Kratochwill, S. Elliott, & M. Gettinger (Ed.), *Advances in school psychology: Preschool and early childhood treatment directions.* Hillsdale, NJ: Lawrence Erlbaum.
14. Cobb, P. (1988). The tension between theories of learning and instruction in mathematics education. *Educational Psychologist, 23*, 87-103.
15. Damon, W., & Killen, M. (1982). Peer interaction and the process of change in children's moral reasoning. *Merrill-Palmer Quarterly, 28*, 347-367.
16. De Corte, E., Verschaffel, L., Hoedemaeders, E., Schrooten, H., & Indemans, R. (1988, April). *Acquiring programming skills in Logo: An exploratory study with sixth graders.* Paper presented at the meeting of the International Association for Computing in Education, New Orleans.
17. De Corte, E., Verschaffel, L., & Schrooten, H. (1990). *Cognitive effects of learning to program in Logo: A one-year study with sixth graders.* Manuscript submitted for publication.
18. Dickinson, D. K. (1986). Cooperation, collaboration, and a computer: Integrating a computer into a first-second grade writing program. *Research in the Teaching of English, 20*, 357-378.
19. Doise, W., & Mugny, G. (1984). *The social development of the intellect.* New York: Pergamon.
20. Forman, E. A., & Cazden, C. B. (1985). Exploring Vygotskian perspectives in education: The cognitive value of peer interaction. In J. V. Wertsch (Ed.), *Culture, communication, and cognition: Vygotskian perspectives* (pp. 323-347). New York: Cambridge University Press.
21. Johnson, D. W., Johnson, R. T., Pierson, W. T., & Lyons, V. (1985). Controversy versus concurrence seeking in multi-grade and single-grade learning groups. *Journal of Research in Science Teaching, 22*, 835-848.
22. Johnson, R. T., Brooker, C., Stutzman, J., Hultman, D., & Johnson, D. W. (1985). The effects of controversy, concurrence seeking, and individualistic learning on achievement and attitude change. *Journal of Research in Science Teaching, 22*, 197-205.
23. Johnson, R. T., Johnson, D. W., & Stanne, M. B. (1986). Comparison of computer-assisted cooperative, competitive, and individualistic learning. *American Educational Research Journal, 23*, 382-392.

24. Lehrer, R., & Smith, P. C. (1986, April). *Logo learning: Are two heads better than one*? Paper presented at the meeting of the American Educational Research Association, San Franciso.
25. Muller, A. A., & Perlmutter, M. (1985). Preschool children's problem–solving interactions at computers and jigsaw puzzles. *Journal of Applied Developmental Psychology, 6*, 173-186.
26. Nastasi, B. K., & Clements, D. H. (1990). *Social-cognitive behaviors and higher-order thinking in educational computer environments*. Manuscript submitted for publication.
27. Nastasi, B. K., & Clements, D. H. (1991). Research on cooperative learning: Implications for practice. *School Psychology Review, 20*, 110-131.
28. National Council of Teachers of Mathematics. (1989). *Curriculum and evaluation standards for school mathematics*. Reston, VA: Author.
29. Perret-Clermont, A. (1980). *Social interaction and cognitive development in children*. New York: Academic Press.
30. Piaget, J., & Inhelder, B. (1967). *The child's conception of space*. New York: W. W. Norton & Co.
31. Sternberg, R. (1985). *Beyond IQ*. Cambridge, MA: Cambridge University Press.
32. Strand, E. B., Gilstad, B., McCollum, P., & Genishi, C. (1986, April). *A descriptive study comparing preschool and kindergarten Logo interaction*. Paper presented at the meeting of the American Educational Research Association, San Francisco, CA.
33. Vygotsky, L. S. (1978). *Mind in society: The development of higher psychological processes*. Cambridge, MA: Harvard University Press.

Effects *with* and *of* computers and the study of computer-based learning environments

Gavriel Salomon

University of Arizona, College of Education, Tucson, AZ 85721, USA

Abstract: The paper discusses the meaning of this book's title emphasizing (a) that effective *computer-based learning environments* are drastically transformed learning settings, and (b) that such learning environments affect *problem solving* in two ways: They affect problem solving during interaction with computer programs and they leave a more lasting cognitive residue as a consequence. Three experimental studies are briefly described that illustrate this point. However, the use of computers in these studies was not accompanied by any significant environmental changes. Such changes came in another project carried out in eighth grade classes. The epistemologies underlying the single-variable experiments (illustrating an analytic approach) and the classroom project (illustrating a systemic approach) are contrasted and discussed, leading to differential implications for the study of computer-based learning environments and problem solving.

Keywords: Intellectual partnership, cognitive residue, learning environment

The issue of "*Computer-based learning environments and problem solving* ", the title of the present volume, needs first some unpacking. Do we mean by the first part of the title no more than the use of computers in regular classroom environments? Or do we imply something wider that goes beyond these common and traditional settings? And when we add "... *and* problem solving" do we mean that to these computer-based learning environments one ought to add the independent quality of problem solving to create a new entity such as "computer-based problem solving"? Or do we mean that computer-based learning environments *affect* (and are affected by) the way learners come to solve problems? And if the latter, what in the process of problem solving is supposed to be affected, why, by what, and in what sense? In what follows I wish to present my understanding of the title of the book, elaborate on it, and discuss the conceptual, practical, and epistemological distinctions that guide me in my explication.

Computers are supposed to be the new wave of the future, the "one in many centuries invention" capable of transforming education in profound ways [7]. Such promises have, however, led to a number of assertions and expectations which, given the wisdom of hindsight, prove to be naive. One of these is the expectation that computers and the software or courseware that can be "run" on them could profoundly affect education. However, the accumulation of research on drill and practice programs, on the effects of programming, even on the use of simulations, has pointed to the unavoidable conclusion: *In and of themselves* computers affect very little [e.g., 6]. And given the proposition that computers are mainly "tools for making, doing, and creating" [21], it becomes apparent that not the computers are the true agents of change, but rather the kinds of learning activities that they afford and invite. The more different these learning activities are from the kind of activities that preceded them, the greater might their impact on classroom learning be.

Computer-Based Learning Environments

But learning activities are not isolated islands in an otherwise unchanging ocean of instructional procedures and settings. Activities are goal-directed, and are carried out within a social and cultural setting which promotes them, sanctions them, and gives them meaning. Seatwork is an activity that emanates from and fits well into a teacher-dominated classroom context in which individual learning and rehearsal of discrete skills is the name of the game. Similarly, computer-literacy classes that are conducted in isolated computer labs emanate from and are congruent with a culture in which isolated skills are drilled away from where they are to be applied (one learns to write with the word processor without having words to process; meaningful words are "processed" in, say, the history class but away from the computer).

The computer, as a tool for making, doing, and creating, does not fit this kind of social and cultural context. The computer, in this sense, functions like a Trojan Horse, for the activities that it affords require profound changes in the learning environment. The very use of computer tools assumes that they serve in more than in an add-on capacity. Indeed, computer tools carry with them implicit assumptions about self-guided exploration and design, even playful activity, team collaboration, integrated curricula, mutual consultation, and teachers' orchestration of activities rather than teacher domination [14]. In other words, to make computer use affect education it cannot just be introduced as an addition to otherwise unchanging classroom practices the way, for example, television could; with its proper introduction *everything* in the classroom, possibly in the school as a whole, changes. If the computers are the trigger they better have something that is triggered by them.

So, here we have the first round of unpacking of the workshop title. Computer-based learning environments are *not* learning environments to which computers have been added. Nor are these environments that appear on computer screens. Rather, these are relatively new environments in which computer-afforded activities have been fully integrated into other activities, affecting them and being affected by them: recitation of isolated facts is replaced by exploration; individual activities are replaced by team work around the computer; the ping-pong-like pattern of communication between single students and teacher is replaced by intensive interaction among students; discipline-based isolated curricula become integrated; and meanings are socially negotiated and "appropriated" rather than handed down ready-made [9]. The question of whether computers are needed to affect such changes is a valid one, and I will deal with it throughout this paper.

Affecting Problem Solving

Turning to the second part of the title -- "... *and problem solving*", the unpacking becomes trickier. What about problem solving? Are we to understand that computer-based learning environments, with the activities that they can promote, are to affect problem-solving *performance* during interaction with the intelligent capabilities of the computer and/or during computer-afforded social interactions? Alternatively, could it not be that such learning environments are expected to promote problem-solving *abilities*? These are two different kinds of effect and we better not confuse them. Elsewhere [18] these effects have been distinguished and were labeled effects *with* computers and effects *of* them, respectively.

Effects *with* computers are those changes in performance that become possible when one is equipped with a tool with which an *intellectual partnership* is created [12, 15]. A computer tool that affords, for example, the activity of building and running a complex multivariate model [8], something that could not possibly have been done without the computerized model-building tool, may well increase the chances that learners engage in higher-order thinking. This becomes possible because the computer tool (e.g., STELLA) shares with learners some of their cognitive load: It carries out low-level computations, it provides memory supports, it offers immediate responses, and -- most importantly -- it can supplant mental processes of simultaneous juggling of interrelated variables that learners cannot possibly carry out in their own heads. By providing all this, it allows learners to generate new hypotheses and test them, to examine the routes they have taken so far, to evaluate their products, and to solve complex problems they have never even tried to solve before. Note that the effects alluded to here totally depend on the presence of powerful computer tools, tools that free learners to engage in mental operations they could not have engaged in otherwise. These are effects that are attained *with* the computer and *during* activity with it.

But there is the possibility of yet another kind of effect -- an effect *of* computer-afforded activity -- that transcends improved performance: It is an effect that pertains to the *cognitive residue* that the partnership with the computer tool might leave behind in the form of improved mastery of a skill or strategy or a better developed ability. From being able to design and test multivariate models of complex systems while using STELLA, learners may come to be better able to think systemically [8]. This does not mean that they could now "run" multivariate models in their heads, but rather that they have acquired some mental strategies for handling such complex life-events. Hearing about the spread of AIDS, they would now ask questions about the many factors that contribute to it, about their interdependence, about self-sustaining and self-correcting feedback loops, and about loops within loops. Similarly, having written essays with the aid of a *Writing Partner* that provides ongoing expert-like guides, young writers would be expected to internalize some of that guidance to make them better able writers when not using that tool any more [23].

The partnership with a computer can be expected to leave some worthwhile cognitive residue *of* it. Such an expectation follows quite nicely from a Vygotskian view of learning and development. The intellectual partnership with computer tools creates a zone of proximal development whereby learners are capable of carrying out tasks they could not possibly carry out without the help and support provided by the computer. This partnership can both offer guidance that might be internalized to become self-guidance [17] and stimulate the development of yet underdeveloped skills, resulting in a higher level of skill mastery. Whether the learners' improved abilities are now assessed with the tool -- how complex are the models they now build? -- or whether they are assessed when on their own, one could see that their intellectual level of performance has improved. The partnership is expected to leave a cognitive residue.

But the distinction between effects *with* and *of* creates a dilemma which has important implications. One side of the dilemma argues that in the presence of powerful tools that can improve the joint performance of learners-plus-computers there is no point in aiming for a cognitive residue. After all, as it is argued [e.g., 13], if people can off-load much of their cognitive activity onto such artifacts as computer tools, what difference does it make whether they become better able as solo problem solvers? Indeed, "Almost any form of human cognition requires one to deal productively and imaginatively with some technology. To attempt to characterize intelligence independently of those technologies seems to be a fundamental error" [10, p. 356]. A similar argument has been made with respect to print literacy: "It is misleading to think of literacy in terms of consequences. What matters is what people do with literacy, not what literacy does to people" [11, p. 15].

The other side of the dilemma remains unsatisfied with the idea that *all* that counts is the better performance of a person-plus-computer. First, one could argue that each partnership-encounter with a computer tool leaves some residue which makes the next encounter more intelligent. After all, students' ability to intelligently use STELLA, the model-building tool, improves from one encounter to the next.

Second, it is argued that computers are not and can never be so ubiquitous as to allow us to relinquish improvements of our solo problem-solving abilities. After all, one could claim, there are countless problems that one would need to solve -- Shall I go to college? What is my opinion of the changes in the USSR? Do I need to repaint my house? How shall I approach my parents, so dissatisfied with my school performance? -- cannot be relegated to any hypothetical partnership with information technology. One would have to rely on one's self and one better be equipped with sufficiently well developed problem-solving abilities to handle such problems intelligently [18].

Third, even the most powerful of partnerships with technology cannot supplant intentionality, the development of well-reasoned beliefs and grounded rationales for choosing one alternative over the other, elements that differentiate "mechanized" from human problem-solving behavior [5]. High level partnerships have their great utility, in and of themselves, but they are no substitute for the cultivation of a cognitive residue in the form of better problem-solving skill mastery.

Emphasizing effects *with* technology as the sole goal versus seeing in them steps toward effects *of* technology has important design implications. If the former is the goal then one would want to design systems that are as clever as one could make them such that the joint performance of person and computer is maximally intelligent. If the latter is the goal then one would want to design systems that make the human partner more intelligent, not the joint system. Let me illustrate. Say you are a hospital director. You would wish to see your physicians come up with the best possible diagnoses. You would, most likely, opt for the best medical expert system you could find, even if it reduced its human partners to no more than data feeders. But consider, on the other hand, that you are the director of a medical school. You want to educate your students to be as good diagnosticians as possible. Thus, you are less likely to opt, for the moment at least, for highly intelligent expert systems but rather for systems that would sharpen the students' diagnostic skills. As Scardamalia, Bereiter, McLean, Swallow and Woodruff [19] have pointed out:

> "In most software design it is presumed desirable to make the software as intelligent as possible and demand as little intelligence as possible from the user. Educational applications, on the other hand, should be aimed at developing the intelligence of the user. Educationally irrelevant burdens should be minimized, but not in ways that deprive students of occasions to develop the planning, monitoring, goal-setting, problem-solving and other higher-order abilities that are important objectives of education" (p. 53).

I wish to propose that what concerns the second part of this book's title applies to the move from effects *with* computer tools that facilitate problem-solving performance during human-computer partnership to effects *of* this partnership whereby improved problem-solving skills and strategies are attained as its consequence.

Empirical Evidence: The Story of Three Studies

Three studies have been carried out to address the question of whether interaction with computer programs that explicitly employ relatively novel expert-like strategies leave measurable cognitive residues in the form of (a) internalization of the strategies, and (b) improved performance that depends on the employment of these strategies. One study was based on the *Reading Partner* - a tool designed to accompany the process of reading for meaning and assist it [17]. The second, the *Writing Partner,* was a similarly designed tool for the purpose of assisting the process of essay writing [23]. A third study was more recently carried out with a new version of the *Writing Partner*, yielding similar results to the ones obtained in the first writing study.

The reading and writing tools were neither designed to teach their users any context-bound specific knowledge nor to provide any drill of such. Rather, they were designed to serve as open-ended tools which also provide expert-like guidance in the processes of writing and reading. In what follows, the tools and the studies conducted with them are briefly described.

The *Reading Partner* entailed a number of texts which were accompanied by metacognitive-like questions based on recent research on the psychology of reading comprehension [e.g., 2, 3] and duplicating expert-like advise pertaining to good reading. While reading, the user encountered metacognitive-like questions as in the following examples: "Can I conjure up an image of the story?"; "Can I predict on the basis of the title what the story is all about?"; "Let me find the key sentences in the text"; "Do I really understand this text?"; "What is it that I don't understand here?".

The study with the *Reading partner* was carried out with 74 thirteen year olds randomly divided into three groups: a text-only group that read eleven texts off a computer screen (to control for reading practice); a text-plus-content-questions group, that read the same texts accompanied by multiple-choice content questions (to control for possible effects of increased attention and focus); and the *Reading Partner* group that interacted with the originally designed tool. Subjects read the eleven texts over three sessions. Reading comprehension was pretested by means of a standard test. Each reading session ended with a self-report questionnaire of effort expenditure, a self-report measure the validity of which as a fair predictor of learning has already been studied [see e.g., 17]. Three posttests were administered two weeks later: (a) subjects were asked to put down in writing the kind of advice they would give a friend "who reads with poor understanding", considered a measure of reading-related metacognitive reconstruction, (b) a second version of the standard reading comprehension test; and (c) a test of far transfer -- an essay *writing* task, administered one month after termination of the study.

A number of important findings emerged. First, subjects in the *Reading Partner* group reported expending significantly more mental effort in the process of interaction with the tool than the other two groups (an important finding in light of the argument that an intellectual partnership with computer tools that is to yield desirable cognitive residues requires mindful engagement; see 15). The difference between the groups accounted for 7% of the effort expenditure variance. Second, the *Reading Partner* group reconstructed reading strategies and reading-related metacognitions significantly better than the other two groups. The difference between the groups accounted for 46% of the variance. Third, the *Reading Partner* group significantly outperformed the other two groups on the reading comprehension posttest. The difference between the groups accounted for 17% of the posttest variance. Fourth, when scores of metacognitive reconstruction were partialled out, the difference in posttest reading comprehension totally disappeared, suggesting that the improvement in reading comprehension was due to the internalization of metacognitions. Moreover, the delayed essay *writing* task, a test of far transfer, put the *Reading Partner* group significantly ahead of the other two groups. The difference between the groups accounted for 13% of the delayed posttest writing task variance. Again, partialling out scores of metacognitive reconstruction erased this difference. It should be noted that nothing in the program explicitly suggested any relationship to writing. The improvement in writing, totally accounted for by the internalization of reading-related metacognitions, suggests that the subjects have not just copied the program's explicit guidance but have acquired a more general *tendency* for self-monitoring.

The first version of the *Writing Partner* was designed in a way that resembled the design of the *Reading Partner*. It was based on a word processor which also forced essay planning and provided metacognitive-like guidance during essay writing, and revising. The tool used interactive questions that followed recent research on the psychology of writing [4]. Examples of the guiding questions were: "Is your essay designed to persuade or to describe?"; "What is the persuasive argument you would want to make?"; "Wouldn't you need an example here?"; "Would a reader with a different opinion agree with this argument?", "Does this paragraph follow from the preceding one?", and the like. Guiding questions were contingent upon the users' inputs.

The first study with the *Writing Partner* was carried out with 60 young adolescents (ages 13-15) randomly divided into three groups: one group was equipped with only a word processor, the second was equipped with the *Writing Partner* which provided guidance without the writers asking for it, and the third group was equipped with the same guidance-providing tool but its guidance was provided only upon the writer's request. Subjects' essay-writing quality was first pretested by expert writing teachers. Then, the subjects - working with one of the three tools - wrote five essays on assorted topics during a two week period. These sessions were followed by a metacognitive reconstruction measure, similar to the one used in the previous study. Two weeks later, an unaided essay writing posttest was administered.

Five important findings emerged from that study. First, the *Writing Partner* group (the one in which guidance was imposed, not voluntarily solicited) wrote significantly better essays while using the tool. The difference between the groups accounted for 35% of the writing quality variance. Second, that group showed a significantly better mastery of writing-related metacognitions than the other two groups. The difference between the groups accounted for 26% of the writing quality variance. Third, the quality of posttest essays of the group with *Writing Partner* entailing imposed guidance was significantly higher than that of the other two groups. Partialling out initial differences, the difference between the groups accounted for 14% of the writing posttest variance. Fourth, and most importantly, when scores of metacognitive internalization were partialled out by means of ANCOVA, the difference in posttest writing quality totally disappeared. This finding suggests that the improvement in writing quality manifested by the *Writing Partner* group was due to the internalization of metacognitions, the kind exposed to during the interactive writing with the tool. Finally, correlational analyses showed that whereas prior writing ability accounted for posttest writing quality in the control and voluntary guidance groups, in the *Writing Partner* group it was effort expenditure, amount of exposure to the expert-like guidance and quality of training essays that accounted for the outcomes.

A new version of the *Writing Partner* was recently designed by us, the guidance of which was far more contingent upon students' inputs, and which offered other guiding and self-diagnosing features such as an "I'm stuck" pull-down screen. A study comparing its effects with those of a word processor clearly showed that (a) quality of essays written with it was superior to the those written by the word processing only group (accounting for 21% of the variance), and (b) quality of pencil-and-paper essays written a while later showed the same advantage (the difference between the groups accounting for 16% of the writing quality variance).

Together, the three studies support the argument that computer tools that embody certain guiding qualities enable students to enter into intellectual partnerships with them. More importantly, though, when sufficient mental effort is expended by the tools' users, such programs can leave a generalizable cognitive residue in the form of improved self-monitoring which facilitates performance later, away from the tool's guidance. It should be noted, of course, that all three studies were carried out in regular classrooms with their regular classroom teachers. It should, however, also be noted that in the three studies no attempt was made to integrate the use of the computer tools into regular classroom activities or curricula. In this sense, the use of the computer tools was an add-on to otherwise unaffected classroom practices.

Moving out From the Laboratory Into a New Classroom Environment

The three studies described above support the contention that cognitive effects *with* computer tools can lead to desirable effects *of* them. But what have these studies to do with computer-based learning environments, as explicated in the first part of this paper? After all, genuine computer-based learning environments, with all the curricular, social, interactional activity, and teacher-role changes that they entail, are not just a simple, linear upscaling of the laboratory conditions studied in controlled classroom experiments. The move from lab to whole novel classroom environments is a qualitative, much-encompassing change. It entails far more than the changing of one variable, keeping all other things as equal as possible. The following illustrates this change.

Working with an eighth grade teacher in a middle class suburban school, we wanted to introduce the use of computer tools into the study of World War I. We could, of course, have continued to teach that chapter in history the way all other chapters are taught, only adding computers for the writing of reports, perhaps even using some ready-made database. But this would render the computer no more than an unnecessary add-on, the effects of which would be miniscule at best. We decided to use the computer and the tools it makes available as triggers for change, restructuring the whole learning environment. So, students started working in teams of three, each with its computer, playing the role of one or another country participating in the War. Computers were linked to each other to allow communication within and outside the classroom. As there were four parallel classes there were, by necessity, four "Italian", British", German", French", "Russian", etc. teams. Each team "prepared" for war, constructing its own database to record the human resources, materiel, strong and weak points, grievances, goals, etc. of its own country and that of its "allies" and "foes". One team served as "historians" announcing each day's events, and sending all teams scrambling to their computers to update their databases, draw maps, write memos, confer through the communication network with their respective allies. Books, atlases, and encyclopedias were heavily used, secrets were traded, and troops were moved on the large cardboard map on the floor. Increasingly more and more diversified use was made of the computers in the numerous capacities that its tools afforded, ranging from word processing to database construction, from communication to graphing, and from spreadsheet construction to model building. Problem-solving performance in the teams appeared to have improved markedly throughout the project, aided by the support that the computer tools and the social collaboration afforded. The project took a few weeks, culminating in a Versaille treaty, negotiated and "signed" during a solemn mock-conference in the school's auditorium.

A number of post-project measures were taken. A transfer measure of problem-solving ability (ability to organize and handle a large and confusing collection of information) showed a large and significant difference between the project classes and no-project ones: participation in the project accounted for 23% of the variance on the information-organization and handling measure. Another, attitudinal questionnaire, showed large pretest to posttest changes in the so-called experimental classes, and between them and the so-called control classes in terms of perceived self efficacy, excitement in learning history, and in feelings of accomplishment (but *not* in the perception of history as a science relevant to life today).

Were the computers the main agent that made the difference in performance, perception, and ability? Such a question would appear to be a legitimate one, but upon closer examination it turns out to be unanswerable. Does the yeast make a difference in baking a cake? Of course it does, but the baked cake, although impossible without the yeast, contains much more than yeast. The yeast is useless in the absence of the other ingredients. The same applies to the project. It would be quite unthinkable to carry out such a project without the computer tools and the activities they afforded, but the project entailed changes that far transcended the sheer introduction of computers. *Most everything has changed -- curriculum, activity, engagement in learning, instruction, social interaction, teacher's role, modes of student evaluation.*

Such system-wide changes in the learning environments need to be studied. But the way we have experimentally studied the cognitive effects of the reading and the writing tools is not the way to study the effects of the project. Too many assumptions underlying the experimental study of one or another computer-related variable are violated when it comes to the study of changes in whole computer-based learning environments. For example, in the experimental setting we manipulate one or two variables, assuming that all other "independent variables" remain unchanged. Not so in the case of studying a system's changes: Everything is allowed to change. Moreover, while in the experimental studies we can easily separate independent from dependent variables, this becomes close to impossible in the study of changes in whole learning environments. The status, for example, of students' mental engagement in the task, while dependent upon the availability of computer tool and team-based exploratory activity, itself affects the use of the tool and the way the team operates. Circular relations replace linear ones.

Similarly, in the experimental setting we assume that variables are independent of each other and that they interact to produce effects. In the study of systems, on the other hand, we assume nothing of the kind. On the contrary, we assume that variables and events not only affect each other reciprocally [e.g., 22], and appear as "clouds of correlated events" [20], but they also *define* each other [1]. Indeed, exploratory team work around the computer derives its specific meaning from its perceived purpose of serving the anticipated final treaty conference. And the meaning attributed to, say, the ongoing construction of a database is part and parcel of the meaning attributed to the competitive nature of the project, which itself derives its meaning from the role-playing that takes place.

Elsewhere [14, 16], I argued that we might benefit from entertaining a *systemic*, rather than an experimental-analytic paradigm for the study of system-wide changes. *A systemic paradigm uses whole systems as its unit of analysis, assumes that variables and events are inseparable from each other, and that they are in constant flux.* It is not the most suitable paradigm for the study of linear causality, but it is suitable for the study of how the *configuration* of variables and events changes over time. How, for example, do the relations among perceived self efficacy, excitement, perceived social pressure to learn, intensity of computer use, amount of social interaction, prior knowledge, and problem-solving performance and ability *jointly* change over the course of a project?

It is time now to return to the title of the book and its meaning. The two approaches to research - the analytic and the systemic - deal with it in different ways and thus maintain a division of labor between them in the study of computers and problem solving. Each approach fits better a particular combination of phenomena to be studied, questions to be answered, and methodologies to be employed [16]. The analytic approach is best suited to test hypotheses that pertain to causal and functional relations between clearly identified input and output variables. In that context, the workshops' title implies questions such as "In what specific ways does activity *X* with computer program *Y* affect problem-solving ability *Z*, for how long, for whom, and under what specific conditions?" This was the case in the first three experimental studies described earlier. On the other hand, the systemic approach is better suited to study how whole learning environments change when the computer is treated more as a trigger for massive change and less as a well-controlled, single addition to an otherwise unchanging environment. It is the changing configuration of interrelated variables that is of interest. In this context, the title of the book implies a question such as "how does the introduction of the computer, and the activities, social processes, curricula, etc. that its introduction affords, affect and become affected by the changes in learners' abilities during and following these activities?" Here, one is willing to forgo the rigorous testing of hypotheses concerned with the amount of problem-solving variance that can be attributed to the computer, *other things kept as constant as possible*, and focus more on the generation of hypotheses pertaining to the changes that are allowed to happen in the learning environment as a whole, *other things allowed to freely change in it*.

The three experiments described earlier treated the question of computer-based learning environments and problem solving in an analytic way and thus could tell us what changes in problem-solving ability can be achieved under controlled conditions with stand-alone computer tools. Such research could help us test hypotheses that emanate from a Vygostkian theory and from other theories such as that of Bereiter and Scardamalia [4] or of Salomon, Perkins, and Globerson [18]. On the other hand, the field study described treated the question of computer-based learning environments and problem solving in a systemic way and thus could tell us what happens with problem-solving ability and a host of related variables when everything is allowed to change and how the environment and the variables of interest in it affect and are affected by each other. No question about it, the two approaches complement each other as each of them sheds a somewhat different light on the issue we are concerned with here.

Summary

I began this paper with an attempt to unpack the title of this book, arguing that computer-based learning environments entail far more than the employment of computer programs, wondrous as they may be. The computer, or rather the activities that its programs and tools afford, are but a cornerstone, a trigger for system-wide changes in the learning environment. I have tried to illustrate such changes with the World War I project we have carried out in eighth grade classes. I also tried to explicate the meaning of the second part of the book's title "... and problem solving", arguing that what is meant by this is that new learning environments might have effects on (a) problem-solving performance *during* a computer-enhanced activity (effects *with* computers), and (b) problem-solving skill and ability the development of which comes as a *consequence* of these activities (effects *of* computers). I illustrated the distinction between the two by three recent experimental studies that showed that both effects are possible and that the latter depends on the former.

However, the combination of the study of effects *with* and *of* computers with the system-wide changes in learning environments that computers can trigger, requires a different research paradigm. One cannot study computers' effects (that is, the effects of the activities their software affords) within whole-system changes the way one can study them under experimental conditions. Hence, I briefly described an alternative research approach - the systemic paradigm - better suited in terms of its underlying assumptions and methodology to what is needed when genuine computer-based learning environments are to be studied. Computer-based learning environments can affect both problem-solving performance during the activities they afford and leave a cognitive residue as a result. But in an environment in which most everything changes, it would be impossible and conceptually unsatisfactory to try to attribute such cognitive changes to one single agent -- the computer.

References

1. Altman, I. (1988). Process, transactional/contextual, and outcome research: An alternative to the traditional distinction between basic and applied research. *Social Behavior, 3*, 259-280.
2. Baker, L., & Brown, A. L. (1984). Metacognitive skills and reading. In P. D. Pearson (Ed.), *Handbook of reading research* (pp. 353-394). New York: Longman.
3. Bereiter, C., & Bird, M. (1985). Use of thinking aloud in identification and teaching of reading comprehension strategies. *Cognition and Instruction, 2*, 131-156.
4. Bereiter, C., & Scardamalia, M. (1987). *The psychology of written composition.* Hillsdale, NJ: Erlbaum.
5. Bruner, J. (1990). *Acts of meaning.* Cambridge, MA: Cambridge University Press.
6. Krendl, K., & Lieberman, D. (1988). Computers and learning: A review of recent research. *Journal of Educational Computing Research, 4*, 367-390.
7. Lepper, M. R., & Gurtner, J. L. (1989). Children and computers: Approaching the twenty-first century. *American Psychologist, 44*, 170-178.
8. Mandinach, E. B. (1989). Model-building and the use of computer simulation of dynamic systems. *Journal of Educational Computing Research, 5*, 221-244.
9. Newman, D., Griffin, P., & Cole, M. (1989). *The construction zone.* Cambridge University Press.
10. Olson, D. R. (1986). Intelligence and literacy: The relationships between intelligence and the technologies of representation and communication. In R. J. Sternberg & R.K. Wagner (Eds.), *Practical intelligence: Nature and origins of competence in the everyday world* (pp. 338-360). New York: Cambridge University Press.
11. Olson, D.R., Torrance, N., & Hildyard, A. (Eds.) (1985). *Literacy, language and learning.* Cambridge, England: Cambridge University Press.
12. Pea, R. D. (1985). Beyond amplification: Using the computer to reorganize mental functioning. *Educational Psychologist, 20*, 167-182.
13. Pea, R. D. (in press). Distributed intelligence and education. In G. Salomon (Ed.), *Distributed cognitions.* New York: Cambridge University Press.
14. Salomon, G. (1990). Studying the flute *and* the orchestra: Controlled experimentation vs. whole classroom research on computers. *International Journal of Educational Research, 14*, 521-532.
15. Salomon, G. (1990). Cognitive effects with and of computer technology. *Communication Research, 17*, 26-44.
16. Salomon, G. (in press). Transcending the quantitative/qualitative debate: The analytic and systemic approaches to educational research. *Educational Researcher.*
17. Salomon, G., Globerson, T., & Guterman, E. (1989). The computer as a zone of proximal development: Internalizing reading-related metacognitions from a reading partner. *Journal of Educational Psychology, 89*, 620-627.

18. Salomon, G., Perkins, D. N., & Globerson, T. (in press). Partners in Cognition: Externding human intelligence with intelligent technologies. *Educational Researcher.*
19. Scardamalia, M. , Bereiter, C., McLean, R. S., Swallow, J., & Woodruff, E. (1989). Computer-supported intentional learning environments. *Journal of Educational Computing Research, 5,* 51-68.
20. Scarr, S. (1985). Constructing psychology: Making facts and fables for our times. *American Psychologist, 40,* 499-512.
21. Sheingold, K. (1987). The microcomputer as a symbolic medium. In R. D. Pea & K. Sheingold (Eds.), *Mirrors of minds: Patterns of experience in educational computing* pp. 198-208). Norwood, NJ: Ablex.
22. Weick, K. E. (1979). *The social psychology of organizing.* Reading, MA: Addison-Wesley.
23. Zellermayer, M., Salomon, G., Globerson, T., & Givon, H. Enhancing writing-related metacognitions through a computerized writing-partner. Manuscript submitted for publication.

Facilitating Domain-General Problem Solving: Computers, Cognitive Processes and Instruction

Richard E. Clark

Educational Psychology, University of Southern California, Los Angeles, CA 90089-0031, U.S.A.

Abstract: Research on the transfer of problem solving between knowledge domains is reviewed. It is concluded that instructional and programming studies do not provide evidence for any essential contribution of the computer to problem solving. Instead, it is recommended that studies focus on the cognitive processes that are required for farther transfer and how these processes can be supported by instruction for students who fail to transfer. Two types of cognitive processes presumed to be active during problem solving are discussed: 1) The selecting of structural features of problem representations and 2) the connecting of features in two or more domains during transfer. The connecting process is discussed in depth and two types of connections are hypothesized, first, horizontal connections between knowledge structures in the form of analogies, and second, vertical connections within knowledge structures in the form of rule-example chains. It is suggested that analogies promote domain-general transfer because they provide horizontal connections between similar knowledge structures in different domains and induce rules that enlarge and extend the links between knowledge. Research focused on the use of analogies in problem solving is offered to support instructional prescriptions intended to facilitate domain-general problem solving.

Keywords: Computer research design, Domain-general problem solving, Metacognitive processes.

Computer Instruction and Computer Programming Effects on Problem Solving

A search for the psychological and educational advantages of newer media has been one of the threads in the historical warp of education [17]. Media such as photography, audio recordings, cinema and television have all been adapted for instructional use shortly after being invented. The computer is only the most recent in a long line of communication media employed for teaching purposes [53]. Like previous media during the period of their ascendancy, computers are the focus of considerable enthusiasm. Many scholars consider the potential contribution of the computer to be far beyond other media for educational purposes for at least three reasons: first, there is a perceived necessity to make study about the operation of computers part of an up-to-date school curriculum in Europe and North America [60]; the second reason for our special interest in computers may be in the expectation that instruction with the computer results in greater learning and problem solving skills than other instructional media [10]; and the third reason for the current emphasis on educational uses of the computer supposedly results from the hope that instruction on computer programming might generalize and enhance our domain-general problem solving ability [19, 54].

Study about the computer seems well established by now in the curricula of many modern secondary schools in Europe and North America [60]. However, it is research evidence about instruction with the computer and studies of instruction on computer programming which are most directly relevant to the topic of this book. If the use of computers to teach and train results in more or better or different learning than instruction with other media (including teachers), then our focus on them is supported. Thus, the discussion turns to an attempt to answer the question: What is the current status of the problem-solving research that examines instruction with computers? Does the instructional use of computers or the learning of computer programming result in problem-solving ability that generalizes across domains of knowledge?

Research on the learning benefits of instruction with computers. Space restrictions prevent a thorough review of research on instruction with computers. However, existing reviews are notable for their lack of compelling evidence for the hypothesis that instruction with computers makes any necessary psychological contribution to learning or to the transfer of what is learned during instruction [11, 44, 55, 19, 18]. To the contrary, evidence in these reviews of research seems to support to the claim that whatever learning advantage one can accomplish with computers can also be accomplished with other instructional media. This conclusion is controversial and not yet generally accepted in the research community. There is a general feeling that some types of instructional simulations (e.g. in science, technology and mathematics) are feasible only with the computer. Clark [11] has suggested that such arguments are based on the economic benefits of computers - not on their learning benefits. He notes that all of the instructional simulations in question could be duplicated in other (non computer) media with the investment of much more time and effort - a practical and an economic benefit but not a learning benefit.

There is also good evidence that in ideal situations, computer-delivered and managed treatments permit us to exercise greater control and reliability over many more instructional variables at less cost and with greater convenience than more traditional alternatives [for example, 10, 11]. Yet, well-controlled studies on the learning or problem solving benefits of instruction with the computer does not support its choice over other, less expensive, instructional delivery media. Might we find a different conclusion in the research on the psychological effects of learning to program computers?

Research on the learning and problem solving benefits of computer programming. It seems reasonable to hypothesize that the learning of computer programming languages might transfer to new and different problem-solving applications. Yet when we review progress to date on these studies, evidence of even limited "domain-specific" learning and transfer is rare. One important line of research in this area explores the hypothesis that adequate instruction in one computer language often will make it easier or quicker to learn yet another computer language [for example, 40, 43]. As Palumbo concludes in his recent review of these studies, "research that specifically addresses thinking skills that closely parallel the programming environment yields the most conclusive result" [47, p. 75]. Other reviewers [14, 54] are even less optimistic. The original hope that the learning of computer programming would sharpen general thinking and problem-solving skills seems unsupported.

Thus, there seems to be a lack of compelling evidence of "domain-general" transfer which is attributable to computer programming expertise [13]. To date, the best evidence supports the conclusion that whenever "domain-general" problem solving or learning to learn skills are acquired, neither computers nor computer-programming skills were necessary aspects of the responsible treatments [11, 47]. Given the weight of negative evidence, the best advice one might derive from a survey of the programming and transfer research is that we should reduce our investment of scarce research funds on transfer studies of programming. It appears that computer programming is no more related to domain-general transfer than the learning of any other analytic problem solving activity [54]. Why then, is there so much interest in the connection between computer instruction and programming and the development of general thinking skills?

The reification of a metaphor. The origin of the confusion over computer use, computer programming skills and the learning of transferrable thinking skills might be found, in part, in the pervasive and sometimes inappropriate use of a metaphor used to understand human cognition. A number of metaphors have served psychology in the past. The mind has been compared to a wax tablet and to the muscular system. Former theories of transfer which resulted from those metaphors were also unproductive. For example, the classical theory of transfer fixed on a muscular metaphor and recommended that generalizable problem-solving and reasoning skills would be developed through the "exercise" of the mind while learning classical subjects such as Latin and Greek.

Norman [45] argues that the computer metaphor for cognition is more difficult to limit than previous metaphors because the architecture of the digital computer was influenced by its designers tacit theories of cognition. Thus, he suggests that we now have a tendency to turn the metaphor around and assume that there is isomorphism between both physical structures and programming processes in the computer and psychological structures and processes. Lohman [41] notes that some go so far as to argue that "similar principles govern the functioning of any system that processes information" (p. 336). For those who advocate a strong computer analogy for cognitive processing, the mind is a mechanism for symbol manipulation [for example, 41]. Thus, in this view, the structural and process differences between the computer and the mind are irrelevant. Some of the more ardent advocates of this model have taken to calling cognitive skills "mindware". In this perspective it is reasonable to assume that writing programs for the computer would not only help the computer to reason but would also "program the programmer". While few would dispute the value of computer models of cognitive processes, it is the inappropriate reification of the models which seems to be at the root of the problem. Norman [45] reminds us that computer models of thought are similar to computer models of the weather yet few of us would expect that a model of violent weather would damage the computer center. Notwithstanding, the value of a model is, in part, its capacity to suggest productive hypotheses about cognitive processing. The reified computer model of domain-general transfer of problem solving may be an inefficient diversion rather than a positive influence.

For those who have conducted research with and on computers and computer programming, there is consolation in the knowledge that researchers in many other areas have achieved nearly identical results. There is wide expectation that science education will promote general problem solving ability yet instructional studies do not support the expectation. For example, Linn and Dalby [40] reached a distressing conclusion about the current results of science instruction research, which they characterized as in a "sorry state". They summarized their concerns in a series of questions such as: "Why do learners stop short of creating well-integrated, self-consistent knowledge structures? Why instead do they acquire larger amounts of isolated, context-dependent, and domain-specific knowledge?" (p. 291). The best evidence to date is that when domain-general transfer of problem solving skills are acquired, the antecedents are instructional methods which teach and compensate for the cognitive processes which are required for domain-general transfer [18, 19]. Insights about transfer will be found in a direct study of the cognitive processes required for domain-general transfer.

What Facilitates Domain-General Transfer of Problem Solving?

The study of transfer has recently focused on task domains such as analogical problem solving [for example, 26, 27, 28], artificial intelligence [for example, 33], concept learning

[5] and science [40]. Along with Gardner [24], we find it useful to characterize current learning and transfer theories not by the tasks they investigate but by the methodologies they employ. In this regard, Gardner suggests that different theoretical and methodological approaches to learning and transfer research can be roughly divided into inductive and deductive camps. It appears that each cluster of theories directs attention at different aspects of transfer and that only one of the clusters directly investigates domain-general transfer. One area of concern that divides these two main approaches closely parallel those that engaged Judd [36] and Thorndike [61] in their arguments about transfer at the turn of the century.

Inductive Theories. In transfer, the more inductive theories [1, 44] tend to often to emphasize how, rather than what knowledge is used during learning and transfer. Inductive theorists focus on the automated learning of experts and the nature and expression of expertise [for example, 7, 8, 56, 32]. Although robust, inductive theories emphasize the relatively "near" transfer of automated knowledge, i.e., sequences of linked productions that form automated, goal-directed steps within specific domains. Research indicates that over time, expertise may produce higher order procedures [for example, 1, 2]. However, these superordinate procedures promote transfer between knowledge domains.

While few studies have explored the extent to which expert procedures transfer outside of knowledge domains, the research on expert problem solving offers some evidence that procedural knowledge may be domain specific [2, 8, 9, 23]. Expert learning and problem-solving procedures have been described as "specific performance routines" [6] or "patterns" that are specific to knowledge domains [2]. Procedural knowledge structures might not facilitate solution of the most novel problems. Lohman [41] has recently argued that it is problem novelty that determines transfer distance requirements. He also suggests that novel problem solving is more associated with intelligence than with expertise. The most novel problems seem to be those which require connections between previously unconnected knowledge domains [28]. If one accepts the logic of Lohman's argument, it is reasonable to assume that the more novel the problem, the greater the likelihood that existing expertise will not be adequate and that problem solving will be correlated with intelligence. One might also speculate that novel problems require the use of conscious, declarative knowledge [52].

also speculate that novel problems require the use of conscious, declarative knowledge [52]. This is the type of knowledge described in the more "top down" theories of learning and problem solving.

Deductive theories. Theories that are more deductive [for example, 23, 26, 27, 34, 52, 64] seem to be more associated with Judd's early [36] argument in favor of domain-general transfer. The focus of investigation in more deductive theories is the "farther" transfer of knowledge. These theorists emphasize domain-general transfer and direct their attention more to what transfers but are less specific about how transfer occurs. The deductive approaches to transfer theory therefore tend to be weaker than the inductive approaches because of the lack of explicitness about the transfer processes. Nevertheless, it is the more deductive approaches that have explored the types of knowledge that are involved in transfer to novel problems and insight. For example, some deductive theorists have examined declarative knowledge and produced promising evidence of transfer between domains [for example, 25, 26, 27, 31, 59].

The role of analogies. Rumelhart and Norman [51, 52] suggest that the knowledge hierarchies which characterize a schema could be duplicated by analogy to form new schema and that "...learning by analogy, [involves] taking one schema and creating another one identical to it except in specified ways" [51, p. 344]. They offer the example of the familiar "slice-of-pie" analogy used by many mathematics teachers as they move students from an understanding of whole numbers to fractions. Presumably, every child's schema for food contains both procedural and declarative knowledge about the definition, utility and (perhaps) even the addition and subtraction of fractions of food. Suggesting that fractions are "like" the pieces of a whole pie implicitly invites students to "copy" the more familiar schema into an "empty" space in cognition and make it available for "editing". The pie and fraction analogy is a good example of domain-general transfer. Knowledge from one schema is transferred to create another and therefore the old knowledge crosses a domain boundary. Within this relatively simple slice-of-pie analogy are at least two implicit hypotheses about the types of knowledge used and the cognitive processes that are engaged when domain-general transfer occurs.

Selecting salient features. First, it seems that analogies require a judgment that essential "similarities" exist between declarative knowledge in the source and declarative knowledge in the "target". Successful similarity judgments require that learners are able to identify or select the most important criteria or structural features that are shared by the source and target schema [26, 27, 28]. In the case of the slice-of-pie analogy, the teacher presumably makes the selection by presenting key concepts and principles in the slicing of pies that are nearly identical to those required for the learning of fractions.

The connecting process. The second hypothesis implicit in the slice-of-pie analogy, is that an analogy is essentially an association or connection between a familiar schema or chunk of declarative knowledge and some problem or knowledge to be learned. Familiar declarative knowledge from a source schema within one domain is "mapped" onto similar information in another domain or copied into an "empty" space to form a new domain. An explication of the cognitive processes and consequences involved in selecting salient structural features of knowledge and making analogical connections is important in understanding how domain-general transfer occurs.

Metacognitive activity during general transfer. Gick and Holyoak [27] reported that in problem solving studies using the Dunker [21] radiation problems, only students with the highest level of general ability spontaneously noticed analogical connections between the problem and the solution. Snow [57] has suggested that fluid ability [35] which presumably represents aptitude for solving novel problems is associated in domain-general transfer of knowledge. There is evidence that learners with higher levels of fluid ability are able to select the critical features of a learning task without assistance [31, 58] but that those with middle level aptitudes may require prompts [27, 59] and that those with very low aptitude levels may require full compensatory models or simulations of the selecting process [16, 58]. In addition, Gray [31] found evidence for cognitive grouping, reorganization and elaboration activities in higher ability students. Analysis of student notes and "teach back" exercises demonstrated that higher aptitude students will often translate text into their own idiosyncratic words and include personal references and elaborations. Lower aptitude students recorded direct quotes without inference. They simply rehearsed information that was presented to them. In short, aptitude differences showed through in the organizational assembly and control activities learners use to adapt to new instruction.

DiVesta and Peverly [20] and Gick and Holyoak [28] have found that when similar examples or limited analogies are used to teach concepts, all subjects seem to have difficulty with domain-general transfer of the concept. DiVesta and Peverly [20] proposed that when concept learning involved the presentation of homogenous examples (encoding constancy) the result was learning which was context dependent. Their research suggests that practice with examples from the same context encourages the use of context as an "orientating" guide which emphasizes certain attributes and relations in the task to be learned. As a result, these attributes and relations become cue-dependent on the context. On the other hand, encoding variability contributes to learning that is more decontextualized and therefore more applicable to novel domains (domain-general transfer). It is possible that students with higher fluid scores would find it more difficult to exercise their skill at domain-general transfer when instruction provides near transfer practice.

Selecting and connecting appear to be high level cognitive processes that seem important in facilitating transfer. Since domain-general transfer is hypothesized to extend beyond the borders of knowledge domains and therefore beyond the influence of expertise, these processes are best conceptualized as "metacognitive" [22]. Both selecting and connecting were described by Corno and Mandinach [16] in their discussion of the development of "self-regulated learning". They related both metacognitive skills to the "transformational" information-processing theory originally suggested by Anderson and Bower [3]. Transformation is a general term which refers to cognitive activity that reorganizes and/or extends knowledge structures to accommodate external learning demands.

Domain-general transfer requires an awareness of the <u>similarities</u> between information located in one area of long term memory (called the "source") and some other bit of information in another area of long term memory (called the "target"). The exact choice of the features of the source which are judged similar to the target are critical in transfer. For example, to judge the formal similarities of "smoke" and "fog", we must know that the criterial features of smoke are that it is "particulate matter resulting from combustion which is suspended in air". In addition, smoke also has odor, color, shape, density, direction, volume and other characteristics which are not criterial to its definition. Assume that fog is defined as "water molecules suspended in air". To judge that smoke and fog are similar, one must notice that their defining attributes are similar, i.e., that both involve matter suspended in air. It is the judgment of such similarities that seems to underlie that part of metacognitive selecting which supports transfer. In order to successfully transfer information, one must select information in the source and target that have the required similarities.

Attention to similarity judgments have been associated with transfer research since Thorndike [61] described the "theory of identical elements". At the core of the theory was the notion that events which share common stimulus properties will be recognized as similar and that the learned response to one event will therefore generalize to the other event. This theory was the origin of our understanding of "near" transfer. It also formed the basis of Osgood's [46] transfer surface theory which was an attempt to explain the role of similarity in both positive and negative transfer.

Studies of the role of similarity judgments in domain-general transfer are aided by the seminal studies of human judgment by Kahneman and Tversky [37]. They have explored the way we use judgment heuristics to make similarity judgments when confronted with a great deal of uncertainty. They suggest that we look both at the obvious similarities and that we also "decompose" items being compared into their most abstract and criterial shared features. Decomposition is related to what transfer researchers have called "decontextualization" [50]. Kahneman and Tversky note our increasing predisposition to judge things as similar when they share a large number of perceptual and familiar features. In these instances, shared criterial features tend to be de-emphasized. For example, we tend to judge robins and sparrows as more alike than robins and penguins. The problem is that when one must choose examples of the concept "bird", robins, sparrows <u>and</u> penguins are all members of the category, even though penguins have many obvious but irrelevant features that differ from the other two instances. Kahneman and Tversky's theory would suggest that for most of us, robins and birds are more "representative" of birds than are penguins.

Negative transfer. The tendency of human beings to make similarity judgments using the representativeness heuristic causes many transfer errors. Effective domain-general transfer requires that we decompose information in both the source and target domain into their most criterial or structural features. It is these features alone which determine their "similarity" for formal purposes. Birds, for example, all have feathers and bear their young in eggs. Many birds fly but this feature is not criterial, although it is a very "representative" trait and is therefore often assumed to be a criterial feature of birds. Whenever the representativeness heuristic leads us to add features that are salient but non-criterial, negative transfer will occur, i.e., the exclusion of the penguin from the category "bird".

Much of the existing research on the learning of categories and concepts is very relevant to our understanding of the typical errors involved in making similarity judgments. For example, the literature on prototype versus best example use in concept identification, makes some of the same predictions of errors in judgment as does the representativeness literature [for example, 30]. The research on transfer contains many examples of problems caused by similarity judgments. In some cases, transfer failures may be caused by subtle differences in the wording of transfer task. A small difference in the order or wording of concepts which are presented for similarity judgments can greatly affect the outcome of studies [63]. This fact might also account for a number of the current disputes in the transfer literature that is focused on problem solving [for example, 27, 59]. In problem-solving tasks, one must choose solutions that correspond to the conditions imposed by the "goal state" [2]. Goal states contain the structural features of a problem category [27, 28]. Structural features include the goal (reason for action), the resources (or enabling conditions), and the constraints (or obstacles). Wording of the goal states in training tasks can affect the similarity judgments made between the training and target problem. If the wording is incomplete, i.e., missing a most criterial feature, such as an important constraint, a more salient but noncriterial feature, i.e., environmental context, may cause transfer errors. Examples of these errors are found in studies by Spencer and Weisberg [59], Glaser and Bassok [29].

Similarity judgments are crucial because they support the next step in the transfer process, the connecting of source and target information. Presumably, one first reaches the judgement that two chunks of information are similar enough to connect and then proceeds to select the type of connection which will be made. The discussion turns next to a description of the nature of the cognitive connections that underlie different amounts of transfer.

Vertical connection by example. As we learn and use concepts and principles, examples and their related "rules" are one of the important units of knowledge hierarchies [2]. It is most likely that we form the rules or definitions of concepts through experience where we select obvious similarities between instances [23, 30]. We transfer these superordinate rules when we make the decision that a new experience is an "instance" of a more general category. Since hierarchies of knowledge may contain a number of levels of rule-example chains [13], this type of connection can be made at a fairly abstract level. This is the type of connection that one would make when confronted with an ostrich for the first time. It has feathers and bears its young in eggs but it does not fly and it is considerably larger than most birds. Successful transfer would require that a learner note that the ostrich has all of the criterial features of birds and that the "unrepresentative" qualities of size and flight are not important. More important is that a successful connection at this point would most certainly be "near", that is, within a knowledge domain and close to the original information about birds. We have concluded that the reason for this near transfer application of rule-example chains is that they are always "vertical" connections in knowledge hierarchies. Examples are always subordinate to rules which are always superordinate to examples. The vertical nature of this type of connection ties it to the context of the schema or domain in which it resides. If houses were domains of knowledge and stairways between floors were connections, one would not expect to easily use the stairs to get to another house. We hypothesize that rule-example connections primarily support learning within but not between domains.

Horizontal connection by analogy. Analogies have received a great deal of attention in the current transfer literature [15, 38, 39]. Many researchers have offered evidence that analogies promote creative problem solving [28], insight [25], an understanding of human brain functions [49], and scientific reasoning [42]. Much of the underlying reason for the recent focus on analogies in higher-order cognitive operations, we suggest, is the nature of the cognitive connection they support. Analogies can be defined as <u>horizontal</u> connections between two similar chunks of information at the same level of <u>abstraction</u>.

Unlike rule-example chains which connect between different levels in knowledge hierarchies, analogies connect two items at the <u>same</u> <u>level</u> <u>of</u> <u>abstraction</u> in a hierarchy. As horizontal connections they may occur between concrete example pairs within a domain, for example, when we suggest that a robin is like an ostrich. A between-domains example would occur when it is suggested that a pressure dial is like a clock face. Horizontal connections by analogy may also be more abstract although they occur within a domain. For example, we can say that the addition of whole numbers and the addition of fractions is analogous. For our purposes, one important aspect of analogies is their capacity to build a bridge between domains of knowledge. Analogical connections prompt the noticing of basic similarities between two domains. When they are provided to students during instruction, analogies prompt the copying of the contents of one schema into an "empty" space within a domain. As a result, new schemas are developed.

An important feature of analogies is that they seem to support "decontextualization" of the information they connect by inducing us to produce a list of their shared attributes. For example, when we suggest that the robin and ostrich are analogous, we imply that they share the same criterial attributes and implicitly invite the <u>subtraction</u> of all unlike attributes. Since robins and ostrichs are more unlike than robins and sparrows, the resulting set of shared criterial attributes are closer to the formal list that defines the category "bird". When robins and sparrows are connected, the resulting list of shared attributes includes things such as their similar size and capacity for flight which do not formally define birds. One would expect therefore, that instruction which uses very divergent examples of the same category would produce a more accurate impression of the criterial attributes of the category than the use of examples that share more non relevant attributes. Divergent examples would also be expected to produce domain-general transfer of the concept than more familiar examples, even though the transfer was within a domain.

One advantage of specifying the type of connection promoted by examples and analogies is to clarify ambiguities in transfer research studies. Treatments which provide examples sometimes promote vertical connections, sometimes horizontal connections within domains and occasionally horizontal connections between domains. In our system, only instances which promote vertical connections are examples. Analogies may have decontextualization consequences which cover the full range from very near to very far. Evidence for transfer qualities of treatments labeled "examples" can be found in the research of Bransford [5]. He has offered evidence that when successive examples of concepts from the same domain are presented, the resulting learning primarily serves nearer transfer. When subjects received examples of a concept selected from successively different domains, the concept more readily transferred farther.

The decontextualization effect of analogies becomes very important when a connection is made between knowledge domains. Simple examples can be found in the common analogies made between the heart and a water pump, the structure of the atom and the solar system or fractions and slices of pie. The strength of horizontal connections between domains is that they commonly induce a general rule made up of the most abstract elements of the common components of the source and target-and therefore more easily promote domain-general transfer.

How Might Instruction Facilitate Farther Transfer?

The discussion of domain-general transfer research has thus far been guided by a question: "What cognitive processes occur as transfer distance increases?". The rationale for the question was the expectation that domain-general transfer learning goals could be achieved only if instruction supported those necessary cognitive processes. Since few students achieve spontaneous domain-general transfer, some instructional support seems necessary. Disputes and important unanswered questions remain in the existing literature. However, for those interested in designing instructional systems which support domain-general transfer, the growing body of studies offers a few useful insights. The remainder of this discussion is an attempt to summarize tentative advice for the design of instructional systems. The suggestions are organized according to the system suggested by Clark [12] that instructional design theories should focus on three classes of variables; 1) learning task; 2) individual differences; and 3) instructional methods. If the current results of descriptive studies were to be organized into those three categories, tentative advice for instructional designers might resemble the headings on the following sections.

Learning task considerations in instruction for farther transfer

Limit learning tasks to declarative knowledge. There are a number of different ways to conceptualize the representations and actions which define declarative knowledge and no agreement on which way is best for instruction [for example, 51]. However, the best evidence supports the claim that composed and automated procedures may not transfer beyond the boundaries of knowledge domains. It is possible that some of the higher-level declarative productions which yield procedures may be available for farther transfer during interdomain problem solving.

Generate specific learning goal statements. This is familiar advice for experienced instructional designers but it appears vital for enhancing a number of aspects of farther transfer research. Goals seem to influence both individual learner actions during mapping and the choice of representational system the designer chooses. For example, some systems for conceptualizing declarative knowledge seem best for representing the processing of textual material [38], others for leading to use mathematical procedures [2], and still others for learning the operation of technological devices [42] and scientific theory [24]. There is also evidence that transfer problems are often influenced by student misunderstanding of goals [29]. Thus, it seems particularly vital in far transfer instruction, very early on in the instructional design process, to follow traditional advice to make a clear statement of the goals and objectives of learning and communicate them to students. In addition to communicating task goals to students, they should be explicitly told that the learning task requires farther transfer.

Carefully identify and define the attributes and relations to be taught. In a number of cases, farther transfer failed because learners fixed on superficial characteristics of the knowledge being transferred. In some instances the wrong attributes or relations were stressed during instruction [for example, 54]. In other cases, individuals had learned inappropriate definitions or uses for existing knowledge [29]. In either case, clear definitions of concepts and principles to be taught will allow for accurate presentations and/or remedial correction of existing declarative knowledge.

Individual difference considerations in instruction for farther transfer

Only a small percentage of the most motivated and intelligent students will achieve spontaneous far transfer. Snow [57] offers three uses of individual differences research for practical application: 1) to select people who can perform some task or set of tasks without training; 2) to compensate for inaptitude through training; and 3) to develop aptitude through training. One must be clear which of these three different educational goals are sought. If selection is the issue, we should note that existing studies of spontaneous solutions to novel and moderately difficult problems indicate that only about five percent of the most capable students succeed without direct instructional support [for example, 27, 28].

Individual differences in specific prior learning of "conceptual entities" that are relevant to the transfer task will influence transfer distance. Studies by Glaser and Bassok [29] and Bassok [4] give important indications of variations in subjects prior experience of categories that may prevent them from "recognizing the structural equivalence" between domains. Bassok [4] speculates that the problem is most difficult when intensive categories (defined by relations between two entities) are mixed with extensive categories (defined by one entity). It seems that different subjects are drawn to familiar "conceptual entities" and have difficulties with unusual combinations of extensive categories. Bassok [4] reports a very high rate of far transfer for problems where barriers to structural equivalence are eliminated. This implies that very specific types of "expertise" or prior knowledge differences between students influence farther transfer and may have to be considered in instructional design.

Instructional method considerations in instruction for farther transfer

Far transfer aptitude deficits may be reduced by instructional methods that support necessary cognitive processing. The number of solvers of far transfer problems increases to about ten percent with a compensating hint or suggestion that familiar knowledge is analogous to a problem being solved [59] and to between twenty to thirty percent when various types of fully compensated transfer aids are provided. Instruction that has targeted support for the cognitive processes of selecting and connecting the structural features of declarative knowledge in two or more domains of knowledge has been particularly successful in promoting farther transfer [4, 13].

To support the selecting process, any instructional presentation form which succeeds in helping students separate structural from superficial elements of knowledge will suffice. Examples are the **highlighting** of key concepts [42] or underlining structural elements.

Connecting involves the "linking" of information rich domains or schema's to distant schemas which are either information rich or poor. Evidence for the usefulness of connecting support can be found in most studies where farther transfer increases as a result of hints, multiple and varied examples and analogies. As with selecting, the presentation form of the help may be less important than whether it achieves its cognitive function. Unlike selecting, successful instructional methods seem to be specific to the amount of information in the domains being connected [38]. Common examples of instructional methods to support connecting between two, information-rich domains are multiple and varied examples; common methods to support connecting between an information-rich and an information-poor domain are analogies. It is also likely that similar connecting functions are served by some forms of illustrated models [42] and simulations.

What can computers contribute to instruction for farther transfer? Given that far transfer is promoted by instructional methods that support the selecting and connecting of structural features that are common to two or more domains of knowledge, there seems to be no feature of the computer that would automatically or exclusively provide such support. Even specific computer-based applications such as programming have no inherent qualities that would, by themselves, either develop or compensate for deficits in ability to select and connect structural features between two or more domains of knowledge. This is borne out by the disappointing results of studies that have investigated the transfer of skills learned in computer programming to other domains [54]. If computer-based activities are to serve any role in promoting farther transfer of knowledge they should be accompanied by or encorporate explicit reference to the structural similarities of source and target domains of knowledge. For example, if programming in a particular language is the source domain, then a target domain to which programming knowledge is expected to transfer should be identified and referred to during programming instruction.

Summary and Conclusion

With the benefit of hindsight, a number of researchers have noted that studies of the transfer consequences of computer programming are reminiscent of other failed theories of learning and transfer. At least three theories which come to mind quickly also hypothesized domain-general transfer of "languages" or media. The ancient "classical transfer theory" suggested that the learning of Greek and Latin exercised the mind and promoted thinking and problem solving skills [24]. More recently there were assumptions that the use of certain electronic media could influence transfer. This may be the origin of some of the interest in the use of computers for instruction in problem solving. A third theory hypothesized that certain "symbol systems" based on the display capabilities of media (such as the capacity to "rotate" objects or "unfold" them from three to two dimensions) had different learning and transfer consequences than more traditional media or symbol systems [14]. The classical theory of transfer has not been seriously entertained for many years. Recent experiments which tested the special medium and symbol system theories have failed to provide compelling positive evidence [10]. It appears that whenever classical curricula or newer media or symbol systems are found to result in more learning or farther transfer, studies may have failed to control for important variables [14]. Yet any of these linguistic, media or symbol systems can be used as convenient task contexts in studies of transfer.

Thus, the meager contributions of the research on computer programming to problem solving and transfer simply do not warrant future investment until a new theory is proposed. In the future, studies of domain-general transfer should focus more on <u>how</u> familiar knowledge is used to solve new problems or supports the development of new domains. An emphasis on the way that instructional examples and analogies promote metacognitive connections seems particularly fruitful. It appears that two examples support transfer only within knowledge domains but that analogies foster both near and far transfer, depending on the way they are constructed and presented. At least two types of metacognitive processes, selecting and connecting, are supported by analogies. In order to achieve transfer, it is hypothesized that one must select different sets of declarative knowledge that have similar structural or criterial features and permit the connecting of at least two different knowledge structures in working memory.

References

1. Anderson, J. R. (1983). *The architecture of cognition.* Cambridge, MA: Harvard University Press.
2. Anderson, J. R. (1985). *Cognitive psychology and its implications.* New York: Freeman.
3. Anderson, J. R., & Bower, G. H. (1973). *Human Associative Memory.* Washington, DC: Winston.
4. Bassok, E. (1990), Transfer of domain-specific problem-solving procedures. *Journal of Experimental Psychology: Learning, Memory and Cognition, 16,* 522-533.
5. Bransford, J. D., Nitsch, K., & Franks, J. (1977). Schooling and the facilitation of knowledge. In R. C. Anderson, R. Spiro, & W. Montague (Eds.), *Schooling and the acquisition of knowledge* (pp. 31-55). Hillsdale, NJ: Lawrence Erlbaum.
6. Brown, A. (1978). Knowing when, where, and how to remember: a problem of metacognition. In R. Glaser (Ed.), *Advances in instructional psychology (Vol. 1)* (pp. 77-165). Hillsdale, NJ: Lawrence Erlbaum.
7. Chase, W. G., & Simon, H. A. (1973). Perception in chess. *Cognitive Psychology,* 55-81.
8. Chi, M. T. H., Glaser, R., & Rees, E. (1982). Expertise in problem solving. In R. Sternberg (Ed.), *Advances in the psychology of human intelligence.* Hillsdale, NJ: Lawrence Erlbaum.
9. Chi, M. T. H., Feltovich, P. J., & Glaser, R. (1981). Categorization and representation of physics problems by experts and novices. *Cognitive Science, 5,* 121-152.
10. Clark, R. E. (1983). Reconsidering research on learning from media. *Review of Educational Research, 53,* 445-459.
11. Clark, R. E. (1985). Confounding in educational computing research. *Journal of Educational Computing Research, 1,* 445-460.
12. Clark, R. E. (April, 1990). *A cognitive theory of instructional method.* Paper presented at the annual meeting of the American Educational Research Association, Boston, MA.
13. Clark, R. E., Blake, S., & Knostman, V. (1989). *Metacognitive processes supporting far transfer: Selecting and connecting.* Paper presented at the annual meeting of the American Educational Research Association, San Francisco, CA.
14. Clark, R. E., & Sugrue, B. M. (1989). Research on instructional media, 1978-1988. In D. Ely (Ed.), *Educational Media Yearbook 1987-88.* Littletown, CO: Libraries Unlimited.
15. Cormier, S. M., & Hagman, J. D. (1987). *Transfer of learning: Contemporary research and applications.* New York: Academic Press.
16. Corno, L., & Mandinach, E., (1983). The role of cognitive engagement in classroom learning and motivation. *Educational Psychologist, 33,* 88-108.
17. Cuban, L. (1986). *A history of instructional technology.* New York: Teachers College Press.
18. De Corte, E. (1990). Towards powerful learning environments for the acquisition of problem-solving skills. *European Journal of Psychology of Education, 5,* 5-19.
19. De Corte, E., Verschaffel, L., & Schrooten, H. (1990). *Cognitive effects of computer-oriented learning.* Paper presented at the Seventh International Conference on Technology and Education, Brussels.

20. DiVesta, F. J., & Peverly, S. T. (1984). The effects of encoding variability, processing activity and rule example sequences on the transfer of conceptual rules. *Journal of Educational Psychology, 76*, 108-119.
21. Duncker, K. (1945). On problem solving. *Psychological Monographs, 58*, (Whole No. 279).
22. Flavell, J. H. (1981). Cognitive monitoring. In W. P. Dickson (Ed.), *Children's oral communication skills*. New York: Academic Press.
23. Gagne, R. M., & White, R. T. (1978). Memory structures and learning outcomes. *Review of Educational Research, 48*, 187-222.
24. Gardner, H. (1985). *The mind's new science: A history of the cognitive revolution*. New York: Basic Books.
25. Gick, M. L., & Patterson, K. (1989). *Constrasting examples, schema acquisition and problem solving transfer*. Unpublished manuscript.
26. Gick, M. L., & Holyoak, K. J. (1980). Analogical problem solving. *Cognitive Psychology, 12*, 306-355.
27. Gick, M. L., & Holyoak, K. J. (1983). Schema induction and analogical transfer. *Cognitive Psychology, 15*, 1-38.
28. Gick, M. L., & Holyoak, K. J. (1987). The cognitive basis of knowledge transfer. In S. M. Cormier & J. D. Hagman (Eds.), *Transfer of learning*. New York: Academic Press.
29. Glaser, R., & Bassok, M. (1989). Learning theory and the study of instruction. In M. R. Rosenwig & L. W. Porter (Eds.), *Annual Review of Psychology (Vol. 40)* (pp. 631-666). Palo Alto, CA: Annual Reviews.
30. Glass, A. L., & Holyoak, K. J. (1986). *Cognition*. New York: Random House.
31. Gray, L. E. (1983). *Aptitude constructs, learning processes, and achievement*. Unpublished dissertation, Stanford University.
32. Greeno, J. G. (1978). Nature of problem solving abilities. In W. K. Estes (Ed.), *Handbook of Learning and Cognitive Processes (Vol. 5)*. Hillsdale, NJ: Lawrence Erlbaum.
33. Hayes-Roth, R., & McDermott, J. (1978). An interface matching technique for inducing abstractions. *Communications of the ACM, 21*, 401-410.
34. Holyoak, K. J. (1985). The pragmatics of analogical transfer. In G. H. Bower (Ed.), *The psychology of learning and motivation (Vol. 19)*. New York: Academic Press.
35. Horn, J. L. (1976). Human abilities: A review of research and theory in the early 1970s. *Annual Review of Psychology, 27*, 437-485.
36. Judd, C. H. (1908). The relation of special training to intelligence. *Educational Review, 36*, 28-42.
37. Kahneman, D., & Tversky, A. (1973). On the psychology of prediction. *Psychological Review, 80*, 237-251.
38. Keane, M. T. (1988). *Analogical Problem Solving*. New York: Halsted Press.
39. Klauer, K. J. (1989). Teaching for analogical transfer as a means of improving problem-solving, thinking and learning. *Instructional Science, 18*, 179-192.
40. Linn, M. C., & Dalbey, J. (1989). Cognitive consequences of programming instruction. In E. Stolway & R.D. Spohrer (Eds.), *Studying the Novice Programmer* (pp. 57-81). Hillsdale, NJ: Lawrence Erlbaum.
41. Lohman, D. F. (1989). Human intelligence: An introduction to advances in theory and research. *Review of Educational Research, 59*, 333-373.
42. Mayer, R. E. (1989). Models for Understanding. *Review of Educational Research, 59*, 43-64.
43. Mayer, R., Dyck, J. L., & Vilberg, W. (1989). Learning to program and learning to think: What's the connection? In E. Stolway & R.D. Spohrer (Eds.), *Studying the Novice Programmer* (pp. 83-111). Hillsdale, NJ: Lawrence Erlbaum.
44. Newell, A., & Simon, H. (1972). Human problem solving. New Jersey: Prentice Hall.
45. Norman, D. A. (1986). Reflections on cognition and parallel distributed processing. In D. E. Rumelhart, J. L. McCelland, & the PDP Research Group (Eds.), *Parallel distributed processing: Vol. 2 Psychological and biological models* (pp. 531-546). Cambridge, MA: MIT Press.
46. Osgood, C. E. (1949). The similarity paradox in human learning: A resolution. *Psychological Review, 56*, 132-143.
47. Palumbo, D. B. (1990). Programming language/problem solving research: A review of relevant

issues. *Review of Educational Research, 60*, 65-89.
48. Phye, G. B., & Andre, T. (1986). *Cognitive classroom education: Understanding, thinking and problem solving*. New York: Academic Press.
49. Pribram, K. H. (1986). The role of analogy in transcending limits in the brain sciences. *Journal of Educational Psychology, 75*, 450-459.
50. Royer, J. M. (1979). Theories of the transfer of learning. *Educational Psychologist, 14*, 53-69.
51. Rumelhart, D. E., & Norman, D. A. (1981). Analogical processes in learning. In J. R. Anderson (Ed.), *Cognitive skills and acquisition*. Hillsdale, NJ: Lawrence Erlbaum.
52. Rumelhart, D. E., & Norman, D. A. (1988). Representation in memory. In R. C. Atkinson, R. J. Herrnstein, G. Lindzey, & R. D. Luce (Eds.), *Steven's handbook of experimental psychology, Second Edition, Volume 2: Learning and Cognition*. New York: Wiley.
53. Salomon, G., & Gardner, H. (1984). The computer as educator: Lessons from television research. *Educational Researcher, 13*, 4-8.
54. Salomon, G., & Perkins, D. N. (1987). Transfer of cognitive skills from programming: When and how? *Journal of Educational Computing Research, 3*, 149-170.
55. Schramm, W. (1977). *Big Media: Little Media*. Beverly Hills, CA: Sage.
56. Simon, D. P., & Simon, H. A. (1978). Individual differences in solving physics problems. In R. Siegler (Ed.), *Children's thinking: What develops?* (pp. 325-348). Hillsdale, NJ: Lawrence Erlbaum.
57. Snow, R. E. (1986). Individual differences and the design of instructional programs. *American Psychologist, 41*, 1029-1039.
58. Snow, R. E., & Lohman, D. F. (1984). Toward a theory of aptitude for learning from instruction. *Journal of Educational Psychology, 76*, 347-378.
59. Spencer, R. M., & Weisberg, R. W. (1986). Context-dependent effects on analogical transfer. *Memory & Cognition, 14*, 442-449.
60. Sugrue, B. M. (1991). A comparative review of European and American approaches to computer-based instruction in schools. In T. Schlechter (Ed.), *Problems and promises of computer-based training*. Norwood, NJ: Ablex.
61. Thorndike, E. L. (1903). *Educational psychology*. New York: Lemcke & Buechner.
62. Tversky, A. (1977). Features of similarity. *Psychological Review, 84*, 327-352.
63. Tversky, A., & Gati, I. (1978). Studies in similarity. In E. Rosch & B. B. Lloyd (Eds.), *Cognition and categorization*. Hillsdale, NJ: Lawrence Erlbaum.
64. Vosniadou, S., & Brewer, W. F. (1987). Theories of knowledge restructuring in development. *Review of Educational Research, 57*, 51-67.

Conceptual Fields, Problem Solving and Intelligent Computer Tools

Gérard Vergnaud

Laboratoire de Psychologie du Développement et de l'Education de l'Enfant, Université Paris V, Centre National de la Recherche Scientifique. 46, rue Saint-Jacques. 75005 Paris, France.

Abstract : The goal of this paper is to present an overview of the theory of conceptual fields, illustrated by the examples of additive structures and multiplicative structures. The main difficulties encountered by students in learning mathematics and other scientific topics are conceptual. These difficulties cannot be analysed if student's conceptions are investigated independently of the schemes that organize their behavior in problem solving situations, and if a concept or a class of situations is examined in isolation. A conceptual field can be defined as both a set of classes of situations, and as a set of interconnected concepts. This paper presents a theoretical analysis of the relationship between explicit knowledge and the implicit operational invariants that underlie schemes. Examples of theorems-in-action and of the relationships between concepts and symbolic representations are presented, as well as ways of using the framework of conceptual fields to devise intelligent computer tools.

Keywords : conceptual field, scheme, theorem-in-action

What is hard problem solving for a 6 year-old may become a routine skill for the same child three years later ; scientific concepts that have become obvious for a good university student have usually gone through a slow process of understanding and several rebuilding steps. The history of science shows that new concepts and new scientific procedures emerge gradually in the process of solving practical and theoretical problems : the concepts of negative number and complex number, function and variable are good illustrations of such progressive elaborations. Most scientists and engineers who use these concepts easily today, are not aware of the conceptual difficulties that scientists have encountered to construct and understand them well. Several classical psychologists such as Piaget [9], Vygotski [14], Bruner [2], also consider that interaction with situations that need to be coped with and problems that need to be solved is a key process in concept formation. Thus the concepts of knowledge-base and problem-solving are dialectically linked : an individual needs a knowledge-base to solve new problems, but his/her knowledge-base stems from his/her activity in solving problems and dealing with new situations.

The theory of conceptual fields is a developmental theory of cognitive complexity. It is based upon the fact that students' competences and conceptions develop through experience and that there are high regularities in the difficulties students have to overcome. For such fields as mathematics, physics and other sciences, these difficulties are mainly conceptual. Thus cognitive tasks, procedures, concepts, errors, linguistic and non-linguistic representations should be classified from a conceptual point of view. The framework of conceptual fields has been developed to provide a framework for better analysis of the process of learning and cognitive development, and to help teachers and curriculum designers. Teachers make decisions all the time : sometimes these decisions are right, sometimes they are wrong, because they rely upon erroneous representations of knowledge, students, and learning. It is useful for teachers to have a more reliable analysis of mathematical situations, the operations of thinking required to deal with them, and the most important errors made by students. This should enable teachers to devise more adequate learning activities, make better diagnoses and evaluations, and provide students with more beneficial explanations. Curricula can also be improved, since present curricula rely upon psychologically weak conceptualizations of knowledge and learning.

This same framework can be used to help produce better computer environments, for instance by relating problem solving and concept formation, since it provides a wider variety of possibilities to reflect the different aspects of the same concept, and can integrate a variety of individual competences. It can also be used to analyse the relevance of different symbolic representations and the possible manipulations of these representations. For instance, one can imagine computer programs and systems that would enable students to tackle problems of varying complexity in a specific field, provided that complexity be carefully analysed, on empirical grounds, and not solely on a priori considerations.

The present paper is divided into four parts :

1. Examples of conceptual fields : chiefly additive and multiplicative structures ;
2. Analysis of the relationships between theoretical concepts such as competence and conception, scheme, concept and theorem, rule and operational invariant ;
3. Analysis of the relationships between concepts, words and symbolic representations ;
4. Potential value of this theory for devising computer tools.

Conceptual Fields

It is a matter of fact that a concept cannot be made meaningful through one kind of situation alone ; similarly a situation cannot often be analysed with just one concept. This implies that the formation of several interconnected concepts needs to be studied concomitantly. There are two ways of approaching a conceptual field : either as a set of situations, or as a set of concepts. These two ways are complementary and therefore useful.

1) A conceptual field is a set of situations, and progressive mastery calls for a variety of interconnected concepts, schemes and symbolic representations.

2) A conceptual field is also a set of concepts, whose meaning and explanatory power stem from their joint intervention in the same situations and schemes.

The meaning assigned here to such terms as "schemes", "concepts" and "symbolic representations" are discussed more extensively in the next section. Readers who are unfamiliar with Piaget's theory, should note that the concept of "scheme" refers to the invariant organization of behavior for a certain category of situations. Piaget also used the word "scheme" for the organization of sensori-motor skills (like swimming), and for intellectual operations (such as the class inclusion scheme).

In order to understand the relationships, filiations and revolutions in the development of a specific conceptual field in students' minds, the relevant elementary cognitive tasks, procedures, concepts and symbolic representations need to be classified. There are many examples of conceptual fields in physics, mathematics and logic. I draw my examples from the fields that I have studied extensively : additive structures and multiplicative structures and will only briefly deal with a few other fields.

Additive structures

There are several meanings of addition and subtraction : some of them are easily grasped by 6 or 7 year-olds, whereas others are still difficult for many 15 year-olds. Additive structures can be described as a set of situations considering that reality is not only a world of objects, properties and relationships, but also and essentially a world of situations to be coped with, tackled and mastered.

But additive structures can also be described as a set of interconnected concepts. Addition and subtraction problems require such concepts as measure, number and sum, as well as set, union, part and whole, state, transformation and relationship, natural and directed number, abscissa and move, unitary operation and binary combination.

It is the organization of situations and concepts that enables researchers to understand continuities and discontinuities in the emergence and development of new competences in the conceptual field of additive structures.

Figure 1 schematizes the six elementary relationships that are necessary to analyse all addition and subtraction problems.

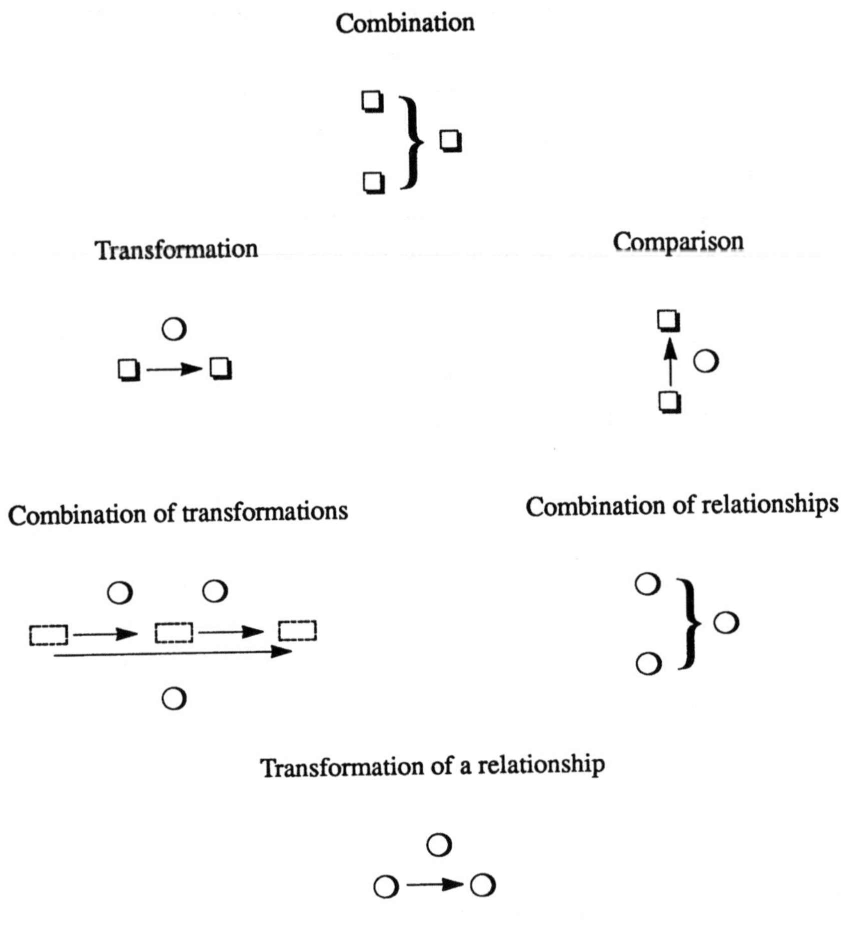

Figure 1. Elementary additive relationships.

It is now well known [3, 5, 8, 10, 12, 13] that the combination relationship (part-part-whole) enables us to generate only two classes of problems: find the whole, knowing the parts (**6 oranges, 5 bananas. How many fruit altogether ?**) and find one part knowing the whole and the other part (**10 children, 6 boys ; how many girls ?**)

In contrast the transformation relationship (state-transformation-state) enables us to generate six classes of problems : find the final state, find the transformation, find the initial state (the transformation being either positive or negative). From a conceptral point of view, the substraction is different when one has to find the initial state (**Robert has just won 5 marbles. He has now 11 marbles. How many marbles did he have before playing ?**), when one has to find the final state (**Veronica had 12 sweets. She eats 4 of them. How many sweets does she have left ?**), when one has to find a negative transformation (**Tom has played marbles with Teresa. He had 9 marbles before playing. He has now 5 marbles. What happened during the game ?**). The same problem with a final state of 13 would give a positive transformation.

The comparison relationship also generates six classes of problems : find the referred, or the relationship, or the referent (the relationship being positive or negative).

It is also recognized that the difficulty of these problems can be contrasted on a partially-ordered scale, and that there are canonical and less canonical procedures to solve them.

It is not so well recognized, in the community of psychologists and mathematics education researchers, that the last three relationships raise interesting cognitive problems at the primary and secondary levels.

The combination of transformations is a very important structure, which provides the best meaning for the addition of directed numbers. **Richard has received 5 dollars from his grand'mother; he has spent 8 dollars on miniature cars. Does he have more or less than what he had before? and how much?**

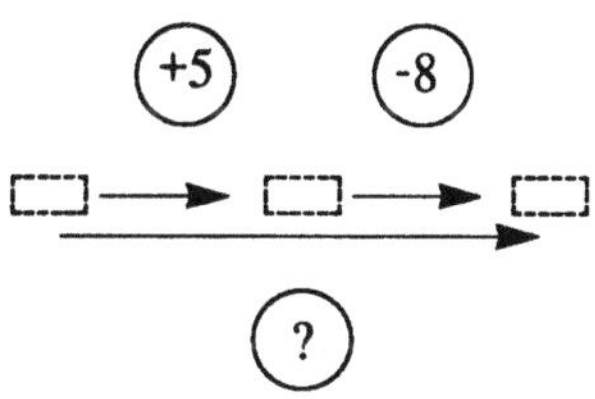

(+5) + (-8) = (?)

There are many different classes of problems, as one may have to find the combined transformation when the two elementary ones are given (problem Richard above), or an elementary transformation when given the other one and the resulting combination of both (problem John below). The value of the givens is an important factor: it can make the same word-problem trivial or totally counterintuitive, as is shown below with the example of problem John.

But, before I explain this example, I would like to make the picture more complete. Relationships also can be combined and decombined: **Janet owed Carolyn 15 dollars, but Carolyn also owes Janet some money because she did not have any money for her meal**

at the restaurant last saturday, and Janet paid for it. Finally Janet says. "You owe me 3 dollars". How expensive was the restaurant?

A relationship (debt, comparison, abscissa,...) can also be modified by a transformation. This is the case of a bank-account for instance.

The last three structures (combination of transformations, combination of relationships, transformation of a relationship) are essential in physics, in algebraic geometry, in accountancy and in many other topics. Some aspects of these structures can be grasped at an early age, some other aspects are difficult for most 9th graders. For instance, many 5 or 6 year olds are able to understand that increasing a quantity by 2 and again by 1 is like increasing it by 3, or that increasing it by 2 and then decreasing it by 2 is the same as no change. Some 6 year olds can even understand that, in a find-the-initial-state problem, one has to invert the transformation. This means that they already understand that transformations can be combined and inversed. It is nevertheless disputable whether operations with directed numbers should be symbolised.

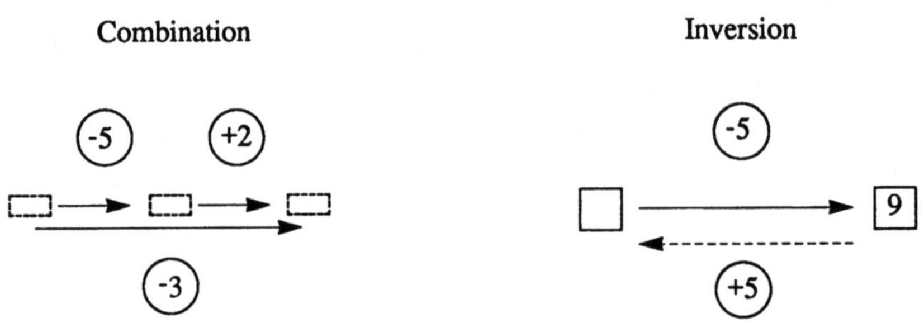

In order to show the discrepancy that can take place in the comparative difficulty of problems, I will now take the example of a more difficult problem, in which one has to find an elementary transformation when given the other one and the resulting combination of both.

John has played two games of marbles. In the first game he won/lost n marbles. Later, he realizes that, altogether, he has won/lost m marbles. What happened during the second game ?

The semantic is the same whatever the value of n and m, and whatever the sign of the transformation (win or lose). Yet the level of difficulty varies considerably for 9 to 15 year-old students.

First game	Altogether	
win 7 lose 7	win 16 lose 16	very easy: 95% success for 6th graders
win 16	win 7	a bit more difficult: 80% success for 6th graders
win 7 win 16 loss 7 loss 16	lose 16 lose 7 win 16 win 7	very difficult: 15% for 6th graders

Why is this so ? Because the mental representation that can be used to tackle the problem in all these cases is conceptually very different.

Irrespective of the numerical values, the canonic representation is the combination and decombination of transformations, i.e. the sum and subtraction of directed numbers.

$$a + x = b \quad \Rightarrow \quad x = b - a$$

whatever the sign and value of a and b may be

But in the "win 7 and win 16" case, one can also use two simpler conceptions:

- the part-part-whole conception $\left.\begin{matrix} 7 \\ \square \end{matrix}\right\} 16$, because the signs are not relevant.

- the state-transformation-state conception $7 \xrightarrow{\bigcirc} 16$, because the first transformation can be identified with an initial state and the altogether transformation with a final state.

. In the "lose 7 and lose 16" case, one can also use the part-part-whole conception ; the state-transformation-state conception is less manageable, because a loss cannot be identified with a state.

. In the "win 16, win 7" case, the part-part whole conception cannot be used, but the state-transformation-state lends itself well $16 \rightarrow 7$, and therefore helps find that the transformation is a lose.

. In the other cases, the first game and the combination of the two games have different signs. Therefore no simpler conception than the combination of transformations can be used, and the subtraction x = b - a results in adding the absolute values. This is totally counterintuitive for most students.

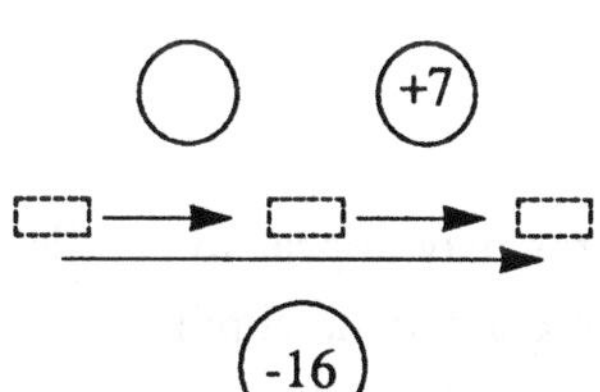

lose 16 - win 7 = lose 23

$x = (-16) - (+7) = -23$

There are other kinds of problems involving the combination of relationships and the transformation of a relationship. The difficulty of reasoning with transformations and relationships of different signs can be traced up to the university level, in physics and in accountancy for instance.

Additive structures offer us a striking example of a conceptual field in which children's primitive conceptions are robust enough to enable them to handle several categories of problems at the age of 4 or 5. Addition can be thought of either as an increase (unary operation), or as the union of two parts (binary combination) ; subtraction can only be thought of as a decrease in the beginning [6]. Students have to overcome these primitive conceptions to tackle other categories of addition and subtraction problems : find the transformation, the initial state, the complementary part, the referent.

They can also use different schemes to solve the same problem, each one exhibiting different properties of the concept. For instance the canonic scheme to find the initial state consists in inversing the transformation and applying this inverse to the final state.

This scheme relies upon the following theorem-in-action

$$F = T(I) \Rightarrow I = T^{-1}(F)$$

F = final state I = initial state T = transformation T^{-1} = inversion of the transformation.

But some children, who cannot inverse the transformation, make a hypothesis on the initial state, apply the direct transformation, find a final state (usually different from the right one) and modify their hypothesis until it is right. The latter scheme is quite rational but is cumbersome for large numbers and is therefore dropped as soon as the ability to inverse the transformation is reached: this ability is dependent on a conceptual change.

The example above of the two games of marbles shows that a linguistic approach is not sufficient and that a mathematical analysis is an absolute necessity. Many university students are still uneasy with problems involving transformations and relationships of different signs, although they have been taught directed numbers, algebra and analytical geometry for years. This shows that the conceptualizing process of additive structures is long and diverse and needs to overcome major epistemological obstacles.

Muliplicative structures

The conceptualizing process of multiplicative structures may appear to parallel the process for additive structures, but this in fact is not true. There are some similarities, but the main conceptual difficulties are very different. Multiplication and division fundamentally arise in problems of proportion, and therefore in four-term relationships. Analysis in terms of states and transformations is not relevant, as there are normally no initial or final states ; dimensional analysis however is crucial. Multiplicative structures involve such concepts as scalar operator, linear function and n-linear function, fraction, ratio, rate and rational number, vector-space, linear combination and linear mapping, and dimensional analysis.

Figure 2 presents a schematic description of the different kinds of relationships and problems involved in simple proportion.

Multiplication

1	a
b	❑

Division I (partition)

1	❑
a	b

Division II (quotition)

1	a
❑	b

General case

a	b
c	❑

Figure 2. Simple proportion relationships.

The following problems illustrate the structure of simple proportion.

Multiplication : **Peter wants to buy 3 cakes. They cost 6 French francs each. How much does he have to pay ?**

Division I (partition) : **Janet has spent 60 francs to buy 12 cakes. How much does 1 cake cost ?** or: **A piece of iron has a volume of 12.3 cubic centimers ; its weight is 96 grams. What is the weight of one cubic centimeter ?**

Division II (quotition) : **Julia has 48 francs in her purse. She wants to buy as many cakes as possible for her birthday. They cost 4 francs each. How many cakes can she buy?** or: **John's father drives at an average speed of 120 km on the motorway. He has 420 km to go. How long will it take him?**

General Case : **How much will it cost to buy 15 miniature cars if 3 cars cost 12 French francs ?** or: **What would be the weight of a big stone of 3.42 cubic meters, if a stone of 425 cubic centimers weighs 1112 grams ?**

There are two very different kinds of multiplicative relationships between the magnitudes or quantities involved, as is shown in the example of the cost of miniature cars.

Miniature cars   French Francs

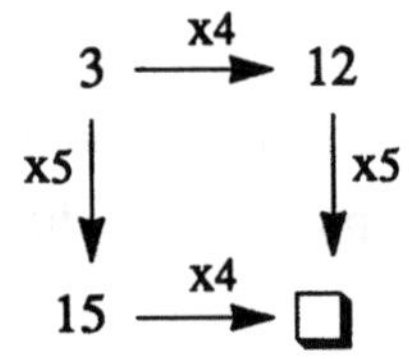

There are also different procedures to solve the problem. These procedures require different theorems-in-action.

1) 15 is five times 3. Therefore the cost will be five times 12 francs. Five is a scalar ratio; it connects two magnitudes of the same kind : 15 cars and 3 cars. The reasoning implies the following theorem-in-action :

$$f(nx) = nf(x) \quad \text{for } x = 3 \text{ and } n = 5$$

If n cannot be found easily, this type of reasoning may be impossible. But the theorem can also be generalized as the isomorphic property of linear functions :

$$f(\lambda x) = \lambda f(x)$$

2) 12 is four times 3. Therefore the cost will be four times 15. Four is the constant coefficient of the function "cost of" ; it connects magnitudes of different kinds : quantities of cars and francs. It is a rate, not a ratio ; it is normally expressed as a quotient of units (francs per car). The reasoning implies the following theorem-in-action :

$$f(x) = ax$$

3) The cost of 1 car will be 3 times less, and the cost of 15 cars will be 15 times more. This again is a scalar reasoning, and two versions of the same theorem are used :

$$f(1) = \frac{1}{n} f(n)$$

$$f(m) = mf(1)$$

4) 15 is 6 + 6 + 3. Therefore the cost is 24 + 24 + 12. In this line of reasoning, the vector 15 cars is seen as a linear combination :

$$15 = (2x3) + (2x3) + 3$$

The same linear combination is applied to the costs :

$$x = (2 \text{ x}12) + (2\text{x}12) + 12$$

The reasoning implies both the additive isomorphism: $f(x + x') = f(x) + f(x')$, and the multiplicative isomorphism of the linear function: $f(nx) = nf(x)$.

The general theorem for linear combinations would be :

$$f(\lambda x + \lambda' x') = \lambda f(x) + \lambda' f(x')$$

Comment : One does not observe any reasoning involving the additive decombination of 12 as 6 + 6. The reason for this is that students do not reason on pure numbers, but on cars and francs : 12 francs cannot be 6 cars + 6 cars.

5) $x = \frac{15x12}{3}$. This is the rule-of-three algorithm. It is different from solution n° 3 in the sense that the multiplication comes first.

Only a few students use it, because it is meaningless for most students, to multiply 15 cars by 12 francs. The theorem actually relies upon a purely numerical theorem of proportions.

$$\frac{x}{b} = \frac{c}{a} \Rightarrow x = \frac{bc}{a}$$

It fails to be intelligible when numbers are thought of as magnitudes or quantities, unless one makes explicit the dimensional analysis :

$$\frac{15 \text{ cars} \times 12 \text{ francs}}{3 \text{ cars}} = x \text{ francs}$$

Figure 3 shows that there are two different ways of combining simple proportions : concatenation and product.

Concatenation — simple cases

1	a	
	1	b
c		d

a=f(1) b=f'(1) d=f o f' (c)

Product — simple cases

	1	b
1	a	
c		d

a=f(1,1) d=f(b,c)

Concatenation — general cases

a	b	
	c	d
e		g

b=f(a) d=f'(c) g=f o f'(e)

Product — general cases

	a	e
b	d	
c		g

d=f(a,b) g=f(e,c)

Figure 3. Combination of simple proportions.

In the simpler cases, there is only one problem that can be solved without any division : find d, given a, b and c. All the other cases require one or several divisions as can be seen in the following examples.

Concatenation : **Robert wants to buy 3 boxes of miniature cars. Each box contains 4 cars. The cars cost 10 French francs each. How much will he have to pay ?**

$a = 4 \quad b = 10 \quad c = 3 \quad$ d is unknown

The solution only requires multiplications.

Given the total cost (120 francs) and asked to find the number of boxes, or the number of cars in a box, or the cost of one car, one would need at least one division to solve the problem.

Product : **Mrs Ronda wants to buy cookies for the week for her 3 children. She needs 4 cookies per child and per day. How many cookies should she buy ?**

$a = 4 \quad b = 3 \quad c = 7 \quad$ d is unknown

The solution only requires multiplications.

When d is not unknown, one may have to calculate a, b or c. These tasks again require at least one division.

The situation becomes more complex when there are no such givens as f(1) or f'(1) or f(1,1). The example below is one of many, as there are a wide variety of problems.

Mrs Turner has bought a box of 100 cookies. She counts that, on the average, 3 children eat 40 cookies in a week. How many days will the box last if instead of 3 children, there are 5 children at home ?

Another kind of difficulty arises when the information is only given in terms of ratios.

Mr Smith has bought a very large fish tank for his den. He already has a fish tank in his kitchen, and he needs 22 liters of water to fill it. How much water does he need for the new fish tank, which is twice as long, three times as high, and one and a half times as wide as the one in the kitchen ?

There are only a few ways of solving this problem. The main conceptual difficulty lies in the trilinear function that relates the variation of volume to the variation in length, width and height. If a student discovers that all you have to do is to multiply 2 by 3 by 1.5 to find the ratio of the volume of the big fish tank to the volume of the smaller one, he implicitly uses a powerful theorem for 3-linear functions

$$f(\lambda_1 x_1, \lambda_2 x_2, \lambda_3 x_3) = \lambda_1 \lambda_2 \lambda_3 f(x_1, x_2, x_3)$$

This is a rather extreme example of a theorem-in-action : although it is not explicit and has never been taught there is no other way to account for the student's sequence of calculations and implicit reasoning.

The field of multiplicative structures has been studied extensively by researchers in mathematics education. The major references can be found in Lesh and Landau [7], Behr and Hiebert [1], Vergnaud [11].

Important mathematical concepts can be derived by students from the study of multiplicative structures, and from the analysis of the solutions they devise, when faced with the variety of situations that we have tried to exemplify. This variety of solutions can be generated by considering the structure of the problem as is shown in the above examples. It can also be generated by referring to different domains of reference (prices, shares, speeds, densities, geometric similarities, etc...) and by different ranges of numerical values (small whole numbers, simple ratios and rates, large numbers, non whole ratios, numbers smaller than one etc...). The concepts of fraction, ratio, rate and rational number are important keys to the whole field of multiplicative structures, but one cannot go very far in analysis, if concepts such as linear function and bi-linear function, dimensional analysis and vector-space are not taken into account.

Other conceptual fields

The framework of conceptual fields is aimed at capturing a wide variety of problems, concepts and cognitive operations in the same "field", which would be misleading to consider in isolation because they are intrinsically tied to one another in the process of conceptualizing, and in the solving of the variety of problems that students may be faced with. It is fairly easy to identify such conceptual fields in physics : dynamics, electrokinetics, calorimetry and thermodynamics... However domains such as the logic of classes and elementary algebra can also be envisaged as conceptual fields, instead of being seen as content-independent conceptual frameworks.

Conceptual fields are distinct from one another but not disconnected. This means that additive structures and multiplicative structures have points in common. Nevertheless the most characteristic features of each field are different. Algebra cannot be disconnected from additive and multiplicative structures ; nevertheless certain features also give algebra its specificity, and make it a specific domain of conceptualization, having its own interesting epistemic filiations and revolutions [4].

Algebra is a powerful tool for solving ordinary problems. It is also a set of new mathematical objects and relationships such as function and variable, equation and unknown, single and multiple solution, dependence and independence of equations, etc. These concepts can be made explicit in the teaching and learning of algebra ; they can also be used implicitly. But it is unlikely that students go very far in the use of algebra whitout conceptualizing these new mathematical objects in one way or another.

Schemes, Concepts and Theorems

There is general consensus among researchers in mathematics education (and among teachers) that students acquire knowledge and develop new competences in mathematics by being faced with new situations, new objects and new properties, by being shown new procedures, by being given verbal explanations and taught how to use symbolic representations. But there is no commonly accepted view on what competences, conceptions and processes consist of. There are even misleading oppositions such as those between procedural and declarative knowledge, analogical and symbolic processes, routine knowledge and problem solving, expert and novice. This prompts me to make my own point of view as explicit as possible.

What are competences and conceptions made of ?

Competences are certainly made of rules, as rules are necessary ingredients to generate action (and more generally behavior). But are rules sufficient ingredients to account for the way students develop increasingly complex competences? I will show below that the answer is no. Conceptions are not made of rules, but rather of propositions (explicit or implicit). These propositions concern the different properties, relationships and transformations of different kinds of objects : the physical quantities of objects like plates, chairs, sweets and miniature cars at the bottom of the architecture, but also very abstract mathematical objects like vector-spaces, topological structures and families of functions, at the top of the architecture.

We need a theory of the interconnections between competences and conceptions. On the one hand, students' conceptions are shaped by their practical experience of situations and therefore by their competences. On the other hand, science and technology give students revised conceptions or even totally new conceptions which enable them to develop more appropriate and more sophisticated competences.

The concept of rule is totally insufficient to account for the way competences are developed.

Even in a very simple rule like :

IF A THEN DO X

one needs to understand how A is conceptualized, how X is conceptualized, and how the properties of X are related to the properties of A. Most rules in the learning of mathematics, science and technology are not mere associations. And even if rules probably exist that stem from reinforced associations without much understanding of the relationship between X and A, some cognitive activity is still required to identify A and the meaning of A from among the different aspects of the environment. It is impossible to avoid the theoretical problem of conceptualization.

It is for these reasons that I have developed the concept of scheme, which Piaget introduced (from Kant) to draw the attention of psychologists to these dynamic and functional totalities as sensori-motor skills and intellectual skills. Piaget was impressed by the "totalities" that gestaltists had discovered in perception, and probably felt the need for a similar idea which would stress the overall organization of action. A more detailed definition of schemes is called for here.

Definition : **a scheme is the invariant organization of behavior for a certain class of situations.**

Because schemes are related to the organization of behavior they do not have the same theoretical goal as schemas, frames, scripts or scenarios. Like algorithms, they are mappings of multidimensional spaces of variables into multidimensional spaces of variables. Sequencing and timing are essential, as is manifest for the scheme of jumping for a champion: his scheme enables him to extract the relevant information over the entire course of the process of running and jumping, and to command the appropriate sequence of movements at the right time.

Even though it is highly invariant, the jumper's scheme is not a stereotype. It varies from one try to the next and the jumper is usually able to take account of slight differences and modify his movements accordingly. I will now give three examples of mathematical schemes.

1. Counting a small set of objects : the child's behavior may vary on different occasions, but there is an invariant organization of gestures (finger, hand, eye) together with the production of the number-words. Two important mathematical ideas underlie this scheme : the idea of bijection between the set of objects and the different sets of gestures, and the idea of cardinality, which is usually illustrated by the repetition or the stressing of the last word : one, two, three, four... four! The first "four" refers to the fourth element, the second "four" to the cardinal of the whole set.

2. Inversing a direct transformation and applying it to the final state : we have seen this example above. It is difficult to see how this scheme could emerge if concepts such as initial state, transformation and final state, and a theorem such as the one above-mentioned $F = T(I) \Rightarrow I = T^{-1}(F)$ were not implicitly contained in the scheme. Therefore, schemes are made of rules, goals and expectations, and operational invariants. There are at least two kinds of operational invariants : theorems-in-action and concepts-in-action. Theorems can be true or false, they are propositions; concepts cannot be true or false, as they are not propositions, but only relevant or irrelevant. Rules of action are propositions: they would not be computable if schemes did not contain theorems. But on the other hand, theorems are necessarily made of concepts. There is a dialectical connection between theorems and concepts, just as there is a dialectical connection between propositions, propositional

functions, and arguments. Propositional functions and arguments would not exist if there was no need for propositions, and there would not be propositions without propositional functions and arguments. The invariant organization of schemes partially rests on the rules that generate behavior, but their foundations consist in the operational invariants that make the scheme relevant and efficient. As these operational invariants are usually implicit, I have termed them concepts-in-action and theorems-in-action.

As schemes are not stereotypes, one must also assume that they have inference potentials. Actions are certainly computed on the spot, depending on the values of the situation-variables. But rules of action themselves can be computed on the spot, in problem solving for instance. Schemes are universals : they address classes of situations.

3. My last example of a mathematical scheme is the script- algorithm that is used to solve equations of the type ax + b = c. The execution of the scheme involves the writing of a script which always has the same organization for certain subsets of values (for instance for $c > b > o$ and $a > o$). I call it script-algorithm because the invariant organization of the scheme does not only come from the script and the sequence of writings : it also comes from such implicit theorems as the conservation of equality when one subtracts the same number from both sides, or when one divides both sides by the same number, and from such implicit goals as the isolation of the unknown on one side of the equal sign.

The scheme to solve ax + b = c equations involves both operations with symbols and operations with concepts and theorems. This is not an exceptional case. There are many schemes whose execution involves a parallel symbolic activity like speaking aloud or subvocally, or drawing a sketch, a diagram or a graph. This means that symbols are a help, and perhaps critical, to the functioning of schemes. This idea is different from the notion that explicit knowledge and implicit knowledge have a different cognitive status, but is connected to it. The relevance of concepts, and the truth and the scope of validity of theorems can only be discussed if they are explicit. Implicit theorems-in-action may certainly be discarded when they are inefficient, but they cannot be proved or disproved. The fact that we develop verbal and symbolic activities while executing a scheme is an important psychological fact ; it shows the need for a theory of the relationship between concepts and symbols.

In conclusion, the above analysis of schemes shows the need to recognize that they are made of rules, goals and expectations, inference potentials and operational invariants. The concept of operational invariant (concepts-in-action and theorems-in-action) is a key concept for a theory of action relying upon a sound theory of representation, regardless of whether this representation is implicit or explicit .

Concepts, Words and Symbolic Representations

Algebra raises the problem of symbols in a way that differs considerably from arithmetic. Symbols are so visibly important in algebra that some people tend to think that algebra is only symbol generation and symbol manipulation. This is of course not true, as a lot of conceptualizing also takes place in the learning and the use of algebra. Nevertheless algebra would not be possible without symbols.

The importance of symbols in mathematics has led some researchers to contend that mathematics is a language. Actually mathematics is knowledge, not a language, irrespective of how essential language and symbols may be in the functioning and the learning of mathematical knowledge.

There are several functions of language in cognitive processes : communicating, representing, helping, controlling, etc... These functions are interconnected, but the representation function is at its core. I will now provide examples to show the efficiency and the limits of symbolic representations.

Let us take the example of the formula for the volume of straight prisms:

$$V = AH$$

V = volume of the prism ; A = area of the base ; H = height.

this formula can be read at different levels :

-calculate V when A and H are given ;

-calculate A when V and H are given ; or calculate H when V and A are given ;

-see the direct proportion between V and A when H is held constant; or between V and H when A is held constant ;

-see the proportion between V and the product AH.

The different ways of understanding the formula are not equally accessible to students and some students stick to the first reading.

Another symbolic system can be used to demonstrate the properties of the double proportion. Figure 4 enables students to perceive and understand the simple proportion between volume and area when height is held constant, and between volume and height when area is held constant.

		Height				
		1	5	10	20	25
Area	1			10		
	7	7	35	70	140	175
	21			210		
	22			220		

Figure 4. Representation of double proportion.

Similarly, Figure 5 enables students to acquire a more solid grasp of the combination of ratios in double proportion situations.

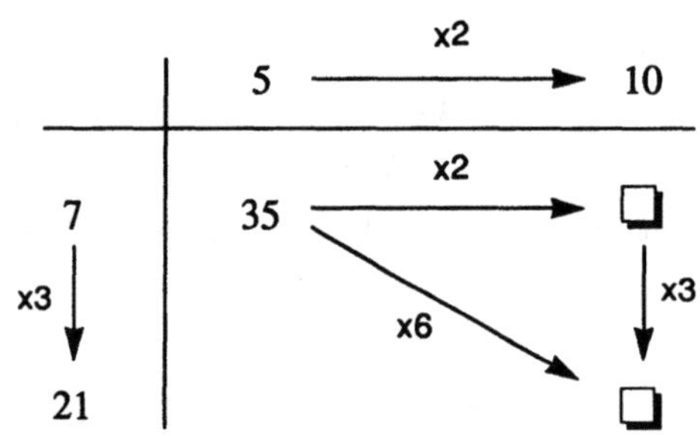

Figure 5. Bilinearity and double proportion.

My next example is the representation of the find-the-initial-state problem, in additive structures.

Maria has just bought a pastry. She paid 8 francs for it. She now has 7 francs. She wonders whether she has lost some change and tries to calculate how much she had before buying the pastry.

Here are several symbolic representations :

a) equation $7 + 8 = \square$

b) arrow diagram $7 \xrightarrow{+8} \square$

c) Euler Venn diagram

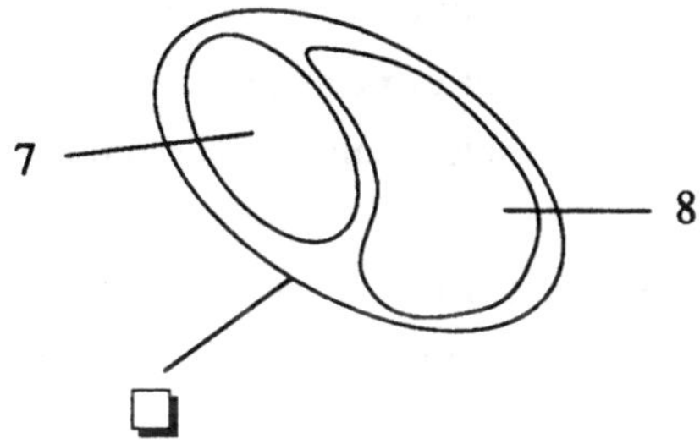

All of these are acceptable but not very useful, since they represent the solution to the problem, and not the problem itself. It is fairly hard to represent the problem and the path from the representation of the problem to the representation of the solution.

a) To deal with the equation, one needs powerful algebraic notions

representation of the problem	$\square - 8 = 7$
algebra	$\square - 8 + 8 = 7 + 8$
	$\square = 7 + 8$
representation of the solution	$\square = 15$

b) In the second case, it is fairly easy to show that one can inverse the transformation that connects the initial state to the final state.

$$\square \underset{+8}{\overset{-8}{\rightleftarrows}} 7$$

c) In the third case, one cannot even represent the problem because Euler-Venn diagrams cannot represent negative transformations. Therefore they are useless.

What emerges from these considerations is that symbols vary in their power to exhibit the properties of mathematical relationships and the relational calculus that enables us to go from the problem to its solution. Euler-Venn diagrams for instance are good tools for part-part-whole relationships and misleading for negative transformations and relationships.

Equalities, formulas and algebraic representations create enormous difficulties at the elementary and early secondary levels. This implies that different kinds of pre-algebraic representations are needed.

When teachers do not have these tools, they try to explain relationships and procedures with words; they also use words to explain other symbolic tools. Words, however may have a different cognitive status. The problem of making explicit the operational invariants involved in schemes is one of the main problems of mathematical education, because language, overt or subvocal, usually accompanies problem solving. The problem of language and thought is examined below.

Words are not concepts, and sentences are not theorems. Words and sentences would be empty shells if there were no operational invariants. On the other hand, explicit concepts and theorems are more than operational invariants : they are public and can therefore be discussed. Operational invariants may only involve the signified side of the coin, whereas concepts and theorems necessarily involve both the signified and the signifier. The core of meaning consists of action and schemes : when we say that a situation or a symbol is meaningless for a student, the connotation is that the student does not know what to do with them. Words have their own importance not only to communicate about objects and their properties, but to think about them and act on them, especially when other people are involved.

A further distinction needs to be made between concepts as tools and concepts as objects. For instance, in problems involving transformations, the initial and the final states are necessarily used as tools to categorize information ; students may be able to use adverbs and the past tenses of verbs to speak of the initial state, but they are not necessarily able to conceptualize the states and the transformation as mathematical objects. Symbolizing plays a part in changing the cognitive status of concepts from tools into objects, but this transformation may create considerable cognitive difficulties, for example when the teacher uses nouns instead of verbs and adjectives to speak of geometrical transformations: reflections, rotations, translations, isometries, homotheties. Verbs and adjectives usually stand for propositonal functions (tools), nouns for arguments (objects). Concepts can be tools and objects. The words used are important.

Finally I will define a concept as a 3-uple of three sets

$$C = (S, I, \mathcal{S})$$

S : the set of situations (or problems) that make the concept meaningful;

I : the set of operational invariants involved in the schemes that subjects can use to treat these situations;

$\mathcal{S}$: the set of symbols (linguistic and non-linguistic) that can be used to represent these invariants, and therefore situations and schemes.

Can the Theory of Conceptual Fields be Useful in Devising Intelligent Computer Tools ?

The main conceptual features of the theory are the following: a few elementary relationships in the same field can serve to generate many classes of problems, including several-step problems. The difficulty of each elementary cognitive task can be described by a partially-ordered hierarchy. Different schemes can be used to cope with the same problem, but their power and generality depend on the conceptual level of the operational invariants upon which they are grounded. Language and other symbolic tools often help students identify relevant information, relevant concepts and theorems, and relevant rules and steps of the problem-solving process. The ultimate goal is to help a student to be able to represent a given situation in a conceptual field, and either activate a relevant scheme to deal with it quasi-instantly, or map this situation into a symbolic representation and then operate inside this representation until the solution is reached.

Computer tools offer a wide variety of possibilities that can be fed by such an approach.

There are reliable ways of generating problems of progressive complexity which give individual students a sample of problems within the scope of their "zone of proximal development" [14].

There are also reliable ways of taking into account the errors and the procedures used by each individual, as they can be identified fairly easily for each class of problems; for instance the most frequent errors made by students in simple and multiple proportion problems are well-known, and there are some identified ways to make them aware of these errors.

There are also well-identified symbolic representations that can be used to help students represent the problem and recognize the status and position of the different pieces of information. There may also be vicariant representations, and therefore ways to translate a problem, a reasoning and a path to the solution from one representation into another.

The kind of interaction that a teacher usually has with students (scaffolding, explaining, monitoring, asking intermediary questions, pointing out relevant properties and relationships...) should also be taken into account as much as possible. This is probably the greatest challenge for intelligent tutoring systems.

Tutoring systems certainly cannot make better choices and decisions than teachers in all facets of teaching activity. But in some, they can be more complete and more reliable. In principle, they can use a more complete picture of the different classes of problems, the different procedures used by students, and the different symbolic representations in a given conceptual field : they can offer for instance powerful possibilities for graph designing. They will probably not be able to interpret students' behavior before a considerable amount of research has been conducted.

References

1. Behr, M., & Hiebert, J. (Eds) (1988). *Number concepts and operations in the middle grades.* Reston, VA, NCTM.
2. Bruner, J. (1983). *Savoir-faire, savoir dire.* Paris : Presses Universitaires de France.
3. Carpenter, T.P., Moser, J.M., & Romberg, T.A. (Eds) (1982). *Addition and subtraction : A cognitive perspective.* Hillsdale, NJ : Erlbaum.
4. Coxford, A. F. (Ed) (1988). *The ideas of algebra, K12 (1988 Yearbook of the National Council of Teachers of Mathematics).* Reston, VA, NCTM.
5. De Corte, E., & Verschaffel, L. (1987). The effect of semantic structure on first graders' strategies for solving addition and subtraction word problems. *Journal for Research in Mathematics Education*, 18, 363-381.
6. Gelman, R., & Gallistel, C.R. (1978). *The child's understanding of number.* Cambridge, MA : Harvard University Press.
7. Lesh, R., & Landau, M. (Eds) (1983). *Acquisition of mathematics concepts and processes.* New York : Academic Press.
8. Nesher, P. (1982). Levels of description in the analysis of addition and substraction word problems. In T.P. Carpenter, J.M. Moser & T.A. Romberg (Eds), *Addition and substraction. A cognitive perspective*, (pp 25-38). Hillsdale, NJ : Erlbaum.
9. Piaget, J. (1967). *Biologie et connaissance.* Paris : Gallimard.
10 Riley, M.S., Greeno, J.G., & Heller, J.I. (1983). Development of children's problem solving ability in arithmetic. In H.P. Ginsburg (Ed), *The development of mathematical thinking.* New York : Academic Press.
11. Vergnaud, G. (1983). Multiplicative Structures. In R. Lesh, & M. Landau, (Eds), *Acquisition of mathematics concepts and processes.* (pp 127-174). New York : Academic Press.
12. Vergnaud, G. (1981). *L'enfant, la mathématique et la réalité.* Bern : Peter Lang.
13. Vergnaud, G. (1982). A classification of cognitive tasks and operations of thought involved in addition and subtraction problems. In T.P. Carpenter, J.M. Moser, & T.A. Romberg, (Eds). *Addition and subtraction : A cognitive perspective.* Hillsdale, NJ : Erlbaum.
14. Vygotski, L.S. (1962). *Thought and language.* Cambridge, MA. : MIT Press.

Part III. Creating Learning Environments

Introduction to Part III

Heinz Mandl
Institut für Empirische Pädagogik und Pädagogische Psychologie, Ludwig-Maximilians-Universität, Leopoldstr. 13, D-8000 München 40

From the viewpoint of cognitive psychology, the learner acquires knowledge and skills in an active manner by interacting with his or her environment and thereby modifying his/her mental structures. Recent instructional approaches emphasize that learning enironments should be designed as to facilitate active and constructive learning processes. This is also true for the design of multimedia learning environments. Concerning the topic of this section, three aspects will be addressed: (a) type of knowledge acquired by means of the learning environment, (b) design of learning environments, (c) integration of learning within the social context.

Type of knowledge

A major part of the contributions in this section deal, certainly not by chance, with the impartation of scientific knowledge - especially with physical sciences knowledge (De Jong & Njoo; Glaser, Schauble, Raghavan, & Zeitz; Goodyear; Mandl, Bollwahn, Hron, Oestermeier, & Tergan; Moar, Spensley, O'Shea, Singer, Hennessey, & Scanlon; Pea; Teodoro). In the framework of computer-based learning environments, physical science topics still appear to be easier to deal with than social science topics. However, it should be noted that the contribution of Glaser et al. is concerned with an economics learning environment. Apart from the impartation of domain-specific knowledge, the authors of this section also emphasize the importance of heuristic strategies in handling learning environments such as strategies of scientific experimentation, designing and interpreting experiments, hypotheses testing, identifying regularities in patterns of data, and reasoning about correlation and covariation in events. Furthermore, Glaser et al. investigate which reasoning and inference activities are employed with some consistency and systematicity from domain to domain, and which activities appear to be used more narrowly in a more limited range of content.

Thereby, cognitive strategies have more bearing. In increasingly acquiring new strategies for problem solving, students are confronted with management problems, for example, how to select among the various problem-solving strategies or how to decide for a change in strategy.

According to Pea, science learning is not that much an issue of mastering the partial skill of being able to manipulate scientific symbol systems and the associated problem-solving skills in dealing with problems, but centrally "constructed" learning by means of scientific conversation - and not a set of mental representational structures: "Science learning consists of entering into the web of social relations and actions that are constituated by various practices, accountabilities and duties that make up the discourse of scientific knowing". In the framework of her MultiMedia Works after-school club for underserved middle-school students, Allen also emphasizes an enlarged knowledge approach: In her learning environments, students learn how to elaborate constructive criticism, to forge community links, collaboration and leadership skills, to search and ask for information help; they learn to use technological skills and a myriad of other skills which are crucial for successful participation in community learning.

Design of learning environments

In contemporary instructional design, there is a clear trend towards creating learning environments, in which learners do not receive ready-made, directly consumable knowledge, but have to create their own knowledge (see De Jong & Njoo). The major part of the contributions in this section deal with exploratory, simulation-based learning (De Jong & Njoo, Glaser et al., Goodyear, Mandl et al., Moar et al., Pea, Teodoro). The contribution of Glaser and coworkers is concerned with three computer laboratories which simulate phenomena in the domains of economics (Smithtown), electric circuits (Voltaville), and the refraction of light through lenses (Refract). In each laboratory students can construct experiments by varying variable values and parameters, take measurements, make predictions about expected outcomes, record data, develop and apply hypotheses, and so forth. A number of authors (e.g., Goodyear, De Jong & Njoo, Mandl et al., Moar et al., Teodoro) articulate the assumption that, if they try to help learners in using exploratory learning environments, there is a need to know about the nature of the exploratory study process. Careful analyses of the exploratory study process will reveal the necessary ingredients of successful exploratory learning. On the basis of a detailed analysis of learning processes performed with a computer-based learning environment, Mandl and coworkers have - in order to support the learning processes - developed a rule-based diagnosis system for identifying misconceptions in a specific domain of physics.

In their SIMULATE Project, De Jong and Njoo evaluate the efficiency of support mechanisms of exploratory learing scratchpads, study-process guidance forms, and ready-made hypotheses. Yet another approach for supporting exploratory learning stems from Moar and coworkers: They focus on five properties computers can process, that is, direct manipulation, memory augmentation, support of qualitative reasoning, and conflict resolution, and the presentation of counter-factual examples. They also describe two ways in which these properties could be used as guiding principles for the design of computer-based learning environments. An interactive simulation, which helps students explore the horizontal motion of objectives is described first; second, the design of a combined animation/modelling system is suggested by means of focusing on issues of memory augmentation and externalisation of visual imagery.

Yet another approach to support exploratory learning has been the conceptual lab developed by Teodoro: By means of his computerized exploratory laboratory, the learner can explore and experiment with concrete or abstract objects, deal with multiple representations and pose assumptions or devise strategies about motion. The conceptual lab is a tool for stimulating verbal and holistic discussions about a formal domain by means of qualitative reasoning. Teodoro endeavours to shift physics instruction to meaningful learning beyond symbols and algorithms. This type of learning requires a novel learning environment called "metabook". A metabook is an integrated package (i.e., written material plus computer software), in which the software is considered to be an integral part of the book, thus allowing for performing what the learner could not perform by means of the book.

Contrary to the previous contributions, Pea and Allen created their learning environments under the perspective of conversation and participation. The goal of Pea's learning environment "Optic Dynagrams" is that students could easily engage in a learning conversation about the conceptual content they are actually investigating. In order to augment the learning conversation, he designed a new kind of activity structures. These activity structures seek to combine the possibilities offered by new media (e.g., accompanying video of optical situations, in order to precipitate inquiry) with new types of conversational structures (i.e., scientific discourse in collaborative groups) and techniques of the domain. Pea focuses on the facilitation of qualitative reasoning in geometrical optics, rather than formal qualitative principles and formulas.

Based on the principle of learning as participation, Allen created the MultiMedia Works studio. In order to facilitate conversation, representation, and socially embedded learning, that learning environment includes computer tools, which are designed to take advantage of specific properties of multimedia representations. In the MultiMedia Works studio both club members and staff define the work by means of participating in the technology design, in the planning of the activities of the group, in the development of multimedia composition, in the planning of the application of their compositions, and in the selection of audience for criticism.

Learning in social context

In creating learning environments, the social framework becomes all the more demanding. Approaches like the situated-cognition approach [2] and the apprenticeship approach [3] focus on the creation of meaningful tasks and working environments rather than on the acquisition of a number of specific skills that students practice in isolation. Proponents of the above-mentioned approaches are very critical towards the type of decontextualized instruction prevalent in schools, because it tends to produce inert knowledge. However, in informal and apprenticeship-type learning, knowledge is acquired in the context of meaningful goals - hence its uses are quite clear rather than obscure [1]. In his contribution, Pea assumes that inquiry-focused discourse is fundamental to learning environments: Learning is not perceived to be information transfer from teacher to learner, but as a process of taking part in community activities, as collaborative sense-making. In the context of the situated-learning perspective, he considers learning to be an ongoing, integral part of active membership communities.

In the context of her MultiMedia club, Allen is concerned with designing and advancing applications of computer tools, which overcome current limitations by means of integrating learning into real world communities beyond the institution of schooling: In her club, learning is organized by the participants outside school. Students and staff members work together in defining the nature of their work. They integrate club-like activities into their curriculum and different subjects traditionally taught separately, for example, languages, history, geography.

Situated cognition and apprenticeship learning are novel and promising approaches. They point a way to redesign traditional forms of learning so students may better acquire expertise and problem-solving skills as well as an improved ability to learn throughout life span [3]. The design of models and their evaluation are, however, still in their infancy. A lot of research and development work must be done in the next years to show the possibilities and limits of situated learning.

References

1. Bransford, J., Goldman, S. R., & Vye, N. J. (in press). Making a difference in people's ability to think: Reflections on a decade of work and some hopes for the future. In R. J. Sternberg & L. Okagaki (Eds.), *Influence on children.*
2. Brown, J.S., Collins, A., & Duguid, P. (1989). Situated cognition and the culture of learning. *Educational Researcher, 18*(1), 32-42.
3. Collins, A., Brown, J. S., & Newman, S. E. (1989). Cognitive apprenticeship: Teaching the crafts of reading, writing and mathematics. In L. B. Resnick (Ed.), *Knowing, learning and instruction* (pp. 453-494). Hillsdale: Erlbaum.

Augmenting the Discourse of Learning with Computer-Based Learning Environments[1]

Roy D. Pea

Institute for Research on Learning, 2550 Hanover, Palo Alto, California 94304 USA

Abstract: Computer tools for learning are often thought of as providing practice in working with symbolic representations. We exemplify a different perspective in which the technology augments the kinds of learning conversations that can take place. Research from the Optics Dynagrams Project illustrates contributions from this perspective. I will describe pre-intervention learning environment characteristics and student learning, our design strategy for new activities and technologies to address problems we observed, and results with a classroom field test of our redesigned learning environment. In the Dynagrams learning environment, small groups of students work with a software simulation of phenomena of geometrical optics. They observe optical situations in the world or laboratory, use dynamic diagramming tools to make predictions and arguments to justify them based on scientific principles, definitions, or prior experiences, and test these predictions in runs of their simulation models. The dynamic diagrams become symbolic vehicles for externalizing student cognitions for peers and teacher, as well as the topic for negotiating group and individual understanding toward physics norms. The pedagogical goal is to have students become better able to engage in appropriate conversations about the conceptual content they are investigating. Such inquiry-focused discourse is a fundamental part of learning environments in authentic practices outside schools; our aim is to examine ways for augmenting such learning conversations in schools.

Keywords: learning theory, science education, classroom discourse.

---

1 We are grateful for support of this research from the National Science Foundation, Grant # MDR88-55582, and from Apple Computer, Inc., External Research. The Dynagrams Project team is made up of Sue Allen, Shelley Goldman, Susanne Jul, Miriam Reiner, Michael Sipusic, and Erik Slavin. The theoretical perspective outlined in this chapter has been deeply influenced by my colleagues at the Institute for Research on Learning.

Introduction

The learning problems our cognitive science research community has studied for several decades center on the use of concepts and on conceptual change, especially in science and mathematics. We have also focused on the acquisition of problem-solving skills and procedures involving representational systems (e.g., graphs, algebraic equations, programming languages) that are the currency of thought for specific content domains. Several responses are common by those concerned with these educational problems. One is to seek out through knowledge diagnosis during problem-solving tasks the conceptions students have which deviate from current scientific understanding. Students are then confronted with problem situations in which these conceptions can be shown to be inadequate, with hopes that such conflict may precipitate conceptual change [e.g., 17, 43, 50, 64, 69]. Another response has been to identify the skills and procedures utilized by competent problem solvers in the domain and to then establish means for training students to use those skills [e.g., 3, 31, 44, 61, 63].

Work with these objectives has made valuable contributions to understanding conditions sufficient for promoting learning. But now we must go on to ask yet harder questions. Do the results of such training carry beyond the school walls? Can experiments establishing transfer of learning results on near-variants of training tasks lead to spontaneous and successful use of such educational innovations after the researcher fades from the scene? These hard indicators have rarely been sought, at the level of social uptake of "solutions" to learning problems that emerge from the research community.

There are foundational issues beyond the common responses of our field which might provide the radical reconstruction of the epistemological eyeglasses with which we view the significant categories of meaning, learning, and knowledge involved in education. What are the characteristics of learning environments that initiate and sustain the learning and use of new concepts and procedures? Our thesis is that these characteristics may be identified by seeking out the properties of *successful* learning environments beyond those provided in current classroom settings.[2] And we are interested in experts, but not so as to just identify their achievements and practices, but to characterize the communities of practice that *give rise* to "experts" and *in terms of which* their achievements have meaning and their disciplines keep dynamically evolving their knowledge claims [e.g., 35, 39]. We believe that current progress in the cognitive sciences of learning may be reformulated and

2 This emphasis on the study of successful learning is a central theme of research work at the Institute for Research on Learning in Palo Alto, California. Among its central predecessors are [8, 40, 50, 55, 62]. Gelman and Brown [21] describe a similar emphasis in studying cognitive competence in the young.

extended by attending to and better understanding this deeper social and situated fabric of cognitive activity.

My method will be to outline aspects of a broader social framework for understanding learning and for investigating how competent problem-solving is established, and to point to alternative characterizations of learning, knowledge, and instructional processes that arise from these considerations. Finally, I will highlight the implications of this social framework for the design of computer-based learning environments, alluding to our Optics Dynagrams Project.

Key concepts of a theory of learning as situated in communities of practice

Lave and Wenger [41], in generalizing the theory of learning as "cognitive apprenticeship" developed by J. S. Brown and colleagues [5, 15], have formulated a situated learning perspective that sees learning as an ongoing and integral part of membership in "communities of practice." Such membership is conceived of as an activity system about which participants share understanding regarding what they are doing and what this means in their lives and for the different communities of practice in which they participate. Learning is conceptualized as a lifelong process integral to *becoming* a member of different communities of practice, and *sustaining* such membership. The construction of personal identities largely involves defining participatory roles in different communities of practice. Persons always are members of multiple communities of practice, which may emerge, change, or disappear during their lifetimes. Allen [1] develops this perspective in her characterizations of the ways in which learning environment design can be conceptualized as providing conditions for the "growing" of communities of practice.

On this view, as in cognitive science, the acquisition of expertise is still viewed as important. But rather than construing expertise primarily as the acquisition of domain facts, problem-solving procedures, heuristics, and metacognition for formal problem-solving, expertise is viewed as a practice of a community. Learning is viewed not only as a relation to problem-solving activities, but in terms of participation as member in the practice of different social communities. For science, such a practice consists of ways of talking and acting (which include many shared goals, concepts, procedures), belief systems about what is interesting or promising about problems, shared views of when it is appropriate to use particular tools, and evolving kinds of sense-making activities that seek to evolve scientific concepts to fit the world (e.g., modelling, theory building, simulations). A community of practice for science includes at its frontiers diverse claims to knowledge, and disputatious means for advancing and resolving such claims, as the success of concepts as resources for resolving new problems is tested [e.g., 36, 67]. Learning then is not perceived as information transfer from teacher to learner, but as a process of participating in the activities of a community, by means of collaborative sense-making in which knowledge functions as a tool to resolve emergent dilemmas [4, 41, 51]. Learning conversations are central to these participations, in the sense I will now describe.

This emphasis on learning through conversations is not intended to *replace* that of learning by other means, such as remembering past experiences when alone and reflecting on the usefulness of one's current knowledge in the face of new problem conditions, or learning by reading and engaging in self-explanations [9]. But conversations *are* a major source of learning resources which have been unreasonably neglected by the cognitive science community in its studies of learning, and yet which, given the pervasiveness of learning through conversations outside schooling institutions, is bound to be critical to achieving successful learning in school settings.

What is a learning conversation?

Learning is fundamentally built up through conversations between persons, involving the *creation* of communications and efforts to *interpret* communications. Creation and interpretation are the reciprocal processes of human conversational action, through which meaning gets established and negotiated [e.g., 24, 32, 51, 60]. Communication is thus not viewed (as it is commonly in educational practice) in terms of one-way transmission and reception of meanings, but as two-way transformational, enabling the progressive construction of meaning through successive turns of action and talk. And conversations are the means by which people collaboratively construct the common ground of beliefs, meanings, and understandings that they share, and also articulate their differences. These conversations also provide the publicly available resources and thus the opportunities for speakers to determine how they were understood, often leading to meaning negotiation and cognitive change. Meaning negotiation takes place using interactional procedures such as commentaries, repairs, paraphrases, and other linguistic devices for signalling and fixing troubles in shared understanding [59].

Learning by participating in the language games and activities of science

In science, as in any communicative exchange, the problems of interpreting speaker meaning are deep. For the novice science learner, the classroom context often radically underdetermines the meaning of technical terms and symbols, and their mapping to the physical world that they are about. In the didactic mode typical of science instruction, few opportunities emerge for resolving the problems that either students or teachers may have interpreting the meaning of their respective talk about science. When science educators write that "it has become a commonplace belief that learning is the result of the interaction between what the student is taught and his current ideas or concepts" [54, p. 211], it is too rarely acknowledged that it is through learning conversations that these differences are most commonly observable and resolvable.

The problem of learning to do scientific conversations is analogous to learning a natural language. During activities in which children participate, adults play language games such as question-answer, of naming and elaboration, through which children "learn how to mean" [6, 27]. Studies of lexical development reveal that through communicative exchanges toddlers engage in what George Miller [47] has called a "spontaneous apprenticeship" with mature practitioners in communities of linguistic practice. Children observe words used by others in contexts and then try out the use of words in contexts, with conversational repair among participants providing opportunities for establishing a working alignment of saying and perceived meaning.

Influenced by these considerations, Hawkins and Pea [30], among others such as Lemke [42], have argued for the need to re-organize science learning environments so that students come to be able to *talk* science, to produce and interpret speech acts involved in participating in scientific activities, rather than just *hear* science. A crucial facet of the practice of science is its rhetoric -- how the discourse of the field is organized, how viewpoints are presented, what counts as an argument and its support, and so on. Science education should result in capabilities to participate in scientific discourse -- to converse about scientific ideas and the scientific aspects of issues and systems generally.

The discourse forms of a discipline can be considered as an example of "language games," the image developed by Wittgenstein in *Philosophical Investigations* [71] for processes by which meaning is communicated and developed. In Wittgenstein's terms, many human activities may be productively viewed as "language games," participation in which can lead to appropriate use of language for the activities involved in those games, through refinement of meaning in contexts of use. Fluency in these practices comes with recognition of membership in a given community of practice [18, 41].

In our work we generalize these observations about *language* to *symbolic forms* more generally, including such representations as diagrams, pictures, mathematical symbols and equations. Just as speakers make speech acts with natural language, they make diagram acts that have analogous interpretive demands in a discourse to those of speech. Such discourse also often involves the use of complex symbolic representational systems in a discourse "workspace" between participants (e.g., diagrams, graphs on a whiteboard, lines of programming code on a computer screen). These representations come to serve as resources that enable speakers to engage in conversations about complex conceptual entities, such as slopes on a graph, or rays of light. They can point to these entities, talk about them to clarify what is meant, and describe how they are connected to other things.

A major part of enculturation in these games of scientific practice is coming to be able to engage in sense-making conversations that use and talk appropriately about how such external representations relate to situations in the physical world and to each other. For science, the key authentic tasks of sense-making in science -- those tasks that are the ordinary activity of the practitioners of a scientific community -- include producing causal accounts including technical concepts, symbols, and models to explain observations of a physical situation, forecasting a future state of a situation given some variation in its properties, and dealing with emergent problems in the course of an inquiry.

This view of learning science relates closely with Hanson's [28] idea about "seeing-as" in his discussion of the philosophy of science. For example, the Copernican revolution taught us to see the sun as a very large stationary object that the earth revolves around rather than a smaller object rotating around the earth. Learning to interpret and use scientific ideas involves learning of ways of "seeing-as" in two important senses. Students learn to see scientific notations such as diagrams and equations as symbols that represent aspects of the conceptual entities of science. They also learn to see systems and events as instances of scientific structures and principles. We emphasize that "seeing-as" is a central part of what communities of practice do, and that the process of learning to see and talk about things in the ways that one's pertinent community sees them is a major aspect of community socialization.

The meaning of representations such as words and diagrams in a community becomes evident through their use and the reshapings of their meanings through commentary by other participants of learning conversations. As such, meaning is dynamic and in continual flux because its use is coupled to the particularities of each new situation, not a static proposition mentally represented in a truth-functional calculus.

Learning through processes of appropriation and interpretation

We now believe that two central mechanisms underlie these processes of enculturation in scientific practice, or what Lave and Wenger [41] would call increasing participation in the "community of practice" affiliated with talking science.

One mechanism is *appropriation*. Appropriation has two sides--appropriation for use and appropriation of what one takes another to mean. Leont'ev (a colleague of Vygotsky whose work [68] established a pioneering theory of the social construction of knowledge), characterized learning as the "appropriation" of cultural tools. For Leont'ev, the biological language of Piaget's interactionism is replaced with the sociohistorical language of "appropriation." Newman, Griffin and Cole [49] apply this concept of appropriation to cognitive change from schooling. They note how:

> For Leont'ev, the objects in the learner's world have a social history and functions that are not discovered through the learner's unaided explorations. The usual function of a hammer, for example, is not understood by exploring the hammer itself (although the learner may discover some facts about weight and balance). The learner's appropriation of culturally devised "tools" comes about through involvement in culturally organized activities in which the tool plays a role....He emphasizes the fact that they [children] cannot and need not reinvent the artifacts that have taken millennia to evolve in order to appropriate such objects into their own system of activity. The learner has only to come to an understanding that is adequate for using the culturally elaborated object in the novel life circumstances he encounters. The appropriation process is always a two-way one. The tool may also be transformed, as it is used by a new member of the culture; some of these changes may be encoded in the culturally elaborated tool [p. 62-63].

Newman et al. apply Leont'ev's concept of appropriation to problems of cognitive change in schools. They observe: "Just as the children do not have to know the full cultural analysis of a tool to begin using it, the teacher does not have to have a complete analysis of the child's understanding of the situation to start using their actions within the larger system." [p. 63]. They see the fact that a given activity by the child can have multiple interpretations (for example, those of the child and of the teacher) as what makes cognitive change possible, through the negotiations of meaning about that situation that arise out of conversations: "While in the ZPD [zone of proximal development] of the activity, the child's actions get interpreted within the system being constructed by the teacher [p. 64]."[3]

Several features of the processes of appropriation should be evident from the description already: (1) One must come to acquire appropriate moves in the context of the activity itself (including its social and material environments); (2) in the context of the

---

3 In the language acquisition literature, this attribution process is described as one in which adults do "rich interpretations" of the talk young children produce. These interpretations may be viewed as *creating* richer meanings in the interaction through the processes of meaning attribution [6]. In the science learning conversations with teacher and student, the teacher may appropriate the student's talk as arising from understandings that the student does not have, but which such appropriation may help create, in the sense that the joint meaning they have constructed in the space between them is then viewable as an appropriate move in the game to which the learner has contributed.

interactive activity, there are different interpretations possible of the actions - a learner or teacher's self-interpretation (how what I am doing means to me), and a learner or teacher's appropriation of other-action to mean what it is taken to mean (which may diverge from the self-interpretation); and (3) with growing participation, one comes to anticipate and produce possible next moves in the game.

Meaning negotiation is the other central mechanism for the social construction of the meaning of what was expressed, and of events that are the target of sense-making activities among conversationalists. Its structure consists of reciprocal acts of *interpretation* between speakers. The resources for meaning negotiation are quite diverse, and include: requests for clarification or elaboration, gestural indications of misapprehension, explicit paraphrasings of what-may-have-been-meant to test for understanding, and explicit commentaries.

Ethnomethodologists such as Garfinkel [19], Garfinkel and Sacks [20], Schegloff and Sacks [60], and Mehan and Wood [45] have highlighted the importance of *indexical support* for such meaning negotiation. With indexical support, speakers opportunistically use the resources of the physical world to clarify what they mean, given the ephemeral nature of spoken language. Their words are "indexed" to referents in a situation, such as words or symbols on a whiteboard or computer screen. Such indexing is critical for establishing a shared semantics of representations, referential mappings between situations and formal symbols depicting world entities. Herbert Clark and colleagues [10, 11] have referred to this achievement as establishing a "common ground". Roschelle and Behrend [57] have emphasized the fundamental indexing roles of action on and gesture towards computer screen representations in constructing shared knowledge in collaborative problem-solving with the "Envisioning Machine," a computer microworld for exploring concepts of velocity and acceleration.

These processes of appropriation and meaning negotiation need to take place in the context of authentic activities that arise from *participating in a community of practice* for science learning. This means engaging in inquiries that require sense-making conversations using the technical concepts and procedures of science, and tasks such as prediction, observation, and explanation. During such inquiries, the meanings of representations for learners such as words for technical concepts and diagram components are continually remade through their *use* and *commentary* on their use, through creation and interpretation.

In summary, competency in the language games of science is co-produced by the participants' actions of appropriation and interpretation in authentic tasks for a community of practice. The teacher's role is to model inquiry, provoke inquiry oriented to students' conceptual change from pre-existing alternative conceptions of the subject domain, and serve to *represent* a community of scientific practice.[4] In the case of our physics students from "expert classrooms," we saw achievement of competency, but it was in a community of practice focused on training for test performance, not oriented to physics inquiry and understanding [53]. How might we change this with the design of the Dynagrams learning environment?

Tools for augmenting learning conversations

J. S. Brown [4] has characterized the epistemological shifts that emerge from a focus on successful learning, including the importance of implicit (tacit) knowledge in understanding[5], the significant process of developing rather than merely acquiring formal concepts, and the central nature for successful learning of the social context, activity[6], features of the material environment, and learner improvisation and exploration. He argues that learning technologies must become capable of drawing on these social, collaborative, constructive, and situated elements of human learning.

By contrast, computer tools for learning are often thought of as especially well-suited to providing solitary practice in the skills of working with externalized knowledge representations (e.g., geometric proof statements; algebraic equations; physics formulae) that it is the student's task to master. Given the need for appropriation activity to make meaning in human discourse, it seems unlikely that computers can be effective agents for directly teaching the language games of science (or any subject, for that matter). Appropriation requires taking a student's utterance, providing it with an interpretation, and engaging with the student in a negotiation that results in a closer approximation to norms of scientific meaning. The intentionality and membership in communities of practice required for this interaction is unavailable in computers in principle. This strengthens the motivation for designing computational resources that can provide things for teachers and students to

---

4 While teachers can rarely literally reproduce all the details of authentic science activity in their classrooms, they can model authentic practice by engagement and reflection on real exploration of topics occasioned by inquiry activities. Lampert [37] and Schoenfeld [61] have experienced success with such practice in mathematics education. We hope that one consequence of the teacher professionalism movement will be to make science teacher participation in communities of practice for science among colleagues outside school more common than it is today.

5 See [70] for related theory on computers and cognition.

6 A focus on learning by doing was central in Dewey's [16] seminal work on education, and in Bruner's [7] influential formulations of an activity-centered, inquiry approach to learning. The perspective on situated learning under development by Brown, Greeno, Pea and others at IRL [4, 5, 15, 25, 52] places greater emphasis that those earlier works on both the *social theory* in terms of which learning-by-doing is framed [41] and the *fine structure of human interactions* through which the collaborative construction of meaning for specific subject matter learning takes place [this chapter, 56, 57].

talk about, and for students to talk about with each other, rather than providing instruction through computers directly.

Here I develop the stance on learning technologies that conceptualizes computer tools as enabling *augmentation* of learning conversations that can take place either between learners, or between learners and more proficient users of the targeted knowledge or skills. In related work at IRL, Roschelle [56] calls this "designing for conversations."

The Optics Dynagrams Project

Science learning research from the Optics Dynagram Project exemplifies our approach for designing tools to augment learning conversations. "Dynagram" is our shorthand for "dynamic diagram," a central kind of symbolic representation in the software we have created as a communication medium for learning conversations about geometrical optics. Visual representations such as diagrams play a far more important role in the reasoning and problem representation processes of scientists than educational practices and learning theories now acknowledge [46]. Diagrams are important symbolic forms for representing concepts and conceptual relations, and provide, in the arguments of many researchers, a "language of thought" that exploits the visual processing capabilities of the human mind [38]. From our perspective, diagrammatic representations also provide conversational artefacts that better enable learners and teachers to become similarly connected to the conceptual content of these representations, and to negotiate differences in beliefs about how such diagrams representing world states will behave under various changes in the world they are about.

The pedagogical objective is to have students become better able to engage in appropriate conversations about the conceptual content they are investigating. These inquiry-focused conversations include such activities as making conjectures, designing experiments to test them, revising conjectures in light of observations of experiments, and so on. Before characterizing how we recrafted the learning environment and technologies to support these aims, let us review what we found upon examining the teaching and learning practice of several "exemplary" physics classrooms.

The shape of conversations in "expert classrooms" before Dynagrams

With the aim of identifying successful learning practices, we selected two classes for study that were taught by highly experienced physics teachers in high schools widely-recognized as producing an unusually high number of scientifically-oriented student graduates (one in New York, one in California).

In one study, we analyzed videorecordings of a teacher's optics lessons in an introductory physics classroom in a science-oriented high school in New York City (henceforth "NY school"), and videos of individual students as they attempted to represent and solve optics problems at a chalkboard using diagrams, equations, and words. Widely considered one of the best U.S. science high schools, it has over a dozen physics faculty and a department chairperson. Geometrical optics was taught for three weeks during the second semester of a required first year introductory course on physical science. Classroom observations and teacher conversations led us to image formation[7] as a particularly challenging and difficult topic, one in which the use and understanding of diagrams is essential. An interview guideline was administered to students immediately following instruction. Each student was asked to draw diagrams at a chalkboard in order to solve basic geometrical optics problems involving image formation with a single lens or mirror. Our goal was to examine student use and comprehension of diagrams as representations for reasoning about optical phenomena, and to document types and likely sources of difficulties during these activities.

For the second study we enlisted the cooperation of the physics faculty of our "CA school", where we have videotaped all optics lessons given by an award-winning physics teacher, and interviews with his students while individually attempting to represent and solve optics problems at a chalkboard using diagrams, equations, and words. These interviews also incorporated use of a simple hands-on laboratory apparatus (light source, converging lens, screen, ruler). We had the student represent and solve optics problems at the chalkboard with diagrams (and equations, when remembered). Then the student was asked to predict what would happen, and why, when the physical apparatus was used to create various optical phenomena. Finally, we asked the student to reconsider the design of the diagram used to justify a prediction if it was disproven by the physical apparatus. (We are now working with this teacher for our observations of the impacts of Optics Dynagrams on teaching and learning activities and outcomes.)

The teachers each predominantly worked by introducing key concepts with definitions, and offering demonstrations of these concepts perhaps common to the students' everyday experiences. Expositions of these situations were then presented using the various technical demonstration apparatuses and the symbolic representations of optics, such as ray

7 Research by Goldberg and McDermott also indicated severe problems in understanding image formation from a plane mirror [23], and of real images formed by a converging lens or concave mirror [22] among college-age introductory physics students both before and after instruction.

diagrams. Preeminent in these discussions was the highlighting by the teacher, and questions by the students, about what they would be held accountable to in quizzes and tests. These tests placed great emphasis on facts and definitions of geometrical optics, rather than their use for prediction and explanation. Labs took place but did not involve students in using these representations for prediction and explanation, or negotiation of their meanings.

Given the learning resources provided by the teacher, we asked whether students could appropriate for their own flexible use the various representations used by the teacher to reason about and explain optical phenomena. Were they able to participate in the language games of geometrical optics? These included technical concepts such as "refraction," ray diagrams, direct or pictorial experience with optical events from everyday situations, and algebraic equations (in NY) used to reason about the quantitative relations among the focal length of a mirror or lens, the size and distance of the object, and the size and distance of the image of the object.

We found that the dominant role of teacher's lectures and demonstrations as the students' resources for learning led to the following results [53]:

- fragile use of symbolic representations including diagrams, equations, and technical terms for reasoning about simple optical situations, both in using the representations to model a situation, and for reasoning about how light would behave in that modelled situation;
- learning was viewed by students as the memorization of definitions of technical terms and states of the diagrams that could be expected to appear on assessment tests;
- students lacked a scientific model of image formation as a point-by-point mapping from object to image governed by laws of reflection and refraction;
- in terms of learning conversations, we see there were few opportunities for students to map the meaning of diagrams or technical terms onto physical situations and apparatus, for making predictions, or modelling the physical situations to explain optical phenomena; and
- virtually no activities involving student groups and the teacher allowing for meaning negotiation for either the linguistic, diagrammatic, or equation representations central to reasoning in the domain.

Students had achieved some competency in a community of practice, but it was in test-taking and surface memory for physics facts, not a flexible capacity to engage in reasoning with the representational resources and techniques we expect as part of scientific practice.

Key aspects of the Optics Dynagrams learning environment

Our thesis is that we can foster learning by augmenting the learning conversations that take place with the Dynagrams learning environment. To do so, we sought to create *powerful affordances* for students to acquire competency in the language games of geometric optics.

Our design involved both technological and social dimensions. To augment learning conversations, we designed new kinds of *activity structures*, with careful attention to the supporting features of the physical and social environments for students' actions and conversations. These activity structures seek to combine: The affordances of new media (e.g., easy creation of diagram graphics and their direct manipulation with a mouse input device; ability to make diagrams dynamic; accompanying video of optical situations for precipitating inquiry), new kinds of conversational structures (scientific discourse in collaborative groups with teacher guidance rather than didactic interactions), and specific opportunities for developing conceptual understanding and techniques for the domain. We largely focused on promoting qualitative understanding of relations in geometrical optics, rather than formal quantitative principles and formulas.

In terms of conceptual understanding, based on problems we found students to have in our prior research, we sought to develop through these activity structures the following notions:

(1) Objects can absorb, refract, reflect, or diffusely reflect light;

(2) A point source emits light rays in all directions;

(3) An extended source is a sum of many point sources, and an object diffusely reflecting light can be viewed as such an extended source;

(4) An extended image is a point-by-point mapping of the object;

(5) Light travels in straight lines infinitely far but decreasing in intensity;

(6) Ray sprays may diverse, converge, or go parallel;

(7) The local mechanisms by which light rays are propagated at surfaces - Snell's Law and the Law of Reflection;

(8) The properties of how a lens bends light depend on its shape and refractive index;

(9) Structure and function of the eye in image formation;

(10) Image properties are often dependent on the position of observer (eye);

(11) Determination of focal point of lens.

Later we provide an example of a Dynagrams learning conversation in which a student group is working on the challenges involved in (1), (6), (7), (9), and (10).

In terms of techniques, we worked to provide activities that would:

- Encourage sense-making with a simulation model through actions of explanation, prediction, modelling, design, and troubleshooting;
- Foster diagram interpretation, diagram use as a qualitative reasoning tool (e.g., to define shadows, find image location, find lines-of-sight for mirrors), and diagram use as indexical support for sense-making arguments and narratives;
- Require continual mapping across diagram representations and physical situations with hands-on materials; and
- Establish collaborative inquiry, including competing conjectures, meaning negotiation, troubleshooting and repair, and refinement of language of description and explanation.

Furthermore, as we have considered science learning by analogy to language learning, we have seen that learning conversations with diagrams need to have three basic properties:

- They need to allow for the *production* of speech acts by learners that incorporate uses of technical concepts, diagrams, and other representations. This called for technology design of a readily-accessible expressive medium for students to use for frequently composing meanings in reasoning during sense-making activities about optical situations (see below). Dynagrams provides an "intelligent" graphical design language for learners to construct ray diagram models of optical situations.
- They need to take place so that learners do their own *interpretations* of the situation. Students need regular opportunities for the social construction of meaning through negotiation around the terms, models, conceptual entities, and causal narratives which they are using to express their beliefs and conjectures with Dynagrams activities. This may mean accepting what has been said, challenging other students' productions with their own, questioning the meaning of terms and actions in such productions, or seeking to account for discrepancies between predictions and experimental results.
- The conversations need to be *sense-making*. The social goal of a learning group is to try to make reasonable causal narratives, attempting to use their ideas and those introduced by the teacher's challenges to provide accounts of observed and predicted phenomena in authentic contexts of problem formation and problem-solving. Note that such sense-making conversations are predominantly synthetic/design tasks seeking congruence/fit between world experience and causal model, rather than analytic tasks seeking verification of hypotheses through problem-solving.

In the Optics Dynagrams Project, small groups of students work with a software simulation, or microworld, that supports the construction and running of graphical models of simple optical situations related to interaction of light and matter. Dynagrams treats the phenomena of reflection and refraction, and deals with absorbance (allowing exploration of topics involving shadows) without any relation to temperature. Learners engage in inquiry cycles of prediction, testing, observation, and explanation.

The 2-D optics simulator we designed and created at IRL [34] allows users to easily create and manipulate one or more scenes made up of optical entities such as spherical, triangular, and rectangular objects (that have assignable properties--materials; reflecting, absorbing, refracting). One may also emit single light rays, or ray sprays over an angle range, from one or more point light sources. Users may also create geometrical entities such as tangent lines, grids, and angles, and measure distances and angles.

We have used the Dynagrams simulator to create a set of challenge activity structures of increasing complexity (e.g., single to multiple light sources for making shadows; single mirrors to multiple mirrors and periscopes; simple lens refraction to a coin-in-pool situation) for small group, sense-making activities.

Students observe real-world optical situations (or their video depictions), use our dynagramming tools to build "scenes" that make predictions and arguments to justify them based on scientific principles, definitions, or prior experiences. The dynagrams bypass many difficulties students have in constructing paper and pencil or chalkboard diagrams. By composing dynagrams representations, students in a group can each graphically express predictions and then use these representations as indexical support for narrative explanations of light behavior in the situations they have modelled. Since the simulator knows how light rays depicted will propagate in the situation students have modelled, they can then run their simulation models and discuss how well each of their graphical conjectures fit the actual results. Through learners' creation and interpretation of these representations in sense-making activities, the dynamic diagrams become symbolic vehicles for expressing students' conjectures about light behavior, and the topic for negotiating group and individual understanding of technical language, concepts, procedures, and skills.

Building on prior work [29, 33, 48], Clement [13] distinguishes between four major types of knowledge used in science: observations, empirical laws summarizing observed regularities, explanatory model hypotheses that introduce theoretical visual models (such as molecules, waves, light rays), and formal quantitative principles. He argues that science educators have overassociated "real" scientific thinking with only empirical laws and formal quantitative principles. Our aim for the students' work with our Dynagram challenges -

which include hands-on work and simulation model building of the physically-observable optical situations - is to have students engaged in building an explanatory model of diverse optical situations using the ray model of light provided by the optics simulator. Clement [13] has argued for the importance to such model construction of having students attempt to give explanations and argue about them in large or small group discussions:

> The complex, tacit, nonobservable, and sometimes counterintuitive nature of scientific models means that misconceptions or "bugs" will be the rule rather than the exception during instruction, requiring critical feedback and correction processes. This means that the learning of complex, unfamiliar, or counterintuitive models in science requires a kind of learning by doing and by construction and criticism than by listening alone [p. 377].

The kinds of critical feedback, correction, and criticism he describes here are integral discourse practices of a scientific community, and build upon the conversational resources students bring with them to the science classroom, but which they rarely have opportunity to utilize in typical didactic instruction.

The shape of learning conversations with Dynagrams use

We have only now just completed a four-week field-test of the Dynagrams environment in a physics classroom in California. The teacher has been regularly involved in the co-planning with our research team of activities using Dynagrams in his classroom.

From pilot teaching experiments with three small groups of physics-naive high school students that took place in Summer 1990, we have begun to see that the kinds of learning conversations we had hoped for with Dynagrams are occurring. Through inquiry activities and "social gedanken experiments," students in their small groups are using, questioning, and refining the meaning of terms in optics such as point source, extended source, reflecting, refracting, absorbing, ray paths, object materials, intersecting rays, object point, image point, image location, real and virtual image, index of refraction. Students are exploring new relations among concepts - such as that a point source emits light in *all directions* not only for a few "special rays," that lenses *bend* light - and new properties of entities (e.g., rays of light propagate in straight lines). There is an excited and enthusiastic use of the dynamic diagramming tools, and mapping activities back and forth between modelling with Dynagrams and hands-on experimenting with optical equipment and light sources. The students engaged in extensive predictions and explorations, far beyond those we have anticipated in the setup of our activities. And after only a few days, groups had achieved a better grasp of the mechanisms of image formation than our pre-Dynagram classroom students had achieved, one girl coming to recognize and teach to others in her group the meaning of a virtual image as including the perspective of the eye. In our previous work, we have found this concept problematic for teachers, not only students.

Studying bids and outcomes of meaning negotiation activities

It may be clarifying to briefly describe some of the ways in which we will be looking at our data from our Dynagrams classroom experiment. We have videotaped for close analysis two groups of three students for each day of optics activities, and the teacher's work setting up and consolidating classroom inquiry activities for the duration of the field-test. Additional outcome evaluations were provided by other data from the classroom as a whole, including frequent homework results, and results from an extensive pre/post test involving paper and pencil reasoning about a broad range of optical situations that tap reasoning processes and concepts described in previous literature on students' pre-instructional alternative conceptions of optics [e.g., 2, 22, 23, 26].

For purposes of analyzing the small group process data, we will be identifying the kinds of meaning negotiation that are central to competent participation in a community of practice for reasoning about geometrical optics. What specific aspects of practice do we find have their meaning negotiated? What are the terms and actions whose meaning is offered up by learners during use that make them available for commentary by others and subsequent cycles of meaning negotiation? And what results emerge in the uptake of repairs by learners who are making moves in these learning conversations?

An Example of Dynagrams Learning Conversations

In this section, I will provide an example that illustrates the critical kinds of meaning negotiation processes we see taking place in the discourse of the Dynagrams classroom, and the kinds of learning that can take place as a result of such discourse. I will focus on a major case of meaning alignment that occurred during the second of four weeks in the Dynagrams classroom where introductory physics students were studying geometrical optics.[8] The group is grappling with the concepts involved in understanding image position for a plane mirror. While they start out a class period with a diversity of views for what "image position" means, at the close of the period they share a very different perspective, which appears robust in the individual reasoning profiles of the students in the group a month later[9].

Some background is required to present the example. During the previous day, the teacher had provided one of his favorite full-class demonstrations. At the front of the classroom is a long plane mirror, with an object in front of it. A number of students each gets a long string. Each student then hands one end of the string to the teacher who is standing behind the mirror. Each student then sits down at his or her chair, and the teacher asks them, one at a time, to follow their line of sight to the image they see in the mirror, and then say where the teacher should hold the string above the mirror such that it is aligned with the image they see. What the class determines with this demonstration is that the various strings, one from each student, intersect at a point that is directly behind the mirror and the same distance behind the mirror as the object in front of it. This hands-on collective demonstration provides an important reference experience for the group as they embark during the next day on their inquiry activity with the Dynagrams simulator.

At the start of the session, the three students (*A, *B, and *C; *T refers to the teacher) are engaged in a collaborative sense-making activity where they construct this diagram with the Dynagrams simulator. In this Dynagram screen display, the target represents the observer's eye in the situation, the twinkle-like object is a point light source that is "on," a ray spray of five rays has been directed at the plane mirror, with the diverging ray spray reflecting off of the mirror surface:

8 I would like to thank Sue Allen and Michael Sipusic of the Dynagrams Project team for their identification and analyses of these sequences of learning discourse from the Dynagrams fieldtest work. For simpler reading of the transcript for our purposes here, I have followed the transcription convention of using // in successive lines to demarcate overlapping speech junctures in the participants' talk. Pertinent contextual actions such as pointing or referent identification are enclosed in parentheses.

9 Reports on individual student performance are in preparation, and not described in this chapter.

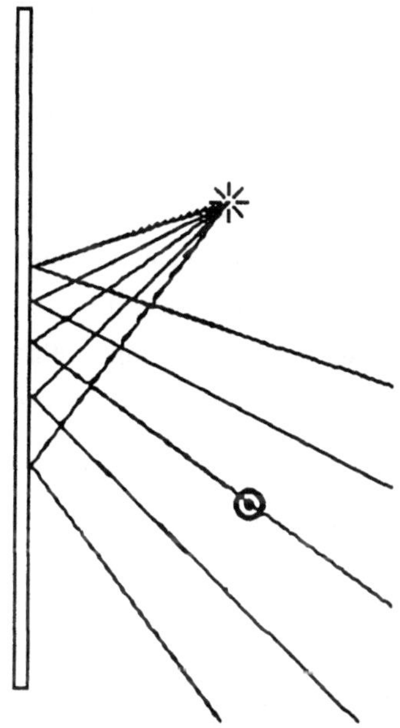

• In this first conversational sequence, we see the students offering multiple interpretations -- all plausible -- of the meaning of the teacher's question: "Where is the image formed by the plane mirror?"

(1) *T: Now, how does this help you find... or what does it help you find out about the image?

(2) *B: About the image?

(3) *T: Yeh. Where is the image according to what you've got here? (the simulator screen).

(4) *C: It's all around//.

(5) *A: The //image according to what we have *here* is on the mirror (points to paper diagram like one they have built on their simulator screen). Yeah, but it is really behind // the mirror.

(6) *B: //Yeah, with the //image.

(7) *T: //OK, but hold it, it can't be on and behind at the same time.

(8) *B: Well, I don't understand this behind thing//

(9) *C: //Well, neither did I

(10) *A: Well, like you said that it appears to be behind the mirror. But, like according to like, these (points to screen - rays bouncing off surface of the mirror), it gets *on* the mirror. Now you see, none of these rays go behind the mirror (points to screen - rays at surface of mirror).

We thus see three interpretations: In (4), *C sees the image as "all around" (as the rays are), in (5), *A sees the image to be "on the mirror" (at the reflective turn of the rays in the diagram) and yet "really behind the mirror" (as they saw yesterday in the string demonstration), while in (8), *B doesn't understand this "behind thing." And in (10), *A notes the hard topological constraint that "none of these rays go behind the mirror." The teacher's role will be to give them a clearer sense of what it means for the image "to be" in some location -- note he is already indicating an unacceptable contradiction between these different interpretations. And the students' collaborative work to negotiate the meaning of these terms will contribute importantly to this objective.

• In a second sequence, we find that this conundrum of "behind the mirror" contradicts *B's beliefs about image formation:

(11) *B: Yeah, like see, like I, I don't really understand this behind the mirror thing.

(12) *B: It's, it's like you look in a mirror, and the only reason that you see anything is because the light reflects off the mirror and you can see an image. Right? (*T nods)

(13) *B: But if there wasn't any light does that mean that the image wouldn't be there? Or would you just not be able to see it?

Her deep philosophical question reveals serious work to figure out what an image "is," to determine the relationship between the rays reaching an observer from an object and the existence of the image.

• In the next sequence, the teacher work through a number of cycles trying to bring the students to see what *he* means by image position. He repeatedly returns to the fundamental idea, conveyed in causal narrative, that our eyes trace back the reflected rays and infer that at the end of those light rays' intersection, behind the mirror, is the location of the image of the light source. The students' diverse contributions to this interchange, including sentence completions with the teacher and noddings, begin to reveal signs of alignment with the teacher:

(14) *T: Alright, now let's go back and pretend that this is a light bulb (points to light source).

(15) *B: OK.

(16) *T: OK. Like the little flashlights.

(17) *B: Right.

(18) *T: Now//

(19) *B: So --

(20) *T: The light is going to be shooting out of that (points to the light source) in all directions, just like we saw here.

(21) *B: Uh huh.

(22) *T: And it is going to be coming towards your eye, right?

(23) *B: Right.

(24) *C: Yeah.

(25) *T: What do our eyes tend to do with the light that they get? They tend to say, "Hey, they started some, someplace."

(26) *B: Uh huh.

(27) *T: OK, so they sort of trace back (triangulating convergent gesture) to say "Oh! at the other end of that light ray --

(28) *B: //Uh huh (nodding).

(29) *C: //Uh huh (nodding).

(30) *T: -- is -- the source." OK? //

(31) *B: // is the source. So --//

(32) *T: // Now, I'm standing out here (points to target in diagram representing observer's eye) and seeing these light rays (points to rays diverging from reflection off of the mirror).

(33) *A: Uh huh.

(34) *T: What are my eyes gonna do with them?

(35) *A: They will think //that they came from --

(36) *B: //-- they came from here (points to surface of the mirror).

(37) *C: //Trace back from here (points to the target-eye icon in diagram) to the mirror.

(38) *T: From there? (*T looks at *C, 'there' marked by skeptical intonation, *A,*B, *C then all turn to his gaze)

(39) *B: Well, they are gonna, you don't see them here, do you? (points behind mirror, which has no rays behind it) and //you are going to think that they came from the mirror. You're not going to look to *here* (points behind the mirror).

(40) *C: //Cause they can't come in back of the mirror (*T starts to physically withdraw from the circle, shaking head)

(41) *A: Well no (*T bends back into group). You are going to think that they are all going to meet....they are all going to meet up back *here* somewhere (points behind the mirror).

(42) *B: Right.

(43) *T: They are?

(44) *B: //That's what you are going to *think* .

(45) *A: //That's what you are going to *think* .

(46) *T: That's what I *think* they are going to come from, huh?

(47) *B: Right.

(48) *T: And, and that's what we are talking about with this *image*, see.

(49) *B: OK.

(50) *T: We are thinking "Where did these light rays appear to come from?" They appeared to come from -- //

(51) *B: //-- somewhere *here* (*T points behind mirror after 'here')

(52) *A: Back *here//*

(53) *T: //back *here.*

(54) *C: //Yeah. Hold it .//

(55) *T: //They *didn't* (shaking his head) We *know* they didn't --//

(56) *B: //Right.

(57) *C: //Yeh, OK.

(58) *T: -- but that is where they *look like* //they came from.

(59) *C: //Yeah, because you would think that they just go straight (pointing gesture with hand "through" the mirror on diagram, as if a ray).

(60) *T: Does that help clarify?

(61) *B: Yeah, that totally... I understand (nods her head).//

(62) *T: I know that you have been fighting that.

(63) *B: Uh huh.

(64) *B: Now I understand.

(65) *T: I've got a whole classroom of people in third period that is fighting it too (shaking his head).

(66) *B: I can see *now* what you mean (*T withdraws from group).

(67) *T: OK (nodding).

Note here that, interestingly, it is *A in (41) who attempts to bring the conceptual work of the group into alignment, but *T must still refine *A's claim that "they (the rays) are all going to meet up back *here* somewhere" (pointing behind the mirror), by questioning his phrasing in (43), and getting the better response in (44-45) from *A and *B, sung out in unison, of "that's what you are going to *think*".

• But the meaning negotiation work is by no means complete. There are suggestions that the teacher is aware of this because in (55) and following, *T is still working on the distinction between how we *think* the light rays appeared to come from behind the mirror, but "they *didn't*, we *know* they didn't." At this point, the teacher uses the simulator as a resource for them to use to disambiguate "trace back" -- he constructs a version of the diagram they have been working with but with fewer rays:

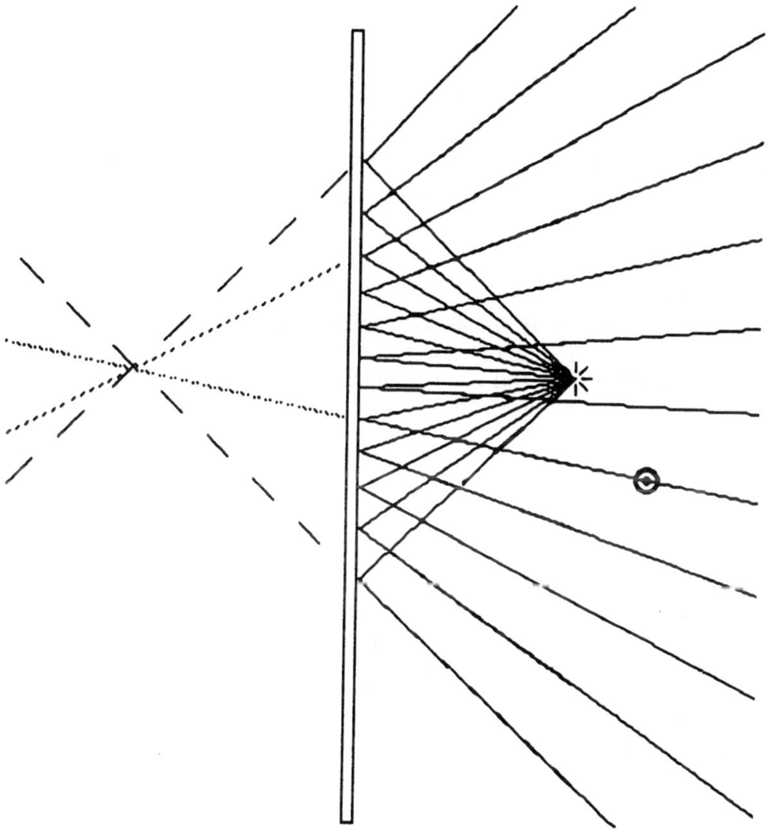

The rays reflect automatically and then the teacher and students add the "dumb lines" included in the diagram, which can then be seen to meet at a point (i.e., those traced back behind the mirror, not dynamic rays sprayed from the source). Here are their observations:

(68) *A: That's the intersection. All the lines meet there... (points to image point)

(69) *C: And it's like even across from the other one (noting symmetry with source on other side of mirror)

The students thus notice the two important features of the geometry of the situation: that the lines all meet at a point, and that the point is symmetrically placed behind the mirror with respect to the object.

• In the next sequence, we see that while *A and *B now are using "behind" in the teacher's sense, *C realizes that she is having troubles (which is true, since she is conflating the source and the image). Now let's see what happens after the teacher leaves:

(70) *C: You guys. I can see why like, we said that they look like they would be coming from back there (gestures ray paths from in front to behind mirror), but what does that have to...are we saying that they *are* coming from back here?

(71) *B: No, we are saying that that is what you *think* when you look in the mirror.

(72) *C: Oh man! You mean like when you look in a mirror, you think like//

(73) *B: //Like, if you//

(74) *C: //like you were behind the mirror (nervous laugh) or something?

(75) *A: Well, we know that, so we don't think that. But remember, like...remember when he, yesterday, like in the classroom when he had the mirror, and then he put the image behind the mirror and you had the line, and the string, remember? And then they intersected behind the mirror.

(76) *C: Yeah, that is true.

(77) *B: So.

(78) *A: And that is where the image is supposed to be.

....

(79) *A: So basically...No, if you see... It (activity sheet) says, "Demonstrate how the light can get from the source to the target." Well then it just goes and bounces off... then the target....If you are looking from the target, you can see that it comes from back here. So you just *think* that it is coming like that. So that is like where the light would come from that you would see, I guess (pointing along path from eye target icon to mirror and behind mirror). (8 second pause without any talk as they look at their papers).

(80) *C: That is not like, if you're saying that the source is *there*, right? (points behind mirror, *A is distracted, looking away as she speaks)

(81) *A: What? (*A then orients to her point to the screen behind the mirror). Yeah. You would be *seeing* the source as there.

Note that in (71), *B is now using the crucial characterization (earlier modelled by the teacher) of *thinking* that the rays are coming from behind the mirror, to help *C out in her confusion expressed in (70), as *C seeks to pin down the meaning of the rays "coming from" somewhere. *C seems to interpret *B's (71) use of "think" in the sense of as-in-an-illusion, which one only thinks to be the case. In (72) and (74), *C checks her comprehension by trying out this interpretation with *B, clearly showing us that she sees this check of "You'd *think* you were behind the mirror?" in (74) as patently absurd.

What happens next? In (75), *A, by saying: "Well, we know that, so we don't think that. Remember the intersecting strings...", does the crucial work of first acknowledging the correctness of *C's viewpoint on the absurdity of thinking one is behind the mirror, thus aligning to *C's meaning. Then he continues to offer her an importantly different interpretation of "think," based on their shared experience with the strings demonstration the day before. *C seems to align to *A's meaning in (76) -- "yeah, that is true," and accepts it. Also note one final clarification check by *C to *A in (80), since he is still not crisply describing the difference between where the light *would* come from, and where an observer *would think* it is coming from. *C says: "that is not like, if you're saying that the source is there, right?" (points behind mirror), which *A affirms.

• Finally, in the last sequence, *C resolves her dilemma, and *B summarizes what she has learned. They are reading together the final question from their activity sheet, and drawing their final conclusions:

(82) *B: Why //

(83) *C: "Draw a conclusion why//

(84) *B: //you would think that the image was coming from behind the mirror." And you would *think* that because you would assume that all the light would converge to that point there behind the mirror.

(85) *B: I mean if you were completely stupid and didn't know, didn't know how mirrors worked. You know what I am saying?

(86) *C: Uh huh

(87) *C: Just because you would think it was like a window, or something.

(88) *B: Right.

These concluding sections of discourse are fascinating for several reasons. Note in (79), how *A is now very much aligned with the teacher, both in words and in form of explanation:

- "it bounces" (mechanism)
- "if you were looking from the target" (observer viewpoint)
- "you'd think it was back here" (points to convergence point of rays)
- "you'd just see these rays" (localness of detector)

The second observation is that *B, who earlier was massively confused about "behind," has in (84-85) joined in on the concepts and language used by the teacher, and earlier in the session, by *A. At the conclusion of this session, *B has basically aligned to the same viewpoint as *A. She has managed to transcend the commonsense perspective she had before this learning. She describes the belief that the light is coming from behind the mirror as something one would think (certainly not her!): "if you were completely stupid and didn't know how mirrors worked". Finally, *C makes a very deep metaphoric comment in summing up her understanding in (87), that "it's like a window." *A, *B, and *C seem through this discourse to have become aligned with the teacher and each other.

Summary of the example

We may capture some important general features of learning conversations by way of this example: Students began a classroom session making different plausible interpretations of the things they were seeing, hearing, and remembering (the diagram; the string demonstration the day before; uses of technical terms such as "think,""behind," and "trace back"). We saw important meaning-making activities in the students' talk and action, as they tried to construct the same meaning as the teacher (evidenced by their active listening, nods, agreement, and eventually by their use of the same phrases and explanatory accounts). They raised their own alternative interpretations and questions, clarifying and repairing their understanding during the use of these concepts and terms for inquiry activities. And when he is available to the group, the teacher tries to understand the student meanings and lead them through the use of these terms to new and appropriate understandings more closely aligned to physics norms.

One can imagine that students *could* generate a variety of idiosyncratic meanings for the scientific terms involved - in this case, involving what it means for the virtual image to be behind the mirror. Similarly, one could imagine that students' work in a group would produce discourse at such a vague level that the different meanings individuals hold would not be manifested or resolved in the discourse. But we can see from this example that the teacher's role is critical in serving as a guide to establishing productive inquiry situations, and in providing the kinds of integrative questions that will lead students toward scientific norms and practice. It is significant, we believe, that the students do a major share of the collaborative sense-making themselves. Given the right activities and encouragement, as well as appropriate resources that allow for establishment of co-reference and the building of a common ground of understanding, students will spontaneously do a great deal of impressive collaborative sense-making and meaning alignment, and facilitate conceptual change for each other.

Looking ahead to analyses of Dynagrams learning conversations

While we are just beginning these analyses, we can say with some confidence that analyses of the following aspects of the data will prove fruitful, and provide other critical loci of *appropriation* and *meaning negotiation* activity (see earlier discussion). Such categories highlight important "bids" for meaning presentation in a student's talk, which may then get accepted, rejected, revised, repaired, and so on by others in subsequent turns of talk, and be ignored, or acknowledged, and so on, in that student's subsequent talk:

- Statements of goals and objectives of an activity (e.g., predict a result, explain an observation)
- Selection of means as appropriate fit for achieving a goal in an activity (e.g., Which representational systems are used - lab apparatus, diagram, words, equation?)
- Details of the use of means (e.g., Are the moves in constructing the situation and ray diagram appropriate? Are the moves in using the terms to describe the situation appropriate? Are appropriate moves used to account for the lack of fit of a prediction and observations of its experimental testing?)
- Conclusions that a goal has been achieved (e.g., How do students resolve what counts as "enough" of a fit between their predictions and evidence coming from testing these predictions by running the simulator or observing lab outcomes?)

We will also be analyzing the discourse outcomes of learners' production of these activities, asking how other members of the group or the teacher interpret previous turns of talk, with a special focus on objections, alternative expressions, and other forms of repair that make up the moves of meaning negotiation. Finally, we will look at how and whether students take up these acts of meaning interpretation in repairing their expressions or not.

While we realize that repair and meaning negotiation may have non-local effects (i.e., occurring significantly after their immediate context in time), we will be able to identify what are likely to be locally significant impacts of repair and meaning negotiation during student's collaborative work on Dynagrams challenges.

Results from these analyses will provide a careful longitudinal picture of the ways in which groups of inquiring students, guided by challenge activities and with some teacher coaching, develop through sense-making activities a deeper understanding of concepts, representations, and procedures for scientific conversations about geometrical optics.

Caveats

While we are quite encouraged by learners' sense-making conversations about geometrical optics we have seen in our teaching experiments this summer, and in the recent field test, we can already see some of the challenges that will remain in truly establishing a classroom community of practice that does sense-making in geometrical optics using Dynagrams. As enthusiastic as our teacher has been about the innovations we have designed and to which he has helped contribute, in the familiar context of the classroom, he regularly lapses into his previous routines of demonstrations (using Dynagrams as a new form of lab apparatus) with explanations, and asking and answering his own questions to the students after such demonstrations. It will require an extended effort, focusing more on supporting the revision of the teacher's roles in the school institution, to evolve communities of practice of the kind we hope to establish for science learning. Lampert [37], in her recent work to establish authentic mathematical discourse in the elementary classroom, has characterized the diverse challenges that such a fundamental shift requires.

General Implications for Learning Environment Design

The construction of learning environments is a challenging task that when examined from a social framework becomes all the more demanding. The issue for science learning from this perspective is not so much one of coming to master the component skills of manipulating scientific symbol systems, and the problem-solving skills associated with their use in working on problems. What is most *centrally* "constructed" through experience in scientific activity is the disposition to engage in appropriate scientific conversations, not a set of mental representational structures. Science learning consists of entering into the web of social relations and actions that are constituted by various practices, accountabilities, and duties that make up the discourse of scientific knowing.

I have laid out some of the specific implications for designers of this perspective, and of the charge that computer tools should serve to augment students' sense-making capabilities and their learning conversations. There are both technological and social design goals that must go together to contribute to effective learning that has some chance of surviving beyond experimental treatments in the ecology of communities of practice and institutions. Among these goals are: authentic activity from a community of practice; in-situ role modelling of appropriate activity for a practitioner in the community of practice, and learners' legitimate peripheral participation in that community; opportunities for use of concepts and skills that allow for social meaning repair and negotiation; and the keystone activity of collaborative sense-making through narration -- to provide reasonable causal stories that account for some event with a set of explanatory constructs. While we expect many challenges to establishing conditions for "growth" of such communities of practice in school institutions, we are optimistic that a focus on augmenting learning conversations with computer tools will go a long way toward taking the insights gleaned from successful learning outside school into the classroom.

References

1. Allen, C. L. (this volume). *Multimedia learning environments designed with organizing principles from non-school settings.*
2. Andersson, B., & Karrqvist, C. (1983). How Swedish pupils, aged 12-15 years, understand light and its properties. *European Journal of Science Education, 5*, 387-402.
3. Anderson, J. R., Boyle, C. F., & Yost, G. (1985). The geometry tutor. In A. Joshi (Ed.), *Proceedings of the Ninth International Joint Conference on Artificial Intelligence* (pp. 1-7). Los Altos, CA: Morgan Kaufman.
4. Brown, J. S. (1990). Toward a new epistemology for learning. In C. Frasson & J. Gauthiar (Eds.), *Intelligent tutoring systems: At the crossroads of AI and education* (pp. 266-282). Norwood, NJ: Ablex.
5. Brown, J. S., Collins, A., & Duguid, P. (1989). Situated cognition and the culture of learning. *Educational Researcher, 18* (1), 32-42.
6. Bruner, J. S. (1976). *Child's talk.* New York: W. W. Norton.
7. Bruner, J. S. (1960). *The process of education.* Cambridge, MA: Harvard University Press.
8. Carraher, T.N., Carraher, D.W., & Schliemann, A.D. (1985). Mathematics in the streets and in schools. *British Journal of Developmental Psychology, 3*, 21-29.
9. Chi, M. T. H., Bassok, M., Lewis, M. W., Reiman, P., & Glaser, R. (1989). Self-explanations: How students study and use examples in learning to solve problems. *Cognitive Science, 13*, 145-182.
10. Clark, H. H., & Shaeffer, E. F. (1989). Contributing to discourse. *Cognitive Science, 13*, 259-294
11. Clark, H. H., & Wilkes-Gibb, D. (1986). Referring as a collaborative process. *Cognition, 22*, 1-40.
12. Clement, J. (1987). Overcoming students' misconceptions in physics: The role of anchoring intuitions and analogical validity. *Proc. Second International Seminar on Misconceptions and Educational Strategies in Science and Mathematics* (pp. 84-97). Ithaca, NY: Cornell University, Department of Education.
13. Clement, J. (1989). Learning via model construction and criticism: Protocol evidence on sources of creativity in science. In G. Glover, R. Ronning & C. Reynolds (Eds.), *Handbook of creativity: Assessment, theory, and research* (pp. 341-381). New York: Plenum.
14. Cognition and Technology Group at Vanderbilt (1990). Anchored instruction and its relationship to situated cognition. *Educational Researcher, 19* (5), 2-10.
15. Collins, A., Brown, J.S., & Newman, S. (1989). Cognitive apprenticeship: Teaching students the craft of reading, writing, and mathematics. In L.B. Resnick (Ed.). *Knowing, learning, and instruction: Essays in honor of Robert Glaser* (pp. 453-494). Hillsdale, NJ: Erlbaum.
16. Dewey, J. (1956). *The child and the curriculum/The school and society.* Chicago, IL: University of Chicago Press.
17. Driver, R., Guesne, E., & Tiberghien, A. (Eds.) (1985). *Children's ideas in science.* Philadelphia: Open University Press.
18. Eckert, P. (1989). *Jocks and burnouts.* New York: Teachers College Press.
19. Garfinkel, H. (1967). *Studies in ethnomethodology.* New York: Prentice-Hall.
20. Garfinkel, H., & Sacks, H. (1970). The formal properties of practical actions. In J. C. McKinney & E.A. Tiryakian (Eds.), *Theoretical sociology* (pp. 338-366). New York: Appleton-Century-Crofts.
21. Gelman, R., & Brown, A.L. (1986). Changing views of cognitive competence in the young. In N.J. Smelser & D.R.Gerstein (Eds.), *Behavioral and social science: Fifty years of discovery* (pp. 175-207). Washington, DC: National Academy Press.
22. Goldberg, F. M., & McDermott, L. C. (1986). An investigation of student understanding of the real image formed by a converging lens or concave mirror. *American. Journal of Physics, 55*, 108-119.
23. Goldberg, F. M., & McDermott, L. C. (1986). Student difficulties in understanding image formation by a plane mirror. *The Physics Teacher, 24*, 472-480.
24. Goodwin, C., & Heritage, J. (1986). Conversation analysis. *Annual Review of Anthropology, 19*, 283-307.

25. Greeno, J. G. (1979). A perspective on thinking. *American Psychologist, 44*, 134-141.

26. Guesne, E. (1985). Light. In R. Driver, E. Guesne & A. Tiberghien (Eds.). *Children's ideas in science* (pp. 10-32). Milton Keynes: Open University Press.

27. Halliday, M. A. K. (1975). *Learning how to mean*. New York: Elsevier.

28. Hanson, N. R. (1958). *Patterns of discovery*. Cambridge: Cambridge University Press.

29. Harre, R. (1961). *Theories and things*. London, England: Newman.

30. Hawkins, J., & Pea, R.D. (1987). Tools for bridging the cultures of everyday and scientific thinking. *Journal for Research in Science Teaching, 24*, 291-307.

31. Heller, J. I., & Reif, F. (1984). Prescribing effective human problem-solving processes: Problem description in physics. *Cognition and Instruction*, 177-216.

32. Heritage, J. (1984). *Garfinkel and ethnomethodology*. Cambridge, England: Polity Press.

33. Hesse, M. (1966). *Models and analogies in science*. South Bend, IN: Notre Dame University Press.

34. Jul, S. (in preparation). Systems architecture of Optics Dynagrams software. *IRL Technical Report*, Institute for Research on Learning, Palo Alto CA .

35. Knorr-Cetina, K., & Mulkay, M. (1983). (Eds.) *Science observed*. London: Sage.

36. Lakatos, I. (1970). Falsification and the methodology of scientific research programmes. In I. Lakatos & A. Musgrave (Eds.), *Criticism and the growth of knowledge* (pp. 91-195) Cambridge: Cambridge University Press.

37. Lampert, M. (1990). When the problem is not the question and the solution is not the answer: Mathematical knowing and teaching. *American Educational Research Journal, 27*, 29-63.

38. Larkin, J. L., & Simon, H. A. (1987). Why a diagram is (sometimes) worth ten thousand words. *Cognitive Science, 11*, 65-100.

39. Latour, B., & Woolgar, S. (1979). *Laboratory life: The social construction of science*. London: Sage.

40. Lave, J. (1988). *Cognition in practice*. New York: Cambridge University Press.

41. Lave, J., & Wenger, E. (in press). *Situated learning: Legitimate peripheral participation*. New York: Cambridge University Press.

42. Lemke, J. L. (1990). *Talking science: Language, learning, and values*. Norwood, NJ: Ablex

43. Linn, M. C. (May 1986). *Establishing a research base for science education: Challenges, trends, and recommendations*. Lawrence Hall of Science and the Graduate School of Education, U. Cal., Berkeley.

44. Linn, M. C., & Clancy, M. J. (in press). Can experts' explanations help students develop program design skills? *International Journal of Man-Machine Studies*.

45. Mehan, H., & Wood, H. (1975). *The reality of ethnomethodology*. New York: John Wiley and Sons.

46. Miller, A. I. (1986). *Imagery in scientific thought*. Cambridge, MA: MIT Press.

47. Miller, G. A. (1977). *Spontaneous apprentices: Children and language*. New York: Seabury Press.

48. Nagel, E. (1961). *The structure of science*. New York: Harcourt, Brace, and World.

49. Newman, D., Griffin, P., & Cole, M. (1989). *The construction zone: Working for cognitive change in school*. New York: Cambridge University Press.

50. Osborne, R., & Freyberg, P. (1985). *Learning in science*. Auckland, Australia: Heineman Publishers.

51. Pea, R. D. (November, 1988). Dialogism and mathematical discourse. *Invited address at the Tenth Annual Meeting of the North American Chapter, International Group for the Psychology of Mathematics Education*. DeKalb, IL.

52. Pea, R. D. (in press). Distributed intelligence and education. In D. Perkins & B. Simmons (Eds.), *Teaching for understanding in an age of technology*. Cambridge, MA: Harvard University Press.

53. Pea, R. D., Sipusic, M., & Allen, S. (in press). Seeing the light on optics: Classroom-based research and development of a learning environment for conceptual change. In S. Strauss (Ed.), *Development and learning environments: Seventh Annual Workshop on Human Development*. Norwood, NJ: Ablex.

54. Posner, G. J., Strike, K. A., Hewson, P. W., & Gertzog, W. A. (1982). Accommodation of a science conception: Toward a theory of conceptual change. *Science Education, 66*, 211-227.

55. Rogoff, B., & Lave, J. (Eds.). (1984). *Everyday cognition: Its development in social context.* Cambridge, MA: Harvard University Press.

56. Roschelle, J. (March, 1990). Designing for conversations. *Paper presented at the AAAI 1990 Spring Symposium on Knowledge-Based Environments for Learning and Teaching.*

57. Roschelle, J., & Behrend, S. (in press). The construction of shared knowledge in collaborative problem solving. In C. O'Malley (Ed.), *Computer-supported collaborative learning.*

58. Saxe, G. B. (in press). The interplay between children's learning in formal and informal social contexts. In A. A. diSessa, F. Reif, M. Gardner, J. G. Greeno, A. H. Schoenfeld, & E. Stage (Eds.), *The science and engineering of science education.* Hillsdale, NJ: Lawrence Erlbaum Associates.

59. Schegloff, E.A. (in press). Conversation analysis and socially shared cognition. In L. Resnick, J. Levine & S. D. Behrend (Eds.), *Socially shared cognition.* Washington DC: APA Press.

60. Schegloff, E.A., & Sacks, H. (1973). Opening up closings. *Semiotica, 7,* 289-327.

61. Schoenfeld, A. (1985). *Mathematical problem solving.* New York: Academic Press.

62. Scribner, S. (1984). (Ed.). Cognitive studies of work. *Special issue of the Quarterly Newsletter of the Laboratory of Comparative Human Cognition, 6.*

63. Soloway, E. (1986). Learning to program = Learning to construct mechanisms and explanations. *Communications of the ACM, 29,* 850-858.

64. Stavy, R., & Berkowitz, B. (1980). Cognitive conflict as a basis for teaching quantitative aspects of the concept of temperature. *Science Education, 64,* 679-692.

65. Suchman, L. (1987). *Plans and situated actions: The problem of human-machine communication.* Cambridge: Cambridge University Press.

66. Tiberghien, A. (1985). The development of ideas with teaching. In R. Driver, E. Guesne, & A. Tiberghien (Eds.), *Children's ideas in science.* Open University Press.

67. Toulmin, S. E. (1972). *Human understanding.* Princeton: Princeton University Press.

68. Vygotsky, L. S. (1978). *Mind in society: The development of the higher psychological processes.* Cambridge, Massachusetts: Harvard University Press.

69. West, L., & Pines, A.L. (1985). (Eds.). *Cognitive structure and conceptual change.* Orlando, FL: Academic Press.

70. Winograd, T., & Flores, F. (1987). *Understanding computers and cognition: A new foundation for design.* Reading, MA: Addison-Wesley.

71. Wittgenstein, L. (1967). (Originally published in 1949). *Philosophical investigations.* Oxford: Blackwell.

Scientific Reasoning Across Different Domains

Robert Glaser, Leona Schauble, Kalyani Raghavan, and Colleen Zeitz

Learning Research and Development Center, University of Pittsburgh, Pittsburgh, PA 15260, USA

Abstract: This study seeks to establish which scientific reasoning skills are primarily domain-general and which appear to be domain-specific. The subjects, 12 university undergraduates, each participated in self-directed experimentation with three different content domains. The experimentation contexts were computer-based laboratories in d.c. circuits (*Voltaville*), microeconomics (*Smithtown*), and the refraction of light (*Refract*). Subjects spent three 1-1/2 hour sessions working with each laboratory and took pretests and posttests that assessed their learning. Specific patterns of strategies used in each laboratory depended primarily on the structural form of the discovery task and the nature of the domain. In a situation that required the discovery of correlational regularities, evidence-generation activities, like the heuristic of controlling variables, were primary. In contexts where the regularities were functional rules, evidence interpretation became important. When the rules were quantitative, mathematical and algebraic heuristics were important. Students appeared very sensitive to the task demands of each laboratory, and adjusted their strategies accordingly. Regardless of this adaptation to specific conditions, they learned more as they proceeded from domain to domain, indicating that they were becoming more effective in planning and carrying out experiments, and in formulating and testing hypotheses based on those experiments. The findings suggest that the most generally useful skills for direct instruction may be those for evaluating the kind of problem at hand and for selecting the most appropriate processes and strategies.

Keywords: Scientific reasoning, experimentation, problem solving.

Previous work in scientific reasoning has investigated performance in various discovery tasks with the objective of characterizing the strategic or reasoning processes associated with successful discovery of lawful regularities [7, 10, 15, 13]. Most of these studies have been carried out in the context of one domain of knowledge. However, we have noted as we work in different domains that there appear to be strong influences of the structure and content of the domain on the particular reasoning and inference skills that people employ, similar to the domain-specificity of performance found in studies of expertise [3]. This observation has led us to investigate the similarities and differences in reasoning that students display when they work to discover the principles that apply in three computer laboratories incorporating simulations of different content domains in the physical and social sciences.

Historically, most of the psychological research on scientific discovery has regarded scientific reasoning in one of two ways. Some studies investigate reasoning processes, in particular, strategies of scientific experimentation, such as designing and interpreting valid experiments, hypothesis testing, identifying regularities in patterns of data, and reasoning about correlation and covariation in events. This tradition tends to cast these skills as being rather general reasoning abilities that presumably are applied across content domains [e.g., 6, 14, 17, 18]. Other work emphasizes the content and structural characteristics of domain knowledge as a function of prior misconceptions or as a function of expertise. Within this line of work, the emphasis is on strategies and heuristics that are quite specific to the domain and the task [e.g., 4, 1]. When an individual is perceptive of the features of a problem, these heuristics often become proceduralized, with the consequence that they may be employed almost automatically when particular task requirements elicit them.

More recent research on scientific reasoning is increasingly attending to the relations between domain-general strategies and domain-specific reasoning heuristics. For example, Kulkarni and Simon [9] have reconstructed the reasoning processes employed by Hans Krebs as he discovered the urea cycle. They concluded that some of the heuristics he employed were closely tied to the domain of biochemistry, whereas others were more general strategies applicable to discovery in all domains of science or to other forms of problem solving. In a similar vein, Linn, Clement, and Pulos [12] studied the relation between task content, task structure, and formal reasoning by asking adolescents a series of questions about constructing, criticizing, planning, and analyzing experiments in a variety of laboratory and naturalistic contexts. Linn et al. found effects of both domain-specific task content and task structure on students' application of scientific reasoning heuristics, like the control of variables.

This study continues the investigation of the relations between general and specific reasoning in science. Kulkarni and Simon's conclusions were based on a reconstruction from historical records, and Linn et al. posed a series of questions about experimentation tasks; here we ask subjects to engage in the actual process of experimentation by planning and carrying out an investigation that lasts over several sessions. Our subjects are university undergraduates who are novices in the domains of investigation. Each subject participates in self-directed exploration in three different content domains, providing the opportunity to investigate which reasoning and inference activities are employed with some consistency and systematicity from domain to domain, and which activities appear to be used more narrowly within a limited range of content.

The three computer laboratories used in this study simulate phenomena in the domains of economics (Smithtown), d.c. electric circuits (Voltaville), and the refraction of light through lenses (Refract). In each laboratory, students construct experiments by varying variables and parameters, take relevant measurements, make predictions about outcomes, record and manage data, and develop and revise hypotheses about the laws and principles that apply in the domain. The laboratories share a common interface and an identical set of tools that support the recording, sorting, and graphing of data, and the generation of hypotheses. In addition to these commonalities in design, the task posed to students working in each laboratory was identical: to try to find as many laws and regularities as possible. To discover the laws in the laboratories, one must generate valid and informative experiments, record and organize the data from observations, and then appropriately interpret the data by developing generalizable laws. We refer to these classes of activities as the generation of evidence, data management, and evidence interpretation, respectively.

Smithtown, the laboratory in microeconomics, is illustrated in Figure 1. In Smithtown [15], students first choose a market to explore (markets are goods or services for production and sale, such as coffee, typewriters, and gasoline). They then make changes to one or more independent variables (for example, the price of the good, the population of the town, or the number of suppliers), and if desired, make one or more predictions about the outcomes of their manipulations. The computer displays the effects of these changes on related dependent variables (such as quantity supplied, quantity demanded, and surplus). In all the laboratories, the student can record as much observed information as desired into an online notebook. He or she may also use a spreadsheet-like table package to sort the data in various ways or a graphing tool to graph relations in the data. Finally, the student uses a hypothesis menu to state the laws or principles discovered; the computer evaluates these hypotheses for accuracy and indicates whether the student has generated sufficient data to support the conclusion.

The underlying structure of Smithtown is basically correlational; that is, changes in certain dependent variables covary with changes in independent variables and parameters. The student's task is to describe these correlational relations in qualitative statements, for example,

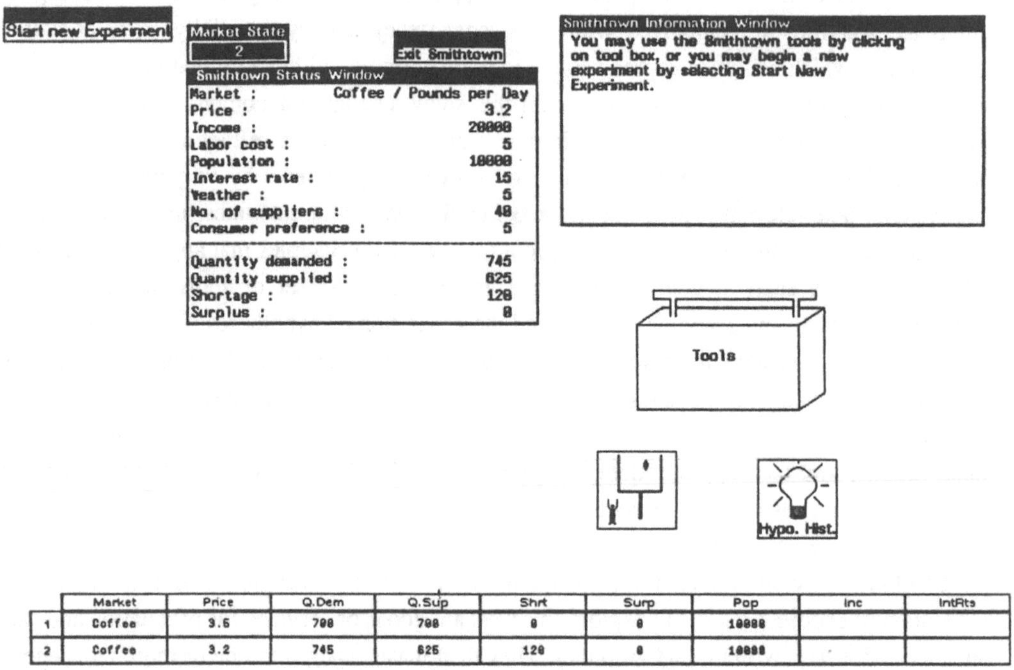

Figure 1: Smithtown - microeconomics.

"As price of tea increases, quantity demanded decreases," a principle in microeconomics known as the Law of Demand. Finding the principles requires finding out which variables are involved in a particular relationship, as well as discerning the general direction of the relationship. Equally important is discovering which variables are not relevant. Discovering these correlational relationships requires the generation of carefully structured patterns of evidence in which extraneous variation is controlled, so that pertinent causal effects can be evaluated in isolation from other candidate causes. Once these patterns of evidence are generated, they are relatively easy to interpret, but it is easy to fall into the trap of generating patterns that are not very informative. For example, a student may succumb to errors like inferring that a variable plays a causal role when in fact other variables also covary with the outcome and therefore may be either solely responsible for it or play a contributing role.

Therefore, strategies associated with the generation of valid and informative evidence appear to be particularly important to successful learning with Smithtown.

Voltaville, the d.c. circuit laboratory, is the second computer laboratory used in the study. As shown in Figure 2, students working with Voltaville first design a test circuit by choosing a voltage source and then inserting resistors of various values, either in series, parallel, or mixed arrangements. Once the circuit is designed, students use a simulated multimeter to measure the values of voltage, current, and resistance in the circuit. Because it is possible to hook the meter to any two points on the circuit by connecting and reconnecting the leads, many measurements can be taken for any of the three variables. As in Smithtown, measurements can be recorded, sorted, manipulated, and graphed. The objective is to find as many of the laws of d.c. current as possible (these are variants of Ohm's and Kirschoff's Laws for series and parallel circuits).

We have mentioned that Smithtown has a correlational structure; in contrast, Voltaville is analagous to the classic rule discovery tasks widely explored in cognitive psychology [c.f., 16, 18]. In rule discovery tasks the objective is to find a rule--in this case, a mathematical formula--that correctly and exactly specifies the relations among all variables in the task.

These rules take on forms like, "V = I times R." or "$R_1 + R_2 + R_3$ = total R." To discover a correlational structure like Smithtown's, one must discern and describe several simple qualitative relations between causes and outcomes. In contrast, to discover the mathematical formulae that comprise the rules in Voltaville, one generates and tests candidate rules until a rule is found that correctly specifies the relation among all the observed values for a set of variables. In Voltaville, unlike Smithtown, it is not necessary to design sets of observations that control for relevant and irrelevant variables. This is because each experiment in Voltaville is fully informative; that is, each experiment yields data illustrating all the laws that apply within that particular kind of circuit. For example, based on only one circuit, a student could, in principle, induce Ohm's Law, as well as Kirchhoff's Laws for Resistance, Current, and Voltage. However, students do not typically discover these laws on the basis of only one case, probably because even though generating informative evidence is not problematic in Voltaville, it can be challenging to identify patterns in the data and interpret the evidence. Because of these structural differences in the two laboratories, evidence generation strategies may be less important in Voltaville than in Smithtown. Instead, in Voltaville, evidence interpretation skills are more fundamental.

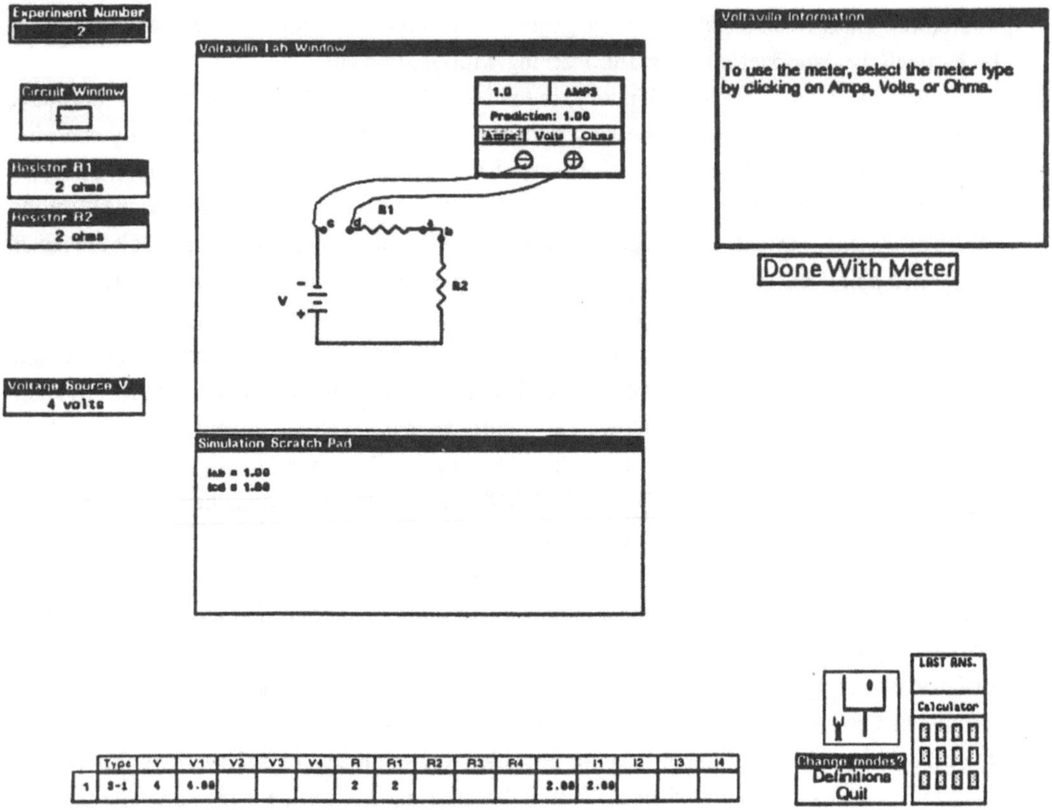

Figure 2: Voltaville - d.c. circuits.

Refract, the third computer laboratory, is illustrated in Figure 3. Students choose a lens shape and lens material to work with and then arrange a point source of light at a fixed distance from the lens and at a particular angle with respect to the lens. The lab includes simulated measurement tools that permit the measurement of angles and distance. Prediction tools support both qualitative and quantitative predictions about the results of experimental manipulations. The laws to be discovered are Snell's Law and the equations specifying the location of images with respect to various kinds of lenses.

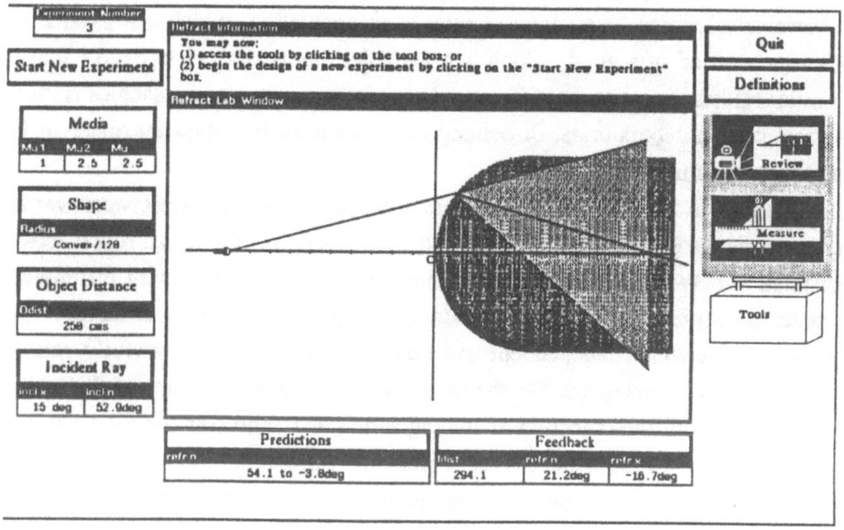

Figure 3: Refract - refraction of light.

Structurally, Refract represents a mixed case. Like Voltaville, it is a rule discovery task, with laws taking the form of mathematical expressions. However, as in Smithtown, one of the challenges in Refract is to identify the particular variables that are implicated when an independent variable or a parameter is manipulated. Therefore, as in Smithtown, accomplishing this objective is facilitated by the systematic generation of evidence in regular patterns. Since not all variables play a role in all laws, systematicity is particularly important in discerning which independent variables are responsible for changes in the corresponding dependent variables. Thus, strategies for generating evidence should play an important role in Refract. In addition, the rules in Refract require somewhat more sophisticated mathematical knowledge than those in Voltaville, and thus strategies for the interpretation of evidence are likely to be important.

Beyond these differences in the structure of the three domains, there are differences in the salience of the structure. As Figures 2 and 3 suggest, the parameters in Refract and Voltaville represent changes in physical objects that can actually be manipulated: lenses made of different materials and shapes, circuits with resistors wired in series or in parallel. Subjects find it intuitively reasonable that changes in these parameters may change the way that variables in the system are related. In contrast, the parameters in Smithtown do not seem qualitatively different from the variables. Income level, interest rates, and population, which are parameters, seem comparable to the price of a good, which is a variable. Students seem to expect that they will all have similar effects. Because the difference between parameters and variables is more salient in the physical science laboratories, discerning this distinction is somewhat easier in Voltaville and Refract than in Smithtown.

In sum, Smithtown has a correlational structure, and the distinction between variables and parameters is particularly difficult to make in this laboratory. Voltaville is a rule discovery task, and the distinction between variables and parameters seems consistent with differences in the physical materials represented in the laboratory. Refract has a mixed structure. Since it is necessary to find out which independent and dependent variables are lawfully related, some correlational reasoning is required. On the other hand, the basic structure of the task is a rule discovery structure, with the objective of finding a rule that expresses the relations among the relevant variables and parameters. As in Voltaville, differentiating between variables and parameters is facilitated by the fact that parameter changes map onto changes in concrete physical materials like lens shape and material. How these structural differences affected exploratory and inference activities in the three laboratories was of major interest in this study.

Method

Subjects

Participants were recruited on a university campus. Since the study required relative novices in the domains of interest, criteria for acceptance in the study were that the candidate be an undergraduate majoring in a nonscience discipline, with no university coursework in either physics or economics. The first twelve applicants who fit these criteria were admitted as subjects, yielding a group of 4 men and 8 women, mean age 21 years (range from 18-25).

Procedure

Sequence of experimental sessions

The study was described to subjects as one concerning learning with computer laboratories. Subjects were told that they would be shown how to use the laboratories and then would spend several sessions working with each lab "as a scientist might" to try to discover as many laws and regularities in the domain as possible.

All participants took a brief test designed to screen for competence in simple algebra and in the ability to make qualitative and quantitative interpretations of numerical data displayed in tables. Subsequently, each subject came to the university from two to three times per week to participate in a total of eleven experimental sessions lasting one and one-half hours each. Total duration of the study was therefore approximately 16 hours for each subject and extended over six weeks.

Subjects were randomly assigned to one of two treatment orders. Because of the time-intensive nature of the study, a completely counterbalanced design was not feasible. Since our task analysis predicted that Refract shared some task requirements with Smithtown and others with Voltaville, it was the best laboratory for studying consistency or transfer of reasoning from the other two labs. Six of the subjects worked for several sessions on Voltaville, and then on Smithtown, whereas for the remaining six, the order was reversed. All subjects worked last on Refract.

Working with the laboratories

Work with each laboratory was preceded by a short pretest (about 20 min) to assess subjects' prior knowledge in the domain. Each pretest included qualitative questions addressing conceptual understanding (for example, a Voltaville pretest concept question might ask whether the brightness of a lightbulb would change if an extra resistor were added at various points in the circuit). In addition, for those domains in which the relations take the form of mathematical expressions (Refract and Voltaville), pretests also included items designed to assess knowledge of and ability to apply these laws (for example, "If the total voltage in a series circuit is 20 volts and the current is 4 amperes, what is the total value of the resistance?"). After the pretest, an interviewer prompted the subject through a standard training and demonstration session (of about 40 min duration) with the appropriate computer laboratory. The purpose was to ensure that the subject understood the activities supported by the laboratory, could operate the computer interface, and was familiar with the discovery tools common to all three laboratories. After this demonstration was completed, the experimenter informed the subject of the task objective: to discover as many laws and regularities as possible. The subject spent the remainder of this introductory session in self-directed experimentation with the computer laboratory. In subsequent sessions subjects continued their exploration. Since the computer laboratories saved each student's activity to a personal file, experiments and records were preserved from session to session, and subjects started off each session with the information and discoveries they had generated in previous sessions. Thus, the study focused on learning that was cumulative over several sessions. The computer records contained a complete trace of all student actions with the laboratories and served as a primary data source for the study.

During the exploration sessions subjects worked individually in the presence of one of three interviewers. The interviewer answered questions about operating the laboratory but avoided directing student exploration. In addition, when appropriate, she prompted subjects to describe what they were thinking, to justify conclusions, and to explain what they were inspecting on the screen. These comments were recorded on audiotape.

Including the introductory sessions, each subject spent three sessions working with Voltaville and Refract. Smithtown encompasses a somewhat larger domain, includes a greater number of goals to discover, and requires more experiments to support each hypothesis. Consequently, each subject spent four sessions working with Smithtown. At the end of the final session on each laboratory, students took a posttest that was parallel in form to the pretest.

Results

We first report student performance in the three laboratories, in particular, with respect to the task characteristics of each laboratory. Next, we discuss the extent to which subjects who work over an extended period with all these laboratories both learn content knowledge and acquire proficiency in the processes of inference and discovery that lead to learning.

Experimentation Activity

We report the results in three general classes of exploratory activity: the generation of evidence, the interpretation of evidence, and data management. These three classes roughly reflect the sequence that subjects engage in when they work with the laboratories; that is, they must first design and set up experiments, then they observe and interpret the results of their experiments, and finally record and further manipulate the data with graphs, tables, or expression columns. Of course in actual discovery, these phases are iterative and do not strictly follow the sequence outlined.

Evidence generation

First we consider the generation of evidence in the three laboratories. Generation of evidence encompasses the amount and breadth of search, the informativeness of search, and the structure of search.

Amount and breadth of search. The problem space comprising the number of possible experiments that can be carried out, sometimes referred to as the e-space [7], can be very large. Furthermore, the informativeness of experiments designed will vary, with some regions of the e-space representing experiments that do not distinguish between rival hypotheses, and other regions representing comparisons that support definitive judgments about a hypothesis. The computer laboratories studied here have e-spaces that are quite large in comparison to those employed in many laboratory tasks. Of the three, Voltaville supports the smallest e-space: it includes three major variables (voltage, with 40 possible values, resistance, with 10, and current, a dependent variable that varies as a function of the values of the other two), and one parameter (circuit type) with eight different levels. In contrast, Refract has two variables (image distance, with 5 values and angle of incident ray, with 7) and two parameters, the relative optical density of lenses, with 4 levels, and lens shape, with 8 levels. In contrast, Smithtown has only one variable, price. However, this variable has an exceptionally large range of values, since it is possible to vary dollar costs in various markets. In addition, Smithtown includes eight parameters (such as income level, population, interest rates, weather, and the like) which also have a very wide range of permissible values that shift the relations among the simple variables. Most subjects find it more difficult to identify the way that parameters work than to discover lawful variable changes [15, 13]. Therefo·· the relative proportion of parameters and variables, as well as its overall larger e-space, make Smithtown the most complex and difficult to master of the computer laboratories. For the same reasons, Refract is of intermediate complexity, and Voltaville contains the least complexity, both in amount and kind of possible variation.

As Table 1 shows, the larger the e-space supported by a lab, the more experiments subjects actually generated. Our subjects appeared to be sensitive to the conditions under which more variation is possible and responded by searching more broadly, thus generating more information. In addition, Table 1 shows that on the average, students made more changes to parameters in Smithtown than in either Voltaville or Refract, and changed variables more frequently in Refract and Smithtown than in Voltaville, a straightforward reflection of the differences in the laboratories' structures.

Informativeness of search. Although each of the computer laboratories permits the generation of many potential experimental combinations, there is for each a much more focused number that comprises the minimal set required to discover all the laws. This minimal amount of evidence varies from a low of only 6 experiments in Voltaville to 20 in Refract and approximately 50 in Smithtown (the number fluctuates somewhat depending on the path of experimentation). Consequently, not only does Smithtown have the largest and most complex e-space whereas Voltaville has the least; in addition, the minimal amount of evidence that must be generated to discover all the laws and relations is greatest for Smithtown and least for Voltaville. As Table 1 shows, subjects typically generate smaller percentages of the minimal required evidence in Smithtown and Refract, a reflection of the larger and more complex evidence patterns required in those laboratories. On the average, students generate only half the evidence required for discovering all of Smithtown's twelve laws. In contrast, they generate all or nearly all of the evidence required to support discovery of the eight laws in Voltaville, even though they may not be able to interpret the evidence to discover them.

Structure of search. Regardless of the amount of evidence they generate, subjects' experimentation can take more or less informative paths. For experiments to be informative, they must reflect patterns compatible with the structure of the laws in each of the laboratories. As discussed above, in Smithtown, laws are qualitative relations, whereas in Voltaville and Refract, they are mathematical expressions. Furthermore, for any one law being explored, most of the factors in Smithtown do not play a causal role, whereas for any one law in Refract and Voltaville, there are fewer noncausal variables. Because of these domain differences, discovering the laws in these three worlds entails structuring experiments in different ways.

In particular, to discover a law in Smithtown, students must generate three price points at two or more levels of a relevant parameter. In contrast, to perceive the covariation among variables that underlies Refract, the relevant comparisons are pairs of observations that differ by only one variable change. This experimentation pattern is less complex than the structure of informative experiments in Smithtown, and there are more alternative paths to solution.

Table 1

*Student Activities That Differed Significantly Across Laboratories**

Activity	Voltaville	Refract	Smithtown
GENERATION OF EVIDENCE			
Amount and breadth of search:			
Mean number of experiments run	15.0	31.0	46.0
Mean number of changes to parameters	6.3	9.1	13.9
Percentage of parameters changed	78.0%	69.0%	38.0%
Mean number of changes made to variables	6.9	22.7	25.2
Informativeness of search:			
Percentage of minimal required evidence	95.8%	80.8%	50.5%
Structure of search:			
Percentage of controlled experiments	30.4%	82.5%	88.5%
INTERPRETATION OF EVIDENCE			
Making predictions:			
Percentage of experiments with predictions	74.9%	73.5%	53.4%
Effects of prior knowledge:			
Number of alternative hypotheses stated	11.7	12.1	17.2
Percentage of goals discovered	88.5%%	58.3%	52.1%
DATA MANAGEMENT			
Percentage of experiments recorded in notebook	91.4%	99.3%	92.1%
Number of tables created	0.9	3.3	2.3

Note. ANOVAs performed on each of these measures are significant, $p < .05$.

However, in both Refract and Smithtown, conclusions depend upon noting regularities in changes from one observation to the next. If the comparisons are not valid, for example, if two or more variables are changed between comparison cases, it will be difficult or impossible to draw definitive conclusions. In contrast, in Voltaville each individual observation produces information that can support one or more valid inferences. To yield meaningful data, there is no need as in the other laboratories to design a set of coordinated experiments that serve as contrasts, because each observation instantiates the rule.

To generate interpretable patterns of evidence, it is necessary in Smithtown and desirable in Refract to follow the pattern of varying only one variable at a time, holding all other variables constant. As indicated, experiments in Voltaville are informative whether one varies one variable, two variables, or many. Our subjects appeared to be sensitive to these task differences. As Table 1 shows, the percentage of experiments in which subjects controlled extraneous variation was very high in Smithtown, and only slightly lower in Refract, but much lower in Voltaville. Note that although students generated controlled experiments much less frequently in Voltaville than in the other labs, they still did so nearly one third of the time, a substantial use of an evidence-generation strategy, given that there is no discernible advantage to using it here.

Evidence interpretation

Generating valid and informative experiments is a necessary but not sufficient condition for discovering the laws in the computer laboratories, since it is also necessary to appropriately interpret and make inferences about the evidence generated. We turn next to strategies in evidence interpretation that can be identified in the three computer laboratories.

Making predictions. Predictions serve both as the products of inferences and as the engines for further inference. However, students make predictions under somewhat different circumstances in the three computer laboratories. Since students perform coordinated sets of observations to discover a law in Smithtown, they tend to work through several successive cases in which their working prediction does not change. In contrast, in Voltaville and Refract, a student typically makes a new prediction for every experiment. For these reasons, as Table 1 shows, students working with Smithtown more frequently omitted making predictions (which are optional in all three laboratories). There is also some indication that the tendency to make regular predictions may have been influenced by the fact that predictions in the different laboratories have different characteristics. We observed from protocols that subjects appeared to find it more satisfying to generate a specific prediction, which was then unambiguously confirmed or disconfirmed by the computer feedback. Specific predictions and feedback of this kind are characteristic of both Voltaville and Refract. In contrast, to make predictions in Smithtown, students work with a menu of phrases to generate predictions like, "Quantity demanded will decrease". These statements are somewhat more ambiguous and less informative than the predictions in Voltaville, which are exact numerical values of variables, or the predictions in Refract, which can be either exact values or sketches indicating where students expect light rays to fall. Students appeared to find the ambiguous prediction statements in Smithtown to be relatively less satisfying, apparently because a mere correlational statement carries less information than an exact predicted value for a variable or a drawing specifying expected location. More informative predictions may encourage more hypothesis-driven search. Although students made predictions less regularly in Smithtown, attempts at inference through predictions resulted in an equal percentage of correct predictions in all three laboratories, averaging three quarters of the time.

Prior knowledge. Evidence interpretation is also influenced by prior knowledge. In general, subjects have experience with buying and selling, and therefore have a great deal of knowledge about consumer and market behavior. As a consequence, they hold a number of expectations about likely causes and effects in Smithtown, which might be correct or false conceptions. Most of our subjects reported that they were much less knowledgeable about the physics domains, in particular, stating that they knew more about economics than electricity, and more about electricity than refraction. These differences in prior knowledge may either help or mislead subjects in deciding where to search for relations, may influence them to be more or less active in search for disconfirming evidence, and may affect their confidence in their conclusions as well as their ability to remember and apply the laws they discover.

Differing prior knowledge could also affect the tendency to check to see whether a candidate law makes "sense" consistent with one's understanding of the phenomenon being described.

Table 1 shows that subjects stated a greater number of alternative hypotheses of all kinds (general and specific, correct and incorrect) while working with Smithtown, in comparison to both Voltaville and Refract. Most of our subjects were not hesitant to try out these tentative conclusions by submitting them to computer evaluation, even if little relevant evidence was available. However, this prior knowledge was a mixed blessing. On the average, subjects discovered a smaller percentage of the Smithtown goal laws than in either Refract or Voltaville. The mean percentage of goals discovered was 52.1%, 58.3%, and 88.5% in Smithtown, Refract, and Voltaville, respectively. Prior knowledge sometimes helps subjects to interpret patterns of evidence, but if prior knowledge is incorrect or only partly correct, it can encourage subjects to distort, ignore, or selectively interpret the evidence that they generate. This finding is a common one in research on scientific reasoning [2].

Data management

The differences among the laboratories also result in differences in how students manage their memory by recording and organizing data. Students appear sensitive to the notion that the utility of a particular analytic tool may vary by content domain. In Smithtown, laws often involve parameter changes that result in function shifts. Consistent with this characteristic, we found that our subjects graphed data more frequently in Smithtown than in the other two laboratories. As mentioned, the Refract laws are moderately complex mathematical expressions. Table 1 shows that in Refract students were particularly likely to use the computer capability for organizing tables, with its spreadsheet sorting and expression-generating functions. With the exception of one relation, the laws in Voltaville are algebraically simple, and thus there was less necessity for students to store or organize data to support the discovery of the relevant laws.

In sum, differences in domain content and structure were associated with differences in task requirements from laboratory to laboratory. These, in turn, were associated with different patterns of student activity in self-directed exploration. The results just described are corroborated by a pattern of intercorrelations run across the relevant activities (those included in Table 1) for all twelve subjects. These correlations reflected no activities in which subject performance was highly related across all three laboratories. Where strong correlations did exist, they were between pairs of laboratories, and they reflected the general structural and task differences already discussed.

In general, the results show that students tended to apply activities appropriate to the structure of the task. There is no simple story about consistency of performance, at least at this level of analysis. In general, our students did not tend to apply activities and processes across the three laboratories. Those relations that did appear were located in the discovery components in which laboratories shared common structural or task requirements.

Learning and Transfer in the Computer Laboratories

What does this pattern of specificity of performance imply for student learning? At least at the level of the overall objective and related subgoals--to discover laws and principles--the tasks posed by all three computer laboratories are the same. Students generate experiments, take measurements, make predictions, record data, and develop and revise hypotheses about the laws that apply. Much work on scientific discovery proceeds from the assumption that subjects differ in the ways that they work toward achieving these subgoals, and our own earlier work attempted to identify patterns of activity in these tasks that account for effective and ineffective learning [13, 15]. However, it appears based on our current results that these patterns are attenuated by content and structural characteristics. These specificities in performance raise two issues about student learning. The first issue concerns the way that students employ their skills to learn specific content knowledge. On this point, our data can be interpreted to suggest that if individuals have differential skills, it is to be expected that some will do better in some domains, and others will do better in others. However, the second issue concerns students' growing capability in learning how to learn. A group of students who possess the identical skills in scientific reasoning may still vary considerably in how and when they apply them. This raises the question of whether students vary in their ability to adapt to fluctuating task demands.

To address this second issue, we performed a number of analyses that focused on student learning across the three laboratories. First explored was whether there are students who are maximally adaptive, that is, students who are uniformly good learners regardless of the content area presented. Second explored, and related to the first, was whether there were relations between amount of learning achieved in the three laboratories. In particular, we wanted to know whether our earlier task analysis of the structural similarities and differences among the laboratories would be verified by results with respect to student learning. Finally, we analyzed whether there were any order effects on learning; that is, whether the amount of learning changed as students progressed from the first laboratory to the second and finally to the third.

Performance across domains

We first considered whether there are students who are generally good or poor learners across all three domains. Amount of learning was measured for each computer laboratory by computing student pre/post test gain scores. Students accomplished significant gains in each of the laboratories. Their gains were relatively higher in Voltaville and Smithtown than in Refract, the most difficult discovery context. Mean gain score for Smithtown was 26.5 percentage points (SD = 15.4) , for Voltaville was 26.3 percentage points (SD = 17.2) , and for Refract was 11.9 percentage points (SD = 17.6) (all of these gains are significant).

However, there was no clear relationship between amount of achievement in one laboratory and amount of achievement in the others. A correlation run on the three gain scores for the 12 subjects yielded only a modest correlation between gains in Refract and Voltaville (r = .27), the two worlds in which the rule discovery structure was shared. (Gain scores were used because both the pre- and post-test scores varied widely. Moreover, students did not gain to ceiling; only one student earned a post-test score of 100% in one laboratory). There was a *negative* correlation between gains in Voltaville and Smithtown (r = -.31), the two labs with different structures, that is, rule discovery versus correlational. There was no meaningful correlation between gain scores on Smithtown and Refract (r = .008), which has mixed properties, so that students who were effective in Smithtown varied in their performance in Refract, and vice versa. In general, then, at the group level, learning in our laboratories appeared to depend to a large extent upon specific structural and task requirements rather than the exercise of generalized reasoning strategies.

This specificity of student performance is further confirmed by the finding that in these complex exploratory situations, no student was clearly the best in all laboratories and no student was the worst. That is, no student was among the one-third with the highest gain scores on all three laboratories (although five students were among the top third achievers on at least two labs). Similarly, no student was among the one-third with the lowest gain scores on all three laboratories (although three students were low achievers on at least two labs).

Learning to learn

The above results appear consistent with the conclusion that some students were better at learning certain kinds of content, whereas others were better at learning other kinds of content. But are students really this specific in their strengths? Data at the group level do imply more general characteristics of performance. When order of learning is considered, we find that the subjects showed more and more learning as they progressed over the three different lab experiences. On the average, there was a mean increase in gain score of 9 percentage points from Lab 1 to Lab 2, regardless of whether subjects began with Voltaville or Smithtown (to evaluate the magnitude of this increase, recall that total gain score for each of these labs was about 25 percentage points). This implies that students may have been achieving familiarity with the overall activity of experimentation and its component processes, including ways of generating evidence, making inferences from this information, searching for regularities, and testing them. That is, beyond learning specific concepts and content, students were also engaging in repeated practice on the skills required for discovering laws using computer laboratories. The findings reported earlier confirm that there are specific relations among content areas, their structural characteristics, and the strategies that lead to successful learning. However, these effects due to the order of learning suggest that in addition to specific skills, there may be a more general set of reasoning skills and strategies associated with learning how to learn successfully in science laboratories.

The existence of more general abilities is further suggested by a comparison of student learning in Voltaville and Smithtown, the first two labs, with their learning in Refract, the final laboratory explored by all students. From the group of twelve subjects, four were identified, for want of a better word, as "improvers." These were the students who made the greatest increase in learning gains when their gain scores on the second laboratory were compared to their gain scores on the first laboratory. Average increase in gain scores among this "improver" group was 33.3 percentage points from the first laboratory to the second. A second group of four made the smallest increase in gain scores, an average increase of -9.5 percentage points. On the third lab, Refract, the improvers gained an average of 19.5 percentage points from pretest to posttest. In fact, three of the four improvers made the largest overall learning gains in Refract, as well. In contrast, the non-improvers' mean gain score was only 5.8 percentage points, indicating that the amount of improvement from Lab 1 to Lab 2 was associated with the amount of learning achieved in Lab 3. Apparently, those students who learned the most about the general objectives and nature of scientific discovery by working with the earlier two labs were able to apply this understanding in Refract. In addition, since Refract is a lab with a mixed structure, it represented an opportunity for subjects to apply the particular relevant skills practiced in both Voltaville and Smithtown.

How did students learn how to learn with the computer laboratories? The process was apparently not one of simply acquiring and applying a set of generally useful strategies; otherwise, there would have been stronger correspondences among the activities that students applied across the laboratories. Rather, this kind of learning appears at the level of overall qualitative patterns of activity like (1) data-driven versus hypothesis-driven search, (2) persistence, (3) focus, (4) use of disconfirming feedback or evidence, (5) applications of prior knowledge, and (6) applications of relevant heuristics and analytical devices.

To identify these qualitative patterns, the protocols of the four improvers and the four nonimprovers were compared. The objective was to characterize how the improvers developed their demonstrated ability to learn better with the laboratories as their experience accumulated. For heuristic clarity, we here describe important characteristics of the performance of one improver, whom we call Allen, and a nonimprover, whom we call Peg (it should be emphasized, however, that improvement was not related to gender). Focusing on these individual cases permits us to demonstrate continuity of approach across the laboratories, and to show how different approaches influenced student learning. Although two particular students are described in this section as examples, the characteristics reported are typical of the improver and nonimprover students, respectively.

Data-driven vs hypothesis-driven search. In general, the search of the improver students was characterized by two sequential phases. In the first phase, which we call the "identify" phase of search, they generated a relatively small amount of data and examined it for a candidate rule or relation. Typically, this phase was relatively brief in contrast to the next one, which we call the "confirm" phase. During the "confirm" phase, the student generated several additional cases in an attempt to either confirm or disconfirm the hypothesized relation. In contrast, the search of the nonimprovers did not appear to have this two-phase character. Instead, their search appeared to be data-driven. That is, they generated experiment after experiment until a possible law or relation was glimpsed. At that point, the candidate relation was nearly always stated as a conclusion, with little additional effort expended on discovering whether it applied in additional cases. When nonimprovers were having difficulty identifying laws, they often defaulted to generating experiment after experiment, with no apparent objective in mind other than a show of activity.

For example, Allen, an improver, typically generated only one or two experiments before stating a tentative hypothesis for evaluation by the computer. Subsequent experiments were then directed at trying to generate more cases in an attempt at confirmation (systematic attempts at disconfirmation were rare among both improvers and nonimprovers). Typically, these strings of experiments culiminated in a restatement of the hypothesis in a general form (that is, one which the student believes applies across several relevant cases). This pattern appeared within all three laboratories. Allen's comments reveal that clear expectations were guiding his search. These comments appeared especially when those expectations were violated. For example, this remark was made immediately after he observed the value for the total resistance in a parallel circuit: "Wait a minute--no, it should be the total resistance. Now wait, something is not working here right. That's a little different from the series one."

In contrast, although Peg generated nearly twice as many experiments as Allen, there was no clear separation in her work between experiments meant to identify tentative hypotheses and experiments meant to confirm hypotheses. As a consequence, within the Voltaville laboratory, she generated far more data than she actually needed to support each conclusion, and spent a relatively smaller proportion of her time trying to interpret the data. Within the Smithtown laboratory, as well, she generated many extraneous experiments; however, many of these were in principle uninterpretable because they did not follow the kinds of patterns that could highlight the lawful regularities in economics. She developed fewer hypotheses overall than Allen did, and unlike Allen, appeared to view hypotheses as conclusions rather than as tentative statements to guide investigation.

Persistence. Improvers appeared to learn when to persist in a line of search and when to abandon it as unfruitful. In contrast, nonimprovers often refused to abandon an idea that had already been directly disconfirmed, or, alternatively, wandered off fruitful paths of search before making much progress.

For example, in Voltaville, Allen worked systematically on all the laws of series circuits before moving on to the more complex parallel cases. In contrast, Peg jumped around from law to law, sometimes remarking that she could not quite recall one that she had just discovered. She also jumped from circuit type to circuit type, that is, from series-2 circuits to parallel-2 circuits, back to series-1 circuits. It is not in principle problematic for a student to move from case to case in this manner. However, within each type of circuit, all the relevant laws are interrelated. For example, Ohm's Law states $V = I * R$, or the value of voltage can be calculated by multiplying the value of the current times the value of the total resistance. In turn, Kirschoff's Laws explain how to calculate the total value of the current, the resistance, and the voltage. Therefore, the variables implicated in these laws overlap. Typically, students who work within a particular circuit type generate and test a number of candidate relations, often fortuitously resulting in the discovery of several laws in close succession. Because of his persistence within a line of search, this was the pattern of Allen's discoveries, whereas Peg made discoveries in a slower, more piecemeal fashion.

Of course, it is not sheer persistence that provides the advantage, but awareness of when to persist with promising lines of search and when to abandon those that are not leading anywhere. As described below under "Use of disconfirming feedback," Peg had particular problems with letting go of ideas that made sense to her, even when she was being directly informed that they were incorrect or misleading.

Focus. Improvers appeared to maintain goal focus over time and in the context of distracting irrelevant information. Nonimprovers often lost sight of what they were doing or even what they had learned. Although they often discovered laws readily, they did not appear to understand them very deeply or to retain them over time.

For example, Allen often reported mulling over perplexing events even between his exploration sessions on the computer. For example, he began his third session with Smithtown by remarking, "There was one thing that I was thinking about last night, and it came to me about the demand..." In contrast, Peg usually had to be reminded of what she had done in the previous session and had to be reoriented to the objective of the work. In contrast to Allen's impressive memory about the details of previous sessions, Peg readily admitted that she often did not recall what she was doing or had done: "I can't remember how I did this now. I can't remember what values I manipulated to get this one. I put in, these are my values, right? Did I do these?"

Use of disconfirming feedback or evidence. As mentioned, the ability to make appropriate use of disconfirming feedback is closely related to the issue of knowing when to persist, with improvers generally showing more sensitivity to cues that they are on the wrong track. Notable in this regard is that while working on Smithtown, both Allen and Peg became stuck on the identical misconception. This misconception is very prevalent in beginning microeconomics students; it consists in failing to distinguish between the effects of variable changes and parameter changes. In particular, students often mistakenly focus on immediate patterns that are in fact the result of higher-order (parametric) relations. When the computer informed him that the actual target relation is a higher-order relation, Allen was able to puzzle through to the correct interpretation. In contrast, after receiving a similar message, Peg devoted an additional 15 experiments to trying to "prove" that changing the average income in Smithtown would *directly* affect the quantity demanded for several goods and services. She justified this strategy by attending to only half of the computer message (the half explaining that she did not have evidence to support this conclusion), and by ignoring the reference to higher-order relations. Thus, she continually misinterpreted the computer feedback as a request for more (as opposed to a different kind of) evidence. "It said that I didn't have enough evidence, but I did. I did an experiment where I increased the income and it (quantity demanded) went down, and I did one where I increased the income...and it increased. I mean, how many more experiments should I do?"

Applications of prior knowledge. Improvers and nonimprovers both rely on their prior knowledge to make sense of new situations, but they use prior knowledge differently. In the reasoning of the improvers, prior knowledge and current evidence are mutually constraining. That is, the students appear to be aware that their qualitative understanding of a phenomenon should be consistent with their interpretation of new data. In contrast, in the reasoning of the nonimprovers, prior beliefs sometimes override the evidence. That is, the student refuses to budge from his or her beginning understanding of the phenomenon. A good example is Peg's expectation that if average income were increased, people in Smithtown would ride the bus less frequently. She interpreted this as "proof" of the following hypothesis: "As income increases,

quantity demanded for bus ride decreases." In fact, although this is in fact the relation observed, the pattern is actually due to a shift in the entire demand curve (that is, the entire relation between price and quantity demanded changes). It is this more complex, indirect relation that is important within the domain of economics. However, Peg's prior knowledge appeared to make it very difficult for her to abandon her belief that there would be a direct relation between changes in average income and changes in the number of bus rides purchased on a given day in Smithtown. In contrast, Allen worked to bring his knowledge into alignment with the data. He began by working from the expectation that changing the number of suppliers should directly affect the quantity supplied for a particular good. His solution to the problem involved discovering shifts in the supply curve, a solution that successfully resolved his qualitative understanding of the context with the data he had generated.

Applications of relevant heuristics and analytical devices. There are analytical devices and heuristics that provide significant leverage within each of the computer laboratories. Within Voltaville and Refract, these are basically mathematical heuristics, for example, the ability to automatically recognize .333 and the fraction 1/3 as the same value, or to think of dividing one variable by another to see if the result yields the third. Within Smithtown, the crucial analytical device is the ability to visualize data graphically, and to conceive of changes in patterns of the data as shifts in demand and supply curves. Students differ widely in their ability to think in these analytical ways. For example, while working in Smithtown, Allen readily induced the schema for shifting curves, whereas even after she had presumably discovered it, Peg admitted, "I'm still not sure about this quantity demanded and the demand thing." Her subsequent experiments indicated that in fact, she had a very slippery grasp of the distinction.

In summary, then, for individual students low or high learning in one laboratory was not directly associated with low or high learning in the others. However, on the average for the group as a whole, practice appeared to result in improvement in students' ability to learn with computer laboratories. The student protocols indicate that there are qualitative differences in students' abilities to learn how to learn with computer laboratories. Part of this learning involves becoming more sensitive to task similarities and differences from domain to domain, and learning how to adapt their experimentation activities accordingly. The implication is that students may develop their skills for determining whether particular skills are appropriate in a particular circumstance, how those skills will be applied, and what other skills need to be developed.

Discussion

Experimentation involves a complex orchestration of activities, and there is typically a great deal of variability in people's performance on these activities. There are mny alternative ways to perform in each, and thus many alternative paths to success in the overall enterprise. Although in general, successful discoverers perform some activities and heuristics more often than those who are unsuccessful or inefficient [e.g., 13, 15], there appears to be no pattern of strategies that guarantees success, as indicated by the lack of consistent patterns of student activity found in this study.

Do these findings imply that experimentation strategies are primarily specific to the domain and structure of a particular task? Many researchers, including ourselves, have investigated what they assumed were scientific reasoning strategies of some generality [7, 8, 15]. We believe that although our current findings complicate this picture to some degree, they do not altogether distort it.

First, it appears that within a particular experimentation task, there are actions that subjects can do or fail to carry out that affect their learning. However, beyond these individual differences, like Linn et al. [12], we also found that the type of task is a second source of performance variance. That is, we found differences in student activity as a function of the particular structure and content of each of the three computer laboratories. In a situation that required the discovery of correlational regularities, evidence-generation activities, like the heuristic of controlling variables, were primary. Where subjects held prior misconceptions, controlled experiments were essential if biases were to be overcome. In discovery situations where the regularities were functional rules, evidence interpretation became more central. When the rules were quantitative, mathematical and algebraic heuristics were particularly important.

However, we also found that over the weeks of the study, our students appeared to improve in their ability to learn from the computer laboratories. We conjecture, therefore, that there may be general strategies that pay off in experimentation, although we now believe they are at a larger grain-size than those we originally studied. First, we suspect that our students learn the rules of the "experiment game," and that those rules result in more effective scientific discovery, at least in situations like these computer laboratories. For example, they learn from experience that discovery is not a five-minute problem like the science problems frequently encountered in school, and that persistence is a sufficient but not necessary condition for success. Students learn that one occasionally encounters conceptual impasses that require the re-evaluation and reformulation of one's approach to the problem. They learn that planful and systematic generation of evidence helps one avoid wasting time by producing data that are uninterpretable.

They learn the importance of attending to counterexamples and disconfirming instances. They become more aware that the laws they are searching for can vary in their complexity, and that they therefore need to both avoid overlooking the obvious relations and become more inventive in generating and testing mathematical heuristics for identifying the complex relations. They learn to defer conclusions until sufficient evidence is available. They learn how and when to record, graph, and manipulate data.

Beyond these skills in learning how to learn in experimental situations, students may also improve their abilities for adapting to the demands and structures of different kinds of problems. Given that different individuals have different kinds of proficiencies, how do they adapt these proficiencies to different kinds of tasks? It may be that with experience and practice, students become sensitive to differences among frequently-encountered kinds of scientific reasoning problems. Thus, one kind of improvement a student might achieve by proceeding through the laboratories would involve inventing or accessing and then fine-tuning the strategies for learning in different kinds of experimentation contexts. A second kind of improvement might involve developing a deeper understanding of the strengths and weaknesses of one's own strategies, including the ability to recognize the occasions and situations when it is appropriate to apply them.

With practice, our undergraduates improved in their ability to learn content knowledge from self-directed exploration. That is, they learned more as they proceeded from domain to domain, indicating that they were somehow becoming more effective in planning and carrying out experiments, and in formulating and testing hypotheses based on those experiments. However, as our work probed into the differential complexity and variance of actual domains of science, we have become increasingly aware of the content and context specificity of effective performance. The activity of scientific discovery depends upon variability in the structural form of the discovery task and the nature of the domain.

What do these findings imply for understanding scientific discovery? For the reasoning of professional scientists, it implies that since most scientists work primarily within the boundaries of their chosen fields and even specialize additionally within those fields, the expertise that develops may be chiefly domain-specific. Sociologists [11] and contemporary philosophers of science [5] confirm that for professionals or journeymen practitioners, acquiring good scientific skills is chiefly a matter of being socialized into the skills of a particular discipline and practice.

The implications are somewhat different if we consider scientific reasoning as part of general education. Given that most students will not become members of any practicing scientific community, and given that it is not feasible to introduce students deeply into the practices and reasoning styles of different topics and domains of science, what do we want students to understand about scientific discovery? The most generally useful message may be that discovery is not a monolithic enterprise where one applies cookbook heuristics described in the standard "scientific method" chapter in most secondary school texts. Actual problem solving in science requires adaptability of reasoning to domain properties. Our point here is not to emphasize extreme domain specificity, nor to discount generally useful strategies. When various discovery contexts are compared, they do have specificities and commonalities. However, when a new exploratory situation is encountered, it is necessary to consider what kind of problem it is, and to apply evaluative and self-regulatory skills to decide which processes and strategies are appropriate to the particular task at hand. That is, the most generally useful heuristics may be those involved in learning to discriminate the discovery and inference requirements of a particular scientific setting.

Thus, the results of this study support the implication that scientific reasoning requires general skills in evidence generation and evidence interpretation. The presence or absence of these skills is associated with success or failure in experimental contexts. However, these general skills take on specific value as they are differentially useful in varying contexts. The implication for early science instruction is that the nature of scientific reasoning of the kind studied here can best be improved when a student is provided with opportunities for practice in identifying situational contexts where scientific reasoning skills can be selectively and adaptively employed.

Acknowledgment

Preparation of this article was sponsored in part by the Center for the Study of Learning (CSL) at the Learning Research and Development Center of the University of Pittsburgh. CSL is funded by the Office of Educational Research and Improvement of the U. S. Department of Education. Refract, Voltaville, and Smithtown were created in association with Jaimie Schultz, Mayura Channarasapa, Valerie Shute, and Peter Reimann. In addition, the authors acknowledge the assistance of Phyllis Akbarzedek, Jenny John, and Robin Wingard.

References

1. Carey, S. (1985). *Conceptual change in childhood.* Cambridge, MA: MIT Press.
2. Carey, S. (1986). Cognitive Science and Science Education. *American Psychologist, 41*(10), 112-130.
3. Chi, M. T. H., Glaser, R., & Farr, M. J., Eds. (1988). *The nature of expertise.* Hillsdale, N.J.: Lawrence Erlbaum Associates.
4. Chi, M. T. H., & Koeske, R. (1983). Network representation of a child's dinosaur knowledge. *Developmental Psychology, 19*, 29-39.

5. Giere, R. (1979). *Understanding scientific reasoning*. NY: Holt, Rinehart, & Winston.
6. Inhelder, B., & Piaget, J. (1958). *The growth of logical thinking from childhood to adolescence*. New York: Basic Books.
7. Klahr, D., & Dunbar, K. (1988). Dual search space during scientific reasoning. *Cognitive Science 12*(1), 1-48.
8. Kuhn, D. (1989). Children and adults as intuitive scientists. *Psychological Review, 96*(4), 674-689.
9. Kulkarni, D., & Simon, H. A. (1988). The processes of scientific discovery: The strategy of experimentation. *Cognitive Science, 12*, 139-175.
10. Langley, P., Simon, H. A., Bradshaw, G. L., & Zytkow, J. M. (1987). *Scientific discovery: Computational exploration of the creative processes*. Cambridge, MA: The MIT Press.
11. Latour, B. & Woolgar, S. (1979). *Laboratory life: The social construction of scientific facts*. Beverly Hills, CA: Sage Publications.
12. Linn, M. C., Clement, C., & Pulos, S. (1983). Is it formal if it's not physics? (The influence of content on formal reasoning). *Journal of Research in Science Teaching, 20*, 755-770.
13. Schauble, L., Glaser, R., Raghavan, K., & Reiner, M. (1991). Causal models and experimentation strategies in scientific reasoning. *The Journal of the Learning Sciences, 1*, 201-238.
14. Shaklee, H., & Mims, M. (1981). Development of rule use in judgment of covariation between events. *Child Development, 52*, 1229-1240.
15. Shute, V., Glaser, R., & Raghavan, K. (1989). Inference and discovery in an exploratory laboratory. In P. L. Ackerman, R. J. Sternberg, & R. Glaser (Eds.). *Learning and individual differences: Advances in theory and research* (pp. 279-326). NY: W. H. Freeman.
16. Simon, H. A., & Kotovsky, K. (1963). Human acquisition of concepts for sequential patterns. *Psychology Review, 70*, 534-546.
17. Tschirgi, J. (1980). Sensible reasoning: A hypothesis about hypotheses. *Child Development, 51*, 1-10.
18. Wason, P. (1960). On the failure to eliminate hypotheses in a conceptual task. *Quarterly Journal of Experimental Psychology, 12*, 129-140.

A Rule-Based Diagnosis System for Identifying Misconceptions in Qualitative Reasoning in the Physical Domain "Superposition of Motion"

Heinz Mandl[1], Jürgen Bollwahn[3], Aemilian Hron[2], Uwe Oestermeier[2] and Sigmar-Olaf Tergan[2]

[1] Institut für Empirische Pädagogik und Pädagogische Psychologie, Ludwig-Maximilians-Universität, Leopoldstr. 13, D-8000 München 40

[2] Deutsches Institut für Fernstudien an der Universität Tübingen, Arbeitsbereich Lernforschung, Bei der Fruchtschranne 6, D-7400 Tübingen

[3] CIM und computerunterstützte interaktive Medien, Mollstr. 40, D-6800 Mannheim 1

Abstract: An automatic diagnosis system for the assessment of conceptual knowledge with a focus on individual misconceptions was developed. The system is a central component of a computer-based learning environment on the physical domain superposition of motion. The diagnosis system is based on a production system approach and has been implemented as a classifier system. Conceptual knowledge manifesting itself in qualitative reasoning on physical problems concerning superposition of motion is represented by means of qualitative rules, the rules representing changes of physical parameters in qualitative terms. Processes of knowledge acquisition are modelled by means of strengthening and discrimination algorithms. The diagnosis sytem is suitable to generate automatically both the correct solution and incorrect solutions corresponding to the learner's correct or incorrect qualitative reasoning. It compares these solutions with the learner's actual solution and thus infers his or her respective conceptual knowledge. The system is able to discriminate between misconceptions resulting in identical problem solutions by means of conflict resolution procedures. To some extent it is also able to cope with ambiguous learner responses and noisy data attributable to unsystematic behavior. The diagnosis system has been implemented in LOOPS and is running on a Xerox 1108 workstation.

Keywords: knowledge diagnosis, qualitative reasoning, misconceptions

Overview

In a project concerning research on knowledge representation and knowledge acquisition in the physical domain of kinematics an automatic diagnosis system has been developed. The diagnosis system is suited to assess individual misconceptions and processes of knowledge acquisition. The system is a central part of an adaptive learning environment concerning the domain "superposition of motion". The environment is intended to foster knowledge acquisition by promoting inductive analog problem solving. It aims at imparting conceptual domain specific knowledge underlying qualitative reasoning. It's focus is on misconception-based tutoring [4]. Learner modelling is based on a production system rationale and is realized by means of a classifier-approach. Individual knowledge as manifested in qualitative reasoning on physical problems concerning "superposition of motion" is modelled with qualitative rules representing changes of physical parameters, e.g., velocity or direction of motion. Processes of knowledge acquisition are represented by means of strengthening and discrimination algorithms.

The diagnosis system generates automatically both the correct solution as well as possible incorrect solutions based on misconceptions. It compares the correct solution with the learner's actual solution and thus infers his or her respective conceptual knowledge. The system is quite efficient because it makes profit of the advantages of classifier systems: the parallel firing of rules as well as the efficient mode of pattern matching. The system is suited to resolve conflicts in case of ambiguous use of misconceptions. The diagnosis system as well as the learning environment are developed in LOOPS on a Xerox 1108 workstation.

The Knowledge Domain

The domain "superposition of motion", a subdomain of kinematics in elementary mechanics, was taken as a knowledge domain. For pragmatic reasons of research the domain was constrained to linear motions of constant velocity.

Superposition of motion refers to the fact that a physical body can carry out different motions at the same time. When described with respect to a fixed reference system the motions superpose to one resulting motion.

Example: A swimmer crosses a river. His/her movement is perpendicular with reference to the moving water. Relative to the reference system "bank" the swimmer is moving in a diagonal direction. The diagonal motion is a superposed motion that is composed by the single motions, the motion of the swimmer and the motion of the river.

From a physical point of view the knowledge domain "superposition of motion" may be described by two concepts, the concept of *relativity* and the concept of independence. The concept of *relativity* means that a motion has to be described relative to a certain system of reference. The concept of *independence* considers the fact that motions are independent from each other: If a body is performing different motions at the same time, these motions superpose without influencing each other.

Novice students often have severe problems in understanding the physical facts concerning "superposition of motion". Empirical studies point to inadequate mental representation of domain-specific knowledge because of the fact that novices incorrectly consider forces as causes of accelerated and unaccelerated motions [1].

Based on relevant literature as well as on results of a rational and empirical task analysis the following misconceptions had been identified as relevant in constituting the knowledge domain.

- *Modified Impetus* misconception (assumption that an object's "own" motion inside the moving reference system is loosing speed after a short time)
- *ActivePassive* misconception (assumption that active motions dominate passive motions)
- *Dependence* misconception (assumption that motions are dependent on each other)
- *DominantMotion* misconception (assumption that a motion is dominant)
- *AbsoluteMotion* misconception (assumption of a steady motion, which is regarded absolute instead of relative to a reference system)
- *CompromiseMotion* misconception (assumption that the complete motion results as the average of velocities and directions of the single motions)
- *ClassicalImpetus* misconception (assumption of a decrease of velocity in a motion considered absolute).

The Learning Environment

As a basis for research concerning the diagnosis component a computer-based learning environment has been developed. The environment represents central aspects of the knowledge domain. It covers problems on superposition of motion referring to the composition and decomposition of motions. It is oriented toward a paradigm of guided inductive analog problem solving. The learning environment comprises about 60 tasks in the domain of superposition of motion. The basic situation of all problem solving tasks refers to the following scenario:

> A mother and her child step onto a overdimensional large conveyor belt at the same time. The conveyor belt moves (with constant speed) toward the right. Immediately after stepping onto the conveyor belt the mother stands still; the child always runs away from its mother in its own constant speed.

According to the child's direction of motion with respect to the direction of motion of the conveyor belt, three basic situations can be distinguished: (1) collinear motions (the child runs either in or opposite to the direction of the belt's motion), (2) perpendicular motions (the child runs perpendicular to the direction of the belt's motion, and (3) oblique-angled motions (the child runs in an oblique angle to the direction of the belt's motion).

The generated tasks are quite similar to one another. Differences between the tasks exist, on one hand, with regard to characteristics of the problem description (e.g., with regard to the speed relation of the child's "own" motion). Differences exist, on the other are relevant for the task solution. For a very elementary task concerning the composition of two single motions, see Figure 1.

Given are qualitative descriptions of the directions of the belt's and the child's motion, the speed relation of both motions as well as the position of the mother, the mother representing the distance, the belt has moved in relation to the environment. The learner's task is to mark the position of the child. Marking is accomplished by mouse-click directly on the screen. Figure 1 shows the final position of the child at the end of a simulation demonstrating the child's (composed) motion relative to the environment.

A physically correct solution is based on the concepts "Relativity" and "Independence". Students wrongly assuming that the motion of the conveyor belt influences the child's motion (Dependence misconception) will predict that the child's motion is deviated in the direction of the conveyor belt. As a result a final position of the child is marked, which is on the right of point g, the actual final position of the child.

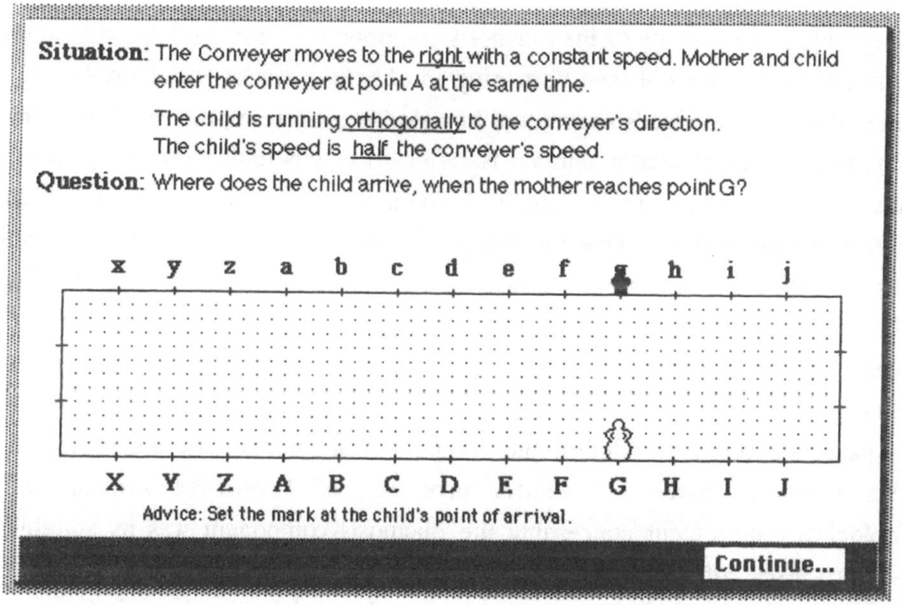

Figure 1. A typical task situation, the grafic showing the final state of a simulation of the child's movement.

The learning environment exists as a fixed sequence of tasks on superposition of motion, the tasks focusing on composition and decomposition of motions. They are sequenced according to their affiliation to one of the basic situations (collinear, perpendicular, oblique-angled). In case of faulty solutions the learner gets tutorial help.

Animated graphic is used to illustrate the reference system dependency of the description of the single motions. Graphical and verbal cues are used to foster the understanding of the underlying physical facts. Tutorial help ensues according to the diagnosed misconception under the control of the diagnosis component.

The Diagnosis Component

Rationale and requirements

The psychological rationale of the diagnosis component is oriented toward assumptions concerning *mental model*-based reasoning in solving physical problems [cf. 5]. It is assumed that an individual's conceptual knowledge concerning the relational structure of a particular subject matter domain is reflected in qualitative reasoning underlying physical problem solving in that domain. Individual problem solutions may be taken as indicators of underlying concepts and misconceptions governing qualitative reasoning.

As a functional requirement the diagnosis component should be suited to diagnose the conceptional knowledge underlying qualitative reasoning of novices in solving problems on "superposition of motion", with a special focus on assessing individual misconceptions. Apart from status diagnosis focusing on the state of knowledge relevant for a particular problem solution the component should also be suited to perform process diagnosis concerning central processes of knowledge acquisition. One psychological requirement concerning the diagnosis component was its suitability to assess processes of restructuring. Restructuring is conceived as fundamental modification of mental representation of knowledge [cf. 9] - here indicated by a reduction of individual misconceptions paralleled by the acquisition of physical knowledge. Another psychological requirement concerning the diagnosis component refers to its suitability for assessing processes of consolidation and generalization of knowledge. A further requirement is to discriminate the underlying conceptual knowledge when in one task two misconceptions lead to identical solutions. Although one may identify in most tasks the one misconception relevant for a particular solution this may not be possible with other tasks. From a technical point of view, the diagnosis component should be suited to cope with ambiguous answers. It should further stand "noisy" data which can be ascribed to unsystematic errors.

Constraints in the development of a diagnosis component suited to cope with the above mentioned requirements result, for example, from the type of tasks that characterize the learning environment. Problem solutions are basically defined quantitatively in terms of the position of the child with reference to the surroundings of the belt. That is why a final identification of misconceptions solely based on quantitative data is not possible.

Other constraints refer to the fact that misconceptions only account for one type of possible errors. There may be other cognitive conditions relevant for explaining incorrect solutions in coping with analog problems on superposition of motion. These are, for example, inappropriate cognitive processes concerning analogical access and analogical mapping [cf. 10]. Incorrect solutions may also be due to action slips related to the omission of a certain parameter of the problem statement as well as to simple counting errors or to a faulty application of a (geo-)metrical rationale like the Pythagoras algorithm.

In the following chapters the approach in modelling qualitative knowledge (with a focus on misconception-based reasoning), in diagnosing conceptual understanding, and in assessing aspects of knowledge acquisition, is outlined in more detail.

Representation of qualitative knowledge

To optain an efficient on-line diagnosis the diagnostic system has been implemented as a classifier system. Classifier systems are production systems, however, contrasted with production systems the pattern-matching is not based on list-comparison but instead on comparison of bit-vectors of fixed length. The advantage is that only a few basic logical operations are needed. Thus, learning algorithms which modify given rules can easyly be implemented. As contrasted to production systems like PRISM or OPS5 classifier systems because of their efficiency and compactness may also be implemented successfully on a PC. For a detailed description of classifier systems see [6] and [7].

The core of the developed diagnosis system consists of a set of rules by which conceptual as well as operational knowledge underlying the correct and incorrect solution of tasks on superposition of motion is modelled. For each physical concept and each misconception up to four rules are used to simulate the effects of conceptual knowledge on the learner's problem solving. A particular set of rules describes the path that leads to a particular problem solutions, thus, providing the basis for inferring the corresponding correct as well as incorrect knowledge.

Qualitative reasoning in solving problems of superposition of motion is described in terms of qualitative representations. In the following illustrations we confine ourselves to two central situational characteristics of tasks on superposition of motion which are important for describing qualitative reasoning: (1) the speed of the child's "own" motion, and (2) the direction of the child's motion with respect to the direction of the belt's motion (right). The speed variable can take the qualitative parameters "constant",

```
[CLASSIFIER Diagnoser Dependence-right
        (IF    ((Level:2 Misconcept:Dependence Direction:right))
               THEN
               (Level:3 ChildVelocity: + ChildDirection: = ))
        (Strength 131)
        (Trace T)]

[CLASSIFIER Diagnoser Dependence-left
        (IF    ((Level:2 Misconcept:Dependence Direction:left))
               THEN
               (Level:3 ChildVelocity:- ChildDirection: = ))
        (Strength 100)
        (Trace T)]

[CLASSIFIER Diagnoser Dependence-ortho
        (IF    ((Level:2 Misconcept:Dependence Direction:ortho))
               THEN
               (Level:3 ChildVelocity: = ChildDirection:-))
        (Strength 100)
        (Trace T)]

[CLASSIFIER Diagnoser ActivePassive-right
        (IF    ((Level:2 Misconcept:Dependence Direction:right))
               THEN
               (Level:3 ChildVelocity: + ChildDirection: = ))
        (Strength 131)
        (Trace T)]

[CLASSIFIER Diagnoser ActivePassive-left
        (IF    ((Level:2 Misconcept:Dependence Direction:left))
               THEN
               (Level:3 ChildVelocity: + ChildDirection: = ))
        (Strength 100)
        (Trace T)]

[CLASSIFIER Diagnoser ActivePassive-ortho
        (IF    ((Level:2 Misconcept:Dependence Direction:ortho))
               THEN
               (Level:3 ChildVelocity: + ChildDirection: = + ))
        (Strength 100)
        (Trace T)]
```

Figure 2. Rules for representing qualitative inferences based on the misconceptions "Dependence" and "ActivePassive". (For reasons of clearness the rules are represented in a readable "symbolic" format, not as typical classifier rules).

"increasing", and "decreasing". The variable of the direction of the child's motion is determined by the parameters "turning to the left", "constant", and "turning to the right".

The modelling mainly refers to changes respectively constancies of the child's motion on the belt with regard to amount and direction.

Figure 2 shows six rules (or classifiers) representing misconception-based qualitative reasoning on superposition of motion.

Underlying misconceptions are the "Dependence" misconception and the "ActivePassive" misconception. The first three rules describe the qualitative conclusions based on the "Dependence" misconception.

Each rule represents one of the three basic situations of the tasks in the learning environment, which are characterized by the three directions - "right", "left", and "perpendicular" - of the child's motion with respect to the direction of the conveyor belt's motion.

Qualitative reasoning on the basis of the "ActivePassive" misconception is described by the other three rules, each concerning one of the three directions of the child's motion.

The first of the rules represented in Figure 3 describes the incorrect qualitative reasoning on account of the "Dependence" misconception, that the speed of the child's "own" motion to the right is increasing while the direction of its motion remains the same. The condition part of the rule refers to the misconception and the direction of the child's motion. The rule's action part refers to the qualitative inference drawn. The symbols "+", "=", and "-" are used for describe changes in direction: "turning to the left", "constant", and "turning to the right". The following rules describe the effects of the "Dependence" misconception in case of a child's motion to the left respectively in perpendicular direction.

The final three rules describe the learner's qualitative reasoning about the child's motion on the conveyor belt under conditions of the "ActivePassive" misconception. Taking the condition "child runs to the right" the assumption based on the "ActivePassive" misconception is that the speed of the child's motion is increasing because of the child's activity, the direction thereby remaining unchanged.

The diagnosis network

Qualitative rules are central for the diagnosis system. There are, however, different types of rules, which altogether build a network structure (Figure 3). The network

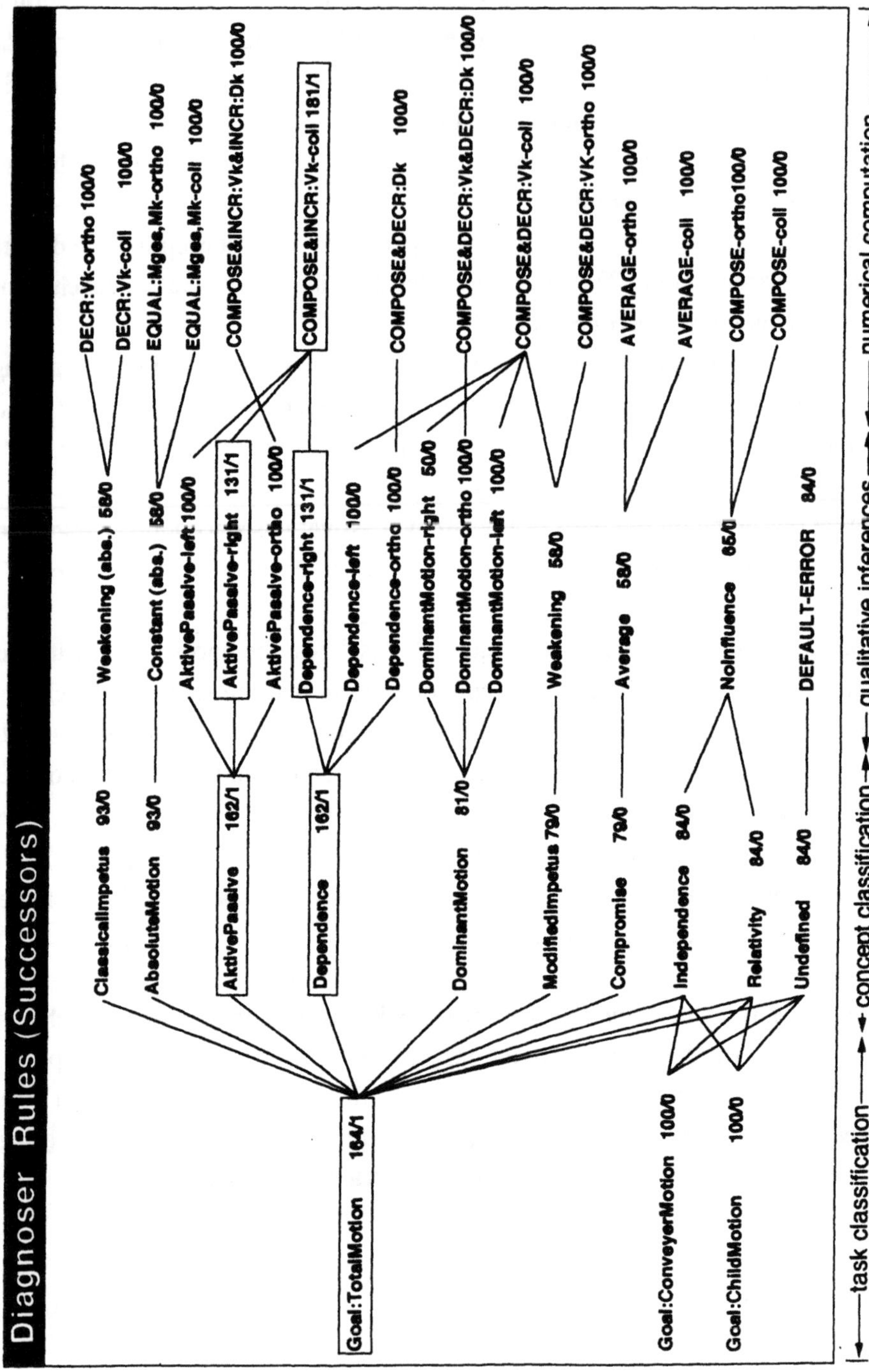

Figure 3. The diagnostic network. The boxed labels indicate potential solution paths.

structure results from the forward chaining of the rules. The nodes represent the rules, the edges the corresponding follow-up relations.

According to the rules' different functions in the diagnosis system, the diagnosis network has a hierarchical structure constituted by four layers: (1) task qualification, (2) concept classification, (3) qualitative inference, (4) quantitative computation. The rules of the first layer classify problems according to whether a composition or decomposition of motion is required. Rules of the second layer serve the purpose of identifying concepts and misconceptions. They are also central for the control of misconception-specific tutorial measures. Rules of the third layer represent qualitative reasoning with regard to the child's motion. The change in the child's motion relative to the physically correct motion is represented as effects of corresponding concepts and misconceptions. These qualitative rules have already been discussed. Rules of the fourth layer are defined as so-called effectors. In opposition to the rules of layer 1 to 3, these rules don't produce any output to enter as new input into the rule system. Effectors instead activate LISP-functions for the quantitative computation of results (positions).

Each new task presented to the learner contains input information for the rule system and leads to the reconstruction of the diagnosis network. Thereby the structure of the network is changed, the rules themselves, however, remain the same. Along with the reconstruction of the network, all possible solutions of a particular task - that may result when different concepts and misconceptions become effective - are calculated on line. The task conditions, the concepts and misconceptions, the qualitative inferences as well as the quantitative computations are inferred automatically by the system from a given task. The knowledge diagnosis is done by means of comparing the system-generated solutions with the learner's task solution; the concepts respectively the misconception underlying an individual learner's task solution are inferred on the basis of the rules used to represent qualitative inferences. This is done by tracing back the path of solution to the corresponding concept or misconception, starting from the effector that produced the same solution as the learner.

Modelling aspects of knowledge acquisition

One central goal of learner modelling was to indicate processes of restructuring, consolidation and generalization. For every task the values (positions) that result under the constraints of the misconceptions and the correct physical concepts are generated. These values are filed in a table, for not having to calculate them anew with every call of the task. If the calculated solution happens to coincide with the learner's present solution, the solution path is traced back, starting from the effector towards the specific misconception. The example of the marked paths in Figure 2 shows that the identification of a misconception cannot always be attained unambiguously. This problem of conflict resolution will be addressed later on. In the following we will go into more detail with respect to the representation of aspects of individual knowledge acquisition.

Departing from the identified paths of solution, the learner's behavior is represented in the network by two standard learning algorithms: strengthening and discrimination of rules.

Modelling with strength values

In order to model strength respectively "fragility" of relevant concepts and misconceptions the paths of solution of the diagnosis network which lead to a solution corresponding to the learner's response are strengthened. Besides, the system notes how often a particular rule lies on a solution path with a final solution equal to the learner's response. Contrary (parallel) to this counter, which is a continuously growing number, the strength value is diminished in the course of time, if a rule is not located in a successful path any more. Changes in strength values are used to model knowledge acquisition in the sense of a restructuring of incorrect knowledge towards a physically adequate representation as well as in the sense of knowledge consolidation.

If the strength value of a rule representing a misconception exceeds a predefined value, tutorial hints and simulations ensue which are specific for this misconception.
The implemented strengthening algorithm is a variation of the bucket brigade algorithm common for classifier systems. The strength values don't beat through the complete path, but are only passed up step by step so that the threshold value of the proper misconception rule in the second layer is only exceeded after multiple strengthening [see 6].

Modelling by means of discrimination of rules

Besides the strengthening of rules, a discrimination algorithm is used to model knowledge acquisition. The discrimination algorithm is assigned to register whether the learner's performance is varying with the tasks' different "surface features". A diagnosed variation indicates, that the learner has not grasped the general principles which allow task solutions in a uniform manner. However, only when the learner's problem solving behavior tends to be uniform with successive tasks with different surface features, a generalization process may be inferred. A generalization process seems to be likely when, at the beginning of a sequence of problem solving tasks, discrimination is needed to model problem solving behavior, whereas at the end of the sequence a consistent problem solving pattern needs to be observed.

The diagnosis system starts with generally defined rules, which are specialized as soon as the learner's behavior is changing. An example for a general misconception rule is:

```
(CLASSIFIER Diagnoser AbsoluteMotion
        (IF((Level: 0   Goal: TotalMotion))
        (THEN(Level: 1   Misconception: AbsoluteMotion)))
```

This rule expresses the hypothesis that for all tasks of calculation of the complete motion the misconception "AbsoluteMotion" is employed.

Let us suppose a learner is solving some tasks on collinear motions correctly, but then makes a mistake, which is typical for the "AbsoluteMotion" misconception, at the first orthogonal task. The system is then assuming that the learner is making this mistake only with orthogonal tasks. The system therefore generates a new rule by specifying the corresponding misconception rule with regard to the task feature "Direction: ortho".

The following new rule is created and added to the existing set of rules:

```
(CLASSIFIER Diagnoser AbsoluteMotion001
        (IF((Level: 0   Goal: TotalMotion  Direction: ortho))
        (THEN(Level: 1   Misconception: AbsoluteMotion)))
```

The rule indicates that the misconception "AbsoluteMotion" is used in situations with perpendicular motions. This new rule can be evaluated just as the other rules by the strengthening mechanism. Rules created by specialization thereby are strengthened with preference, so that the modelling of the learner's behavior in the diagnosis network is done as specific as possible [8].

Conflict resolution

As already described, different paths often lead to the same solution. Because of this reason it isn't always possible to identify the learner's misconceptions unambiguously. Besides, other mistakes have to be expected, because the diagnosis component only covers one category of mistakes (misconceptions).

In case that different paths (i.e. different misconceptions) lead to the same solution, the diagnosis component selects such tasks that can distinguish best between the potential misconceptions. For this purpose a table is elaborated by means of the diagnosis network, in which the resulting solution is calculated and registered for every task and every misconception. The following figure again shows the possible incorrect solutions for a task (Figure 4).

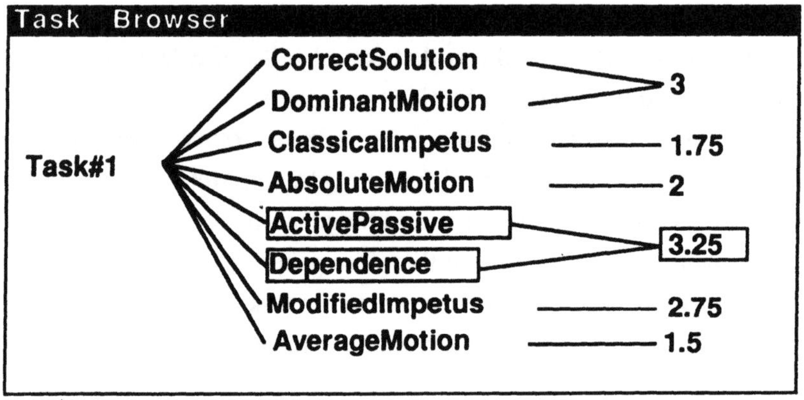

Figure 4. Two misconceptions leading to the same result. The numbers 1, 2, 3 ... correspond to the solutions A, B, C ....

In a case of conflict, a task is selected as successor problem, which is more appropriate than the starting task to discriminate between the possible misconceptions (Figure 5).

Figure 5 shows that the selected successor task leads to different solutions with the "ActivePassive" and "Dependence" misconceptions. It is therefore more suitable for the diagnosis of these misconceptions than the original task.

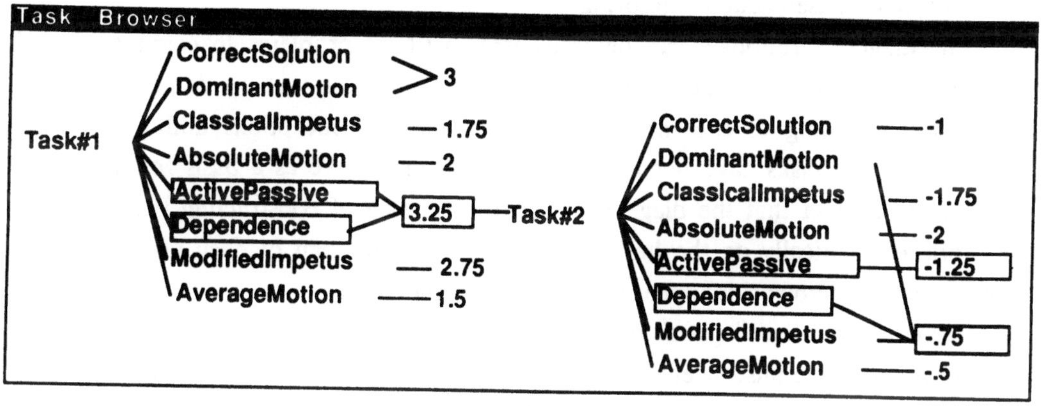

Figure 5. Conflict resolution by using a discriminating second task. The negative numbers -1, -2, -3 correspond to the solutions Z, Y, X.

From the list of given tasks always the first task suitable to discriminate best between the potential misconceptions is selected. Through the directed task selection the different development of strength values is accelerated, thus, making the diagnosis more efficient.

Tutoring based on a diagnosed mistake only ensues when a misconception is identified and the same mistake has shown up again in consecutive tasks. The described procedure is contrary to the principle of immediate informative feedback after appearing mistakes [see 2, 3]. The decision to first identify a misconception relative clearly before tutoring is started, was taken for reasons of improving the system's diagnostic validity.

Final Remarks

The developed diagnosis component offers appropriate opportunities to be successfully applied to the diagnosis of misconceptions in computer-based learning environments. Within the described learning environment on superposition of motion single misconceptions can be diagnosed with sufficient certainty, thus, rendering possible misconception-based adaptive tutoring.

The diagnosis rationale can be transferred without problems to computer-based learning environments in the sense of microworlds, provided that the domain has been structural according to the inherent concepts and misconceptions and appropriate problem solving tasks have been constructed that may serve as a basis for diagnosis. Especially the fact that the diagnosis system is capable of constructing the diagnosis network automatically on the basis of the problem statement of each task makes its application in microworlds appear especially convenient.

Restrictions in the diagnosis of misconceptions result from the fact that so far only the diagnosis of one misconception per task is possible. This may be sufficient with most of the tasks. There are, however, some tasks in which several misconceptions appear in combination. As to the diagnosis network, an explosion of possibly relevant rules would result when combinations of all misconceptions would be considered. One way of avoiding this problem would be the assignment of tutorial help to heal one of the envolved misconceptions. By this, the combination of different types of errors would be disintegrated and single misconceptions could be identified.

Further constraints result from the fact that incorrect solutions concerning problem solving tasks on superposition of motion can't be attributed to misconceptions exclusively. Incorrect solutions may appear due to unsystematic arithmetical errors in the quantitative determination of the solution (e.g. counting errors). With the system described mistakes in trying to find a solution in a purely (geo-)metrical way aren't diagnosable either. Mistakes of this kind were not of interest in the development of the diagnosis component and were therefore not taken into account.

References

1. Aguirre, J., & Ericksen, G. (1984). Student's conceptions about the vector characteristics of three physical concepts. *Journal of Research in Science Teaching, 21*, 439-457.
2. Anderson, J.R., & Reiser, B.J. (1985). The LISP-tutor. *Byte*, 159-175.
3. Anderson, J.R., Farrell, R., & Sauers, R. (1984). Learning to program in LISP. *Cognitive Science, 8*, 87.
4. Becker, L.A. (1988). Computer-aided misconception-based tutoring and exercise generation. Programmed Learning and Educational Technology. *Journal of the Association for Educational & Training Technology, 25*, 67-75.
5. Forbus, K.D., & Gentner, D. (1986). Learning physical domains: Toward a theoretical framework. In R.S. Michalski, J.G. Carbonell & T.M. Mitchell (Eds.), *Machine Learning. An Artificial Intelligence Approach, Vol. II* (pp. 311-348). Los Altos, CA: Morgan Kaufmann.

6. Holland, J.H. (1986). Escaping brittleness: The possibilities of general machine learning algorithms applied to parallel rule-based systems. In R.S. Michalski, J.G. Carbonell & T.M. Mitchell (Eds.), *Machine learning: An artificial intelligence approach, Vol. II* (pp. 593-623). Los Altos, CA: Morgan Kaufmann.
7. Holland, J.H., Holyoak, K.J., Nisbett, R.E. & Thagard, P.R. (1986). *Induction: Processes of inferences, learning, and discovery*. Cambridge, MA: MIT Press.
8. Langley, P. (1987). A general theory of discrimination learning. In D. Klahr, P. Langley & R. Neches (Eds.), *Production system models of learning and development* (pp. 99-161). Cambridge, MA: MIT Press.
9. Rumelhart, D.E., & Norman, D.A. (1978). Accretion, tuning and restructuring: Three modes of learning. In J.W. Cotton & R.L. Klatzky (Eds.), *Semantic factors in cognition* (pp. 37-53). Hillsdale, N.J.: Erlbaum.
10. Vosniadou, S., & Orthony, A. (1989). Similarity and analogical reasoning: a synthesis. In S. Vosniadou & A. Orthony (Eds.), *Similarity and analogical reasoning* (pp. 1-20). Cambridge: Cambridge University Press.

The Provision of Tutorial Support for Learning with Computer-Based Simulations

Peter Goodyear

Department of Educational Research, University of Lancaster, Bailrigg, Lancaster, LA1 4YL, England.

Abstract: This chapter considers a number of key issues concerned with supporting exploratory simulation-based learning through the provision of appropriate tutorial interventions. It focuses on sources of pedagogical knowledge that have the capacity to inform real time decision-making in such contexts. It looks to both empirical studies of learning, and research on the action and thinking of teachers, as potential sources of practical pedagogical knowledge. The chapter has implications for both Intelligent Tutoring Systems (ITS) research and for teacher education.

Keywords: Simulations, Intelligent Tutoring Systems, pedagogical knowledge

Introduction

Other than in distance education contexts, the environment for computer-based learning (CBL) typically involves a threefold interaction: between a computer, one or more learners and a teacher. The role of the teacher can be quite complex, including such activities as:

- selecting appropriate software
- planning its integration with other learning activities
- watching over the learners' use of a program
- judging when and how to make interventions during use of the program
- using the learners' activity at the computer as a window onto their thinking and cognitive development
- summarising, and helping learners reflect on, their experiences using the program
- arbitrating disputes and managing the allocation of computer time between learners

How teachers carry out this role, and how teachers can be helped to *learn* to carry out this role, are topics which have received limited attention in educational computing research. This chapter reports on some work which is primarily concerned with the activity of providing real-time tutorial support for learners engaged in "exploratory" CBL.[1] More specifically, it is concerned with tutorial support for simulation-based learning. It is part of an on-going programme of research which is attempting to understand the nature of tutorial action, and the knowledge on which it draws, for the purposes of both teacher education and intelligent tutoring systems design [13].

This programme of research has a central organising question: what knowledge enables an intelligent teaching agent to make good pedagogical decisions? Some unpacking of assumptions buried in this question is necessary.

Firstly, "intelligent teaching agent" is used as a term sufficiently general to include human tutors (both professional teachers and many others) and also those parts of artificial knowledge-based tutoring systems whose function is pedagogical decision-making. I believe there can be merit in working at such a level of abstraction. I use the terms "teaching", "tutoring", "pedagogical" without any restrictive connotation - in the sense of implying particular forms of educational action such as "knowledge communication" [32].[2] The domain of these terms is all forms of action whose general goal is the facilitation of learning.

Second, decision-making is seen as a key object of enquiry in the understanding of action. I do not restrict "decision-making" to the sense of *considered* choices between alternative lines of action [5], but include also those forms of choice to which little conscious attention is paid and whose flavour may best be captured in production system representations or through phrases like "knowledge-in-action" [22]. The explication of the tacit knowledge embedded in action becomes a key task.

Thirdly, I believe that pedagogical decisions can be evaluated, though such evaluation is highly problematic.[3]

As well as the *forms* of pedagogical knowledge, its *content* and *origins* are of considerable interest. In agreement with Self [27], I see some value in distinguishing between (relatively stable) pedagogical knowledge and (relatively dynamic) knowledge of a *particular* learner, or

---

1 CBL wherein the learner has a relatively high level of control over the distance and direction in which they travel.

2 I am labouring to avoid the assumption that "tutoring", for example, is essentially a strongly and narrowly didactic or instructing activity.

3 And operates on a number of levels - there can be severe methodological problems in identifying good pedagogical decision-makers, on one criterion, and assuming their decision-making is sound in other contexts [3, 19]

set of learners, with whom the tutor is currently interacting. *Excluded* from "pedagogical knowledge" are

- a representation of the current learner(s) - beliefs about their knowledge, misconceptions, learning preferences and other educationally-relevant attributes
- a representation of the on-going interaction with the learner(s)

though I would very definitely *include* the knowledge that supports the construction and maintenance of these representations and the "case knowledge" into which significant aspects of the tutor's interaction experiences are encoded.

Unlike Self, and many other ITS developers, I am less sure about the separation of "domain" and "pedagogical" knowledge and am convinced of the importance of "pedagogical content knowledge" [33]. By this I mean (a) important in the knowledge base of an effective teaching agent are representations of subject matter specialised for pedagogical usage, (b) representations of subject matter entirely disconnected from pedagogical usages are probably rare and rarely called on, but (c) there is such a thing as domain independent pedagogical knowledge.[4]

In this chapter, the focus is squarely on the origins or sources of pedagogical knowledge. From where can an intelligent teaching agent derive knowledge which will support the making of good pedagogical decisions when the agent's goals are concerned with supporting simulation-based learning? I examine two sources: research on learning and research on teaching.

Research on Learning with Computer Simulations

Learning with computer-based simulations can be thought of as a subset of exploratory learning, which is, in turn, a subset of all learning. What holds for "all learning" or even "learning from instruction" [29] or "instructionless learning" [28] may hold for simulation-based learning. However, as we shall see, the knowledge engineering problem is often one of specificity of application. Hence, the more circumscribed studies provide an adequate test. (If one can't make the more specific studies applicable, one has even less chance with the more general studies.) My current purpose is to examine one influential study of simulation-based learning in order to see what pedagogical knowledge might be derived from it. It is not

4 Though see the comments on the slipperyness of domain definitions, below

necessary to provide a comprehensive survey of the literature in this area.[5] Instead, we must focus on the form of statements contained in reports of empirical educational research. The adequacy of this research-based knowledge as a source of practical pedagogical knowledge for an intelligent teaching agent is the primary concern.

A Representative Study: Rivers & Vockell (1987)

Rivers and Vockell [26] carried out three linked enquiries working with high school biology students. The students found themselves in one of three groups. A control group used no computer simulations in their course. A "guided discovery" group used computer simulations with some supporting text that contained advice on the productive use of each program. An "unguided discovery" group used the simulations but without the advice.

An example set of hints for guidance, referring in this case to a predator:prey simulation, is given in the report (p407):

a) It is a good idea to change only one variable at a time
b) The population of the predator and the prey are inter-related
c) A change in one generally has an effect on the other
d) If the deer population becomes too large it will affect the carrying capacity of the environment. Eventually their food will run out and they will die off. (Carrying capacity is the number of deer the environment can support without having the food run out.)
e) The denser the vegetation of the environment, the easier it is for the wolves to capture the deer
f) Each wolf has to find and eat about 25 deer per year to survive
g) Use this program to test your hypotheses. Look for patterns or relationships between populations as you systematically change the variables.[6]

The guidance, in this case, consists of a mixture of concept definitions, propositions about relationships between entities in the model and suggestions of a more strategic nature.

The authors' summary of their research includes the following as a key finding

5 de Jong [17] and Goodyear & Tait [14] offer reviews
6 These statements were presented by the computer program which also ran the simulation

> "...the students using the simulations often surpassed the control students on the pretests for subsequent units, on tests measuring scientific thought processes, and on a test of critical thinking. In most cases the students using the Guided version of the simulations developed these generalized skills more effectively than those using an Unguided version." (p413).

The researchers' outcome tests included customised pre- and post-tests, the BSCS Process of Science Test, the Test of Integrated Process Skills, and the Watson-Glaser Critical Thinking Appraisal.

How may practical pedagogical knowledge be derived from such a study? From a reading of the study, one can begin to engineer some pedagogical heuristics, which might be of the following kind:

IF you're tutoring the learner's use of a simulation program
AND you have a goal of increasing the learner's scientific problem-solving skills

THEN supply them with some statements about (a) the definition of terms used in the simulation, (b) relationships in the model, (c) strategies for orderly hypothesis testing.

To go further in such an engineering task, and we need to go a lot further if we are to support pedagogical decision-making in actuality, we need to analyse and re-specify such statements. The report of the empirical study helps us with some, but only some, of this work.

It tells us, for example, how "scientific problem-solving skills" were operationally defined and tested. We need to decide whether or not, or in what circumstances, we are prepared to agree with such definitions. It may be, for example, that we believe it important to distinguish between espoused and enacted processes of scientific enquiry before defining desirable behaviours for our students to emulate [18]. Or it may be that we want to choose a different trade-off between the acquisition of content knowledge and the acquisition of generalised scientific reasoning skills. In either case, to the extent that our instructional goals differ from those implicit in the study, direct extraction of pedagogical heuristics will be unhelpful.

Of course, the report gives more detail than is contained in the summary statement of key findings. We learn, for example, that only some of the simulation programs used in the study produced the pattern of findings reported. We learn that a change in forms of the Watson-Glaser test used in the study affected some of the results. We learn that some sub-tests of the Watson-Glaser showed no significant differences between groups. Our pedagogical heuristic would hold for "Interpretation" sub-skills (the ability to weigh evidence) but not, it seems, for

"Deduction" (the ability to determine whether specified conclusions necessarily follow from information in given statements or premises) or "Recognition of Assumptions" or "Evaluation of Arguments".

We also need to make decisions about the representativeness we are prepared to accord to the sample. (Should our rule's condition side contain clauses about the age, nationality, ethnic background and socio-economic status of the learner? Does the rule apply only in Biology subject domains? Or sciences generally? Or only in formal education contexts? Or, for reasons bound up with shifting curricula and teaching styles, only to mid-west American high-school students in the early 1980's?)

The study gives us little direct help with refinement of the action side of the rule. What mix of guiding statement types is helpful? What propositions about relationships in the model should be given and which left for discovery? What terms need clarification? Or if the need for clarification is to be determined dynamically, how do we detect the need and detect its satisfaction? Are the strategic pointers (such as statements a and g) valid, sufficient and necessary? Do they need exemplification? How? What constitutes a pattern? What constitutes systematicity? *Is it* always a good idea only to change one variable at a time?

I do not mean to single out this study for special criticism; it is representative of many studies, perhaps of the dominant paradigm in educational research. My intention is to highlight the difficulty of deriving practical pedagogical knowledge from such empirical studies. It *is* possible, by various forms of argument, to derive pedagogical guidelines. The authors themselves attempt this:

"The previous paragraphs indicate that in order to do a thorough job of teaching problem-solving or thinking skills, the programs described in this report are helpful. However, the programs do not cover the entire spectrum of appropriate thinking skills. To teach generalized problem-solving skills effectively, it would be necessary to either (a) restructure the simulations to focus more specifically on the specific skills (such as deduction) ... or (b) combine the programs with other experiences which address these additional skills." (p412).

Analysis of the process by which such guidelines are derivable reveals, perhaps unsurprisingly, the necessity for some underpinning implicit, informal, pedagogical theory. If we look at either (a) how authors derive pedagogical guidelines within their articles, or (b) how teachers, teacher educators or instructional designers derive pedagogical guidelines or build pedagogical knowledge from their reading of such articles, we realise the impossibility of effecting such derivations without the use of some underpinning pedagogical theory.[7] The general problem of how to derive pedagogical knowledge, for use by an intelligent teaching agent, from the educational research literature, thereby hinges upon the explication of the largely tacit pedagogical knowledge of the agent doing the derivation. One cannot see the general problem as merely a recoding of propositions, such as from the quasi-natural language of a research report into a Prolog clause in an Intelligent Tutoring System. That would obfuscate the mediating role of the informal, implicit pedagogical theory in the mind of the system creator.[8]

It is a mistake to see the educational research literature on empirical studies of learning from instruction as a source of law-like statements whose encoding is sufficient to create a knowledge-base that will support pedagogical decision-making in the real world.[9]

I shall now move to consider an alternative source of pedagogical knowledge: studies of teaching.

Research on Teaching

Notwithstanding the limitations of empirical educational research, intelligent teaching agents exist. If our primary goal is increase the numbers of such agents, and/or to extend their powers, we would do well to study their action and the knowledge on which it draws. This assumes that representations of their action and knowledge can be made available to other agents, and can be internalised by other agents, for the purpose of improving their practice.

There are many problems to be tackled in understanding a teaching performance or endeavouring to elicit the teaching knowledge implicated in it [6, 11, 12, 16]. Knowledge engineering in *formal* domains is a non-trivial task. In the fuzzy domains of professional practice, conventional elicitation methods are further strained, or prove inadequate. Representational problems abound. If our goal is to encode pedagogical knowledge in an ITS to support simulation-based learning, for example, the knowledge representation formalisms of contemporary AI are terribly restrictive [13]. But that is not an argument for abandoning the enterprise. Rather, we need to recognise the utility of early explorations in pedagogic

7 I would stress again the informality and largely tacit nature of such personal theory, which might best be thought of as a loosely connected set of pedagogical beliefs.

8 An example drawn from instructional design rather than ITS development, is [24]

9 Some ITS researchers have too simple a view of this problem, saying only that the descriptions provided by educational research are too coarse-grained or informally phrased. While this may be true, the key difficulty resides in the *interpretive* work needed to transform knowledge about learning into practical pedagogical knowledge.

knowledge engineering, through which greater and more specific demands can be made on the tools and techniques of cognitive modelling.

So what *may* be learned from studies of teaching? Clearly the perceived value of what may be obtained, in terms of knowledge which can inform practice, depends in part on the nature of the intelligent agent for whom/which the knowledge is intended. I offer two examples in the remainder of this chapter: the first with implications to be drawn out in teacher education, the second an empirical exercise of our own with implications for ITS design.

A useful study in the area of teacher thinking was carried out by Deborah Blank for her PhD thesis [4]. Blank looked at one-to-one tutoring in a context where learners were engaged in individual work with a computer simulation. 15 teacher:learner pairings were observed and videotaped and the videos were used in a stimulated recall session immediately after the lesson. The focus was on when, how and why tutors chose to make interventions. Beliefs about characteristics of the individual learners accounted for significant amounts of variation in the action of any particular tutor. Variation between tutors was explained in terms of differing beliefs about teaching and learning (differences in informal pedagogical theory). Such differences included: beliefs about the tutor's role in supporting simulation-based learning, beliefs about learners' cognitive and affective characteristics, beliefs about the value of feedback, reflection and self-evaluation, and feelings of self-efficacy. There was some consistency in teaching action taken across teaching sessions, but only at a gross level of description. Questioning was the major form of tutorial intervention - usually with the purpose of checking on the learner's understanding or of encouraging them to reason.

This summary is meant merely to indicate the kinds of statements about practice, and its underpinning knowledge and beliefs, that can be generated through studies such as Blank's. The summary itself is not enough to constitute practical knowledge and conceivably neither is the text of Blank's thesis. The key task to be addressed concerns the forms of representation that can make the bridge between the practice and thought of Blank's subjects and the growth of practical knowledge in other intelligent teaching agents. I would argue that this is a fundamentally different, but more tractable, exercise than might be involved in re-rendering accounts of student learning in order to derive prescriptions for tutorial action.

An Experimental Study

The second study to be reported here took place within the EC DELTA SAFE/SIMULATE project [17]. The goal of SIMULATE, in its current phase, is to deliver a set of requirements and specifications for an authoring workbench. This workbench is specialised for the creation of ISLEs (Intelligent Simulation Learning Environments). An ISLE can be thought of as a computer-based simulation set in a supportive, intelligently adaptive instructional environment. SIMULATE presupposes certain important attributes of simulation-based learning: emphasising its exploratory character, the necessity for active learning and a learner with a high degree of autonomy. It also presupposes that certain kinds of instructional intervention, particularly those which are strongly directive, or primarily concerned with information transmission, will be absent or rare [14, 17].

The goal of this study was to access some of the knowledge and decision-making processes of human tutors engaged in supporting a learner using a simulation program. Given the location of the study, within the SIMULATE project, our primary purpose was to derive some information which would guide the design of ITSs specialised for supporting simulation-based learning, rather than information for feeding into teacher development programmes. Nevertheless, there is scope to generalise some of our findings for such purposes.

We focussed on a simple, quantitative dynamic-systems type simulation in which possession and/or acquisition of specialist domain knowledge was not a salient issue. We wanted to avoid the generation of lengthy student:tutor protocols which were primarily *about* specific subject matters. As far as possible, we wanted to focus the dialogue on topics with more general implications. Strategies for exploring a complex problem-space, or for formulating hypotheses and experiments, are good examples.

A detailed understanding of the actual simulation is not needed to evaluate the account presented below. However, the reader may find it helpful to know that:

a) the model was of selected factors influencing the length of a queue of student hitch-hikers intent on travelling between Lancaster University and the town of Lancaster

b) the model was relatively simple in its structure, containing three independent variables influencing the length of the queue and one negative feedback loop

c) the model was implemented in Stella on the Apple Macintosh [30]. An image of the "Diagram" window is given as Figure 1. This shows the structural view of the model which was available to the learner.[10] Other views, primarily tabular or graphical outputs from running the model, were also available. These were the learner's main method of inspecting quantitative relationships in the model. Figure 2 gives an example graph. The learner has a high level of control over the definition of graphs, tables etc. and can store output from successive runs for later comparison.

10 Further details of the simulation model are given in [15].

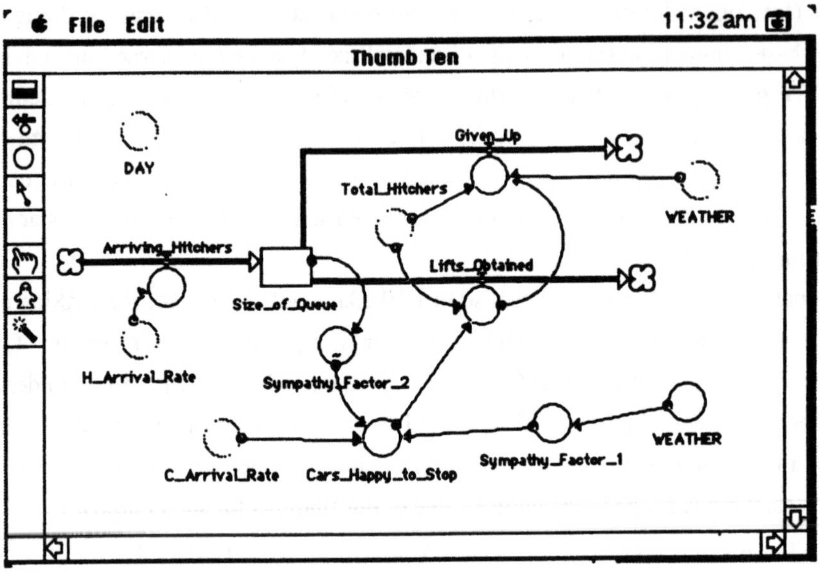

Figure 1. Diagram view of the structure of the model.

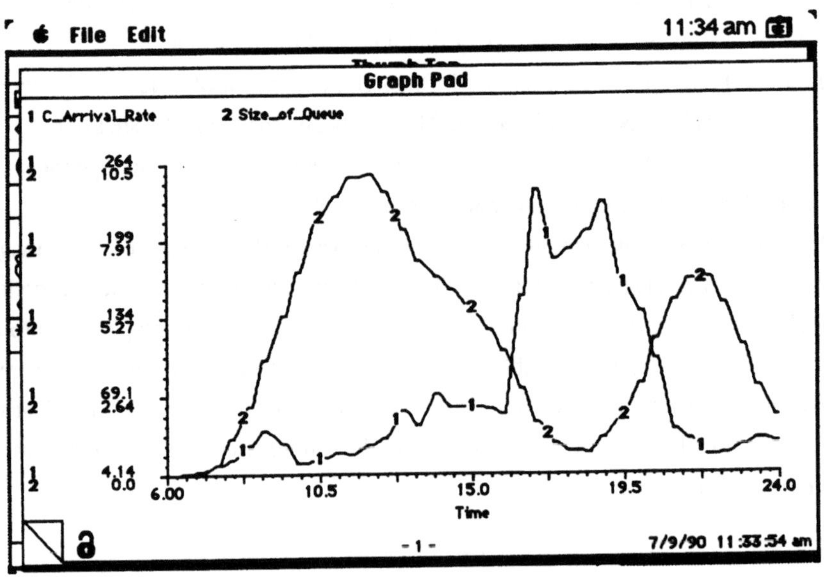

Figure 2. Sample of graphical output.

Eliciting Pedagogical Knowledge

In classical studies of problem-solving within the information-processing paradigm [21], researchers usually have the advantage of possessing a privileged view of the problem-space. They have a detailed representation of problem states, goals and operators on which to map the behaviour (including the utterances) of their study subjects. Analysis of verbal protocols gains considerable leverage from possession of this representation. In most knowledge engineering studies, of the kind used in expert systems development, this privileged view of the problem space is not available. Knowledge engineers typically start out with faint and fragmentary maps of the domain on which to locate their expert's problem-solving behaviour. Understanding teaching action and eliciting pedagogical knowledge is closer to the second than the first situation.[11]

In this study, we used 11 paid volunteer subjects to act as learners exploring the Hitch-Hiker simulation. Detailed analysis of their behaviour can be found in [15]. After some structured practice with Stella the subjects were given a simple briefing, intended to stimulate their exploration of the model. Formulation of an appropriate briefing was surprisingly difficult. In pilot studies, we found that those subjects who were given no guidance, or only very general guidance, about goals for their exploration soon demanded it of the experimenters. If suggestions were too specific, the subject turned the exercise into a look-up task - a qualitatively different experience from model exploration. The briefing we eventually settled on (i) offered some sample questions about relationships in the model, (ii) said we would be asking the learner some *similar kinds* of questions after their exploration of the model and that we would be asking them for a description, in their own words, of the general pattern of relationships in the model. This sufficed, with the 11 subjects in the main study, to promote a relatively purposeful but still exploratory interaction with the simulation.[12] Members of the research team acted as tutors. For part of the time, each subject explored the simulation without tutorial support.

We adopted a number of complementary techniques for observing tutors' problem-solving behaviour and accessing their decision-making [12, 13]. The most fruitful centered on videotaping tutor:learner:computer interactions and analysing either or both concurrent and retrospective protocols. Full details are available in [15] but a brief rationale for the approach should be provided here. A small number of tutor-subjects was used. This might raise

11 James Voss's work on problem-solving in social science shares some of these difficulties [31]

12 The difficulty entailed in providing an appropriate degree of structure for the learning task is worthy of mention because it has such strong implications for the provision of tutorial guidance in exploratory learning situations.

questions of "representativeness". (In what sense are we sampling a population of tutors? How idiosyncratic was the action of our tutor-subjects? How do we know their action was efficacious?) At this juncture, we believe such questions are beside the point. Given the limitations of the state of the art in ITS construction, we can learn a great deal by problematising the performance of almost any passage of teaching action.

We have experimented with placing paired tutor-subjects in a separate room, mainly so that we might elicit concurrent verbal protocols but also to mimic some of the restricted bandwidth of communication between a learner and an ITS (using telephone or computer-mediated links, see [12]). Concurrent protocols are to be preferred over retrospective in so far as the latter are subject to greater transformation through memory effects, post-hoc rationalisation, etc. Paired subjects are preferred in order to provide natural grounds for verbalisation [20]. Strategic interruption can be used to provoke a break-down in the problem-solving process or to allow interviewing methods to unpick the bases of some observed action. The former is useful in that it can allow the subjects introspective access to processes which are otherwise routinized [7]. And while interviewing methods carry the risks of other high-inference techniques, they do allow some access to invaluable accounts which are otherwise unobtainable [11-13].

While this "Wizard-of-Oz" methodology can provide access to certain kinds of cognitive processing and tutorial behaviour that are difficult to study by other means, the nature of the tutor:learner interaction is highly artificial. Furthermore, in the case of computer-mediated communication, the tutor:learner interaction cost is quite high. That is, the cost of making a supporting move - the time it takes to type in a comment or the interruption it causes in the flow of the learner:computer interaction - may frequently dissuade the tutor from acting (though not necessarily from saying something about an intention to act). Consequently, we have drawn heavily on the more highly interactive tutoring that goes on in face-to-face situations, even though this is significantly more difficult to analyse.[13]

Transcripts of selected tutoring segments are available in [15]. Here, I shall attempt to summarise some of the salient issues emerging from the study.

a) A crucial feature was *the role of interaction.* This may seem a commonplace observation, given that we were observing interactive tutoring. The difficulty of studying teachers' thought processes in interactive tutoring means it is a relatively unexplored area (compared with teachers' planning, for example). Where interactive teaching has been studied, it has usually been in *class* teaching situations, where the forms of interaction are relatively impoverished in comparison to the interaction that can be achieved in one-to-one or small group

---

13 Due to both the unavailability of concurrent think aloud data (as opposed to tutorial utterances) and to the added richness of the tutorial interaction (use of gesture etc).

tutoring. Findings from class teaching studies have rarely been able to demonstrate significant "on-line" decision-making [5]. It may be thought that interactive tutoring is cognitively more demanding than "off-line" teaching tasks. It may be so, but the danger is to ignore the fact that interactivity is both a demand on *and a resource for* the tutor's action. Our tutor-subjects did not engage much in detailed tutorial *planning*. Their activity cannot easily be modelled as a purposeful search through a well-defined problem space with the object of attaining a well-defined instructional goal. And to say this is not a *criticism* of their action. The tutoring which they exemplified is unlike a chess game [21]. The tutor is unlike a grandmaster who has a single objective and a clear understanding of goals and operators, facing an opponent whose knowledge of the pieces, moves and purposes of the game is at best rudimentary. In actuality, both parties to the educational interaction engage considerable resources in negotiating and clarifying the nature and purposes of the activity *while it is in progress.*

This resonates with recent studies of tutoring by Sarah Douglas [10], who allocates a key role to processes of failure detection and repair. Knowing that one's action as a tutor or a learner will be at best partially understood, and knowing that there will be further opportunities for clarification, elaboration or rejection of one's action - these things are crucial in determining the character of interactive teaching.

A good example concerns the asking of "open questions". Open questions occur quite frequently in the transcripts. They serve a number of purposes, but a common feature is their ability to provide an appropriate level of constraint on the space of answers that a learner may consider. A successful open question is precise enough to cause the learner to attend to a usefully small subset of all possible answers, without determining precisely the answer that they will give. Failure to ask effective open questions (indeed, *any* open questions) is a frequent criticism of the performance of classroom teachers. Not much is known about successful methods for generating open questions. But what seems clear from our study is that they are significantly easier to generate in interactive situations, where the participants can co-operate in detecting and repairing communication failures. The successful formulation of open questions, or "non-directive guidance" [17] more generally, depends on having a good model of the learner - especially of their beliefs about the nature, purpose and current focus of the task. Open questions fail to the extent that the tutor has a flawed model of the learner. In class teaching situations, the teacher must work with a general or collective model of the learners listening to them. Those learners who differ significantly from the general model will find it harder to interpret the open question. Moreover, they do not usually have much opportunity to clarify their understanding of the teacher's intent through further interaction.

The situation is different in one-to-one tutoring. Here, the tutor is able to work with a mental model of the *actual* learner with whom they are interacting, so the probability of model-based failure is lower [10]. Moreover, the learner can readily initiate a process of clarification or refinement of the meaning of the tutor's utterance. Knowing this, the tutor can be less painstaking in formulating their initial question, since the learner can initiate a conversational repair. We can think of this as a situation in which *interaction cost* is an important influence on the behaviour of intelligent agents [15]. [14] Because the interaction cost is low, the tutor will more readily make ambiguous utterances or perform ambiguous actions. Ambiguity is a necessary attribute of "non-directive" tutorial actions (such as the asking of an open question). In interactive tutoring, the possibility of (a) having a mental model of the actual learner with whom one is interacting and (b) repairing failures in the tutorial dialogue, mean that the use of ambiguous actions in supporting exploratory learning can be a low-risk strategy.

One implication of this, for the design of certain classes of ITS, may be that the knowledge necessary to tutor is not so much knowledge that supports the derivation and elaboration of teaching plans (from goals and pedagogical principles), so much as the knowledge necessary to establish, through interaction, a shared, evolving understanding of the tasks at hand. Some consequences of this suggestion, for further research, can be found at the end of the chapter.

b) A second issue concerns the *salience of domain knowledge* as a constituent of the tutor's expertise and base for action. Very few of the tutors' utterances drew on any special knowledge of the domain. Utterances may have had domain content, but often only as a necessary part of identifying variables, for example. Of course, the accuracy of this conclusion depends on how one defines "domain content". One could define the domain as being "about hitch-hiking". Or one could define it as "about systems and modelling concepts, such as feedback, equilibrium, causality, etc". Under neither of these definitions could the tutors' transcripts be said to contain many references to domain knowledge. But one could also define the domain so that it includes the higher level knowledge acquisition or scientific research skills that play a key part in productive exploration of a simulation [17]. These *were* objects of concern to the tutors. If we choose to define the domain in narrow terms (knowledge about hitch-hiking and the hitch-hiking model), then we can say that relatively domain-free knowledge played an important part in the tutoring dialogues. Since such knowledge should be of quite general applicability, this is an encouraging finding for potential ISLE builders.

---

14 Imagine a situation in which an intelligent agent has two methods by which they can acquire a piece of information. An example might be trying to remember the correct syntax for a rarely used command during the use of a complex computer program. The agent can either (a) expend some cognitive effort trying to recall the correct syntax or (b) try what they believe to be the correct command and see whether they get an error message or the result they intended. In *some* situations, the second is the more efficient strategy. We can say that such situations are ones with a "low interaction cost". Situations with low interaction cost are those in which it is more efficient to try some action (and see what its consequences are) than to try to remember or reason one's way to a satisfactory solution. The interaction need not be human:computer but can be human:human.

The question of how far one can get, in tutoring without specialist knowledge of the domain, is an intriguing one. Cumming and Self, [8] for example, have proposed a separation of "conversation levels" in tutorial discourse; positing that useful tutoring can be done at either or both of "task" and "discussion" levels. Following Anzai and Simon [1] they see a learner as both *doing* some task, and *reflecting* on their task. Tutoring that focuses on the second of these is tutoring at the discussion level. The question Cumming and Self pose, and which we are now investigating, is whether (and to what extent) it is possible to decouple tutoring at the discussion and task levels.

c) A third observation arises from the existence of what appear to be two clear sets of problems for the learner to deal with. There are the substantive problems of learning about the modelled phenomena and there are problems concerned with learning and driving the Stella interface. Tutors did not demonstrate any difficulty in adopting, and maintaining in parallel, two different pedagogical styles. The approach in supporting the learner's exploration of the model was low-key with few interventions. They spent much of the time watching the learner's exploration, intervening in the exploration only when they felt they had constructed a strong warrant for doing so. In contrast, they saw instances of the learner's having trouble with the Stella interface as moments when quick intervention was easily justified. In short, they seemed happy to take a pro-active role where the learner's difficulties were essentially technical/operational ones, but content to take a back seat during extensive periods of model exploration. This accords with observations we have made of face-to-face tutoring of the acquisition of PC-Paint skills in an art class, and with earlier work on Logo tutoring strategies [13].

d) A fourth set of issues concern tutors' identification of problems - in particular, the evidence they use to detect moments when the learner may be in difficulties. Even with relatively experienced users of complex software packages like Stella, the user/learner may make quite frequent slips - in the sense of performing actions other than those intended - through inattention rather than because of a deep misconception. A common example is the "wrong mode" problem, where a command appropriate to one mode is given in another because the learner was inattentive to modality. This class of slip is often easily detected and rectified when a pair of learners or users are working together: a second pair of eyes looking over one's shoulder is useful. What makes the slip obvious to the observer is that it contradicts his expectation about the command or action set from which the user's next move will be drawn. A knowledge of the grammar of the interface is both essential to the observer/tutor and relatively easily provided as part of the knowledge-base of an ISLE or ITS.

Other kinds of difficulty are less straightforward to detect and decisions about appropriate tutorial action more complex to make. The notion of "case-based reasoning" seems applicable here. A significant number of tutoring decisions involved reference to analogous events or situations: both recent and otherwise. In the latter case, a pair of tutors may not have shared experience of the case(s) referred to but nevertheless one of them could provide a sketch description of case features which was (i) interpretable by the other tutor and (ii) usable as a basis for further shared reasoning. Recent work on representations of expertise in both research on teaching and AI [25] are suggestive of some interesting ways forward here.

In this brief summary I have attempted to give a flavour of some issues arising out of the analysis of just one set of teaching performances. Such accounts can be useful in themselves, as stimuli for considering basic design features in ITS research and development, for example. However, much is also to be gained from detailed modelling, through computational representation of such performances and of the knowledge on which they appear to draw. This will continue to be a difficult, problematic and expensive activity for some time to come. Nevertheless, focussing on the practical pedagogical knowledge which is *demonstrably* sufficient for the action of human tutors must be a more direct, and less expensive, route to the goal of ITS construction than one which begins by wandering through the fields of student learning.

Further Research

Some of the work on which we are now embarked is looking at the question of decoupling task and discussion levels in tutoring. We are working with tutor-subjects in a knowledge-elicitation situation similar to that described above, except that certain key kinds of domain-related information are removed from the tutors' view of the simulation. Instead of seeing "Weather" or "Length of Queue", the tutors see only variables labelled "x" or "y". Their knowledge and action are thereby focused on more domain-independent features of the problem-solving situation. Our expectation is that it will nevertheless be possible to elicit many potentially valuable and generally applicable pedagogical strategies.

A second line of research will move from looking at the analysis of *any* teaching performance to the analysis of teaching performances based on some special kinds of expertise. Rather than initiate an elaborate hunt, a la Leinhardt or Berliner [3, 19], for "expert teachers", we prefer observing the *growth* of expertise specific to this pedagogical task. We intend to engage a small number of tutors in a series of tutoring interactions, focusing on the development of their expertise. Our intuition is that such subjects will develop, try and evaluate methods, through engagement in real tutoring tasks, in a manner which will be at least as productive as attempts to derive pedagogical heuristics from alternative sources. And as useful by-products, we will gather data on processes of reflection and on forms for experientially-derived practical knowledge, that may be informative in the consideration of self-improving systems. We expect to be able to make linkages with other recent work on case-based pedagogical reasoning and reflection [9].

Conclusion

This chapter arises out of a concern for the nature of pedagogical knowledge which can be used by an intelligent teaching agent. It advances the notion that such knowledge should not be conceived as some kind of large, declarative knowledge base of propositions, drawn from educational research on learning, which an agent can use to deduce a basis for its next teaching action. Instead, it proposes that we pay greater attention to the performances of existing teaching agents to try to understand the forms of knowledge which enable their action. There are some indications that (i) the relatively generic knowledge used to drive dialogue [2, 23] and (ii) forms of case-knowledge encoding and abstracting from experience of learners and learning interactions, may be of considerable importance in supporting tutorial action. At this stage in the development of ITS research, it may be more important to get such architectural foundations right than to worry about demonstrations of the utility of specific tutoring moves.

Acknowledgement

This work was partially supported by EC DELTA project grant P7061/D1014 SAFE/SIMULATE. I should like to acknowledge the stimulating collaboration of the many partners in that enterprise, but particularly Hans Hijne, Robert de Hoog, Ton de Jong, Robin Johnson, Melanie Njoo, Ken Tait, Michael Twidale, Jos van Berkum and Wouter van Joolingen. Annet van Kouwen and Tamara Ronteltap deserve special praise for their skilful management of many aspects of the empirical work reported here. Useful comments on ideas in this paper have been volunteered by Stefano Cerri, Pierre Dillenbourg, Vitor Duarte Teodoro, Patrick Mendelshon, Peter Reimann, Andy diSessa, Dan Schneider and Tim O'Shea. Nevertheless, the opinions, errors and infelicities of style are my responsibility alone.

References

1. Anzai, Y. & Simon, H. (1979) The theory of learning by doing. *Psychological Review,.86*, 124-40.
2. Baker, M. (1990) Arguing with the tutor. In M. Elsom-Cook (Ed.), *Guided discovery tutoring* (pp 141-170). London: Paul Chapman
3. Berliner, D. (1986) In pursuit of the expert pedagogue. *Educational Researcher,15*, 5-13.
4. Blank, D. (1985) *An investigation of teacher decision-making while tutoring fifth grade students engaged in problem-solving using a computer simulation*. Unpublished doctoral thesis. University of Maryland. MA, USA.
5. Calderhead, J. (1984) *Teachers' classroom decision making*. London: Holt
6. Calderhead, J. (1987) Exploring teachers' thinking. London: Cassell
7. von Cranach, M. (1982) Ordinary interactive action: Theory, methods and some empirical findings. In M von Cranach & R. Harre, (Eds.) *The Analysis of Action*. Cambridge: Cambridge University Press
8. Cumming, G., & Self, J. (1989) Collaborative intelligent educational systems. In D. Bierman, J. Breuker & J. Sandberg (Eds.), *Artificial Intelligence and Education* Amsterdam: IOS
9. Dillenbourg, P., & Goodyear, P. (1989) Towards reflective tutoring systems: self-representation and self-improvement. In D. Bierman, J. Breuker & J. Sandberg (Eds.), *Artificial Intelligence and Education*, Amsterdam: IOS
10. Douglas, S. (in press) Tutoring as interaction: Detecting and repairing tutoring failures. In P. Goodyear (Ed.), *Teaching Knowledge and Intelligent Tutoring*. Norwood, NJ: Ablex
11. Goodyear, P. (1988) Approaches to the empirical derivation of teaching knowledge for ITS design. *Proceedings of ITS88 Conference on Intelligent Tutoring Systems*, (pp291-298). Montreal: University of Montreal
12. Goodyear, P. (1989) Expert systems and intelligent tutoring: Some issues in the engineering of pedagogic knowledge. In H. Oosthoek & T. Vroeijenstijn (Eds.), *Higher Education and New Technologies*. Oxford: Pergamon Press
13. Goodyear, P. (in press) Research on teaching and the design of intelligent tutoring systems. In P. Goodyear (Ed.), *Teaching Knowledge and Intelligent Tutoring*. Norwood, NJ: Ablex
14. Goodyear, P., & Tait, K. (in press) Learning with computer-based simulations: Tutoring and student modelling requirements for an intelligent learning advisor. *Proceedings of the 3rd Conference of the European Association for Research on Learning and Instruction* Oxford: Pergamon Press
15. Goodyear, P., van Kouwen, A. & Ronteltap, T. (1991) Exploring quantitative simulation models, (Technical Report), University of Lancaster, Lancaster, England.
16. Halkes, R., & Olson, J. (Eds.) (1984) *Teacher Thinking*. Lisse: Swets & Zeitlinger
17. de Jong, T. (Ed.) (in press) Computer simulations in an instructional context. *Education and Computing*.
18. Laurillard, D. (1988) Computers and the emancipation of students: Giving control to the learner. In P. Ramsden (Ed.), *Improving Learning: New Perspectives*. London: Kogan Page
19. Leinhardt, G. & Greeno, J. (1986) The cognitive skill of teaching. *Journal of Educational Psychology, 78*, 75-95
20. Miyake, N. (1986) Constructive interaction and the iterative process of understanding. *Cognitive Science, 10*, 151-177
21. Newell, A., & Simon, H. (1972) *Human Problem-Solving*. Englewood Cliffs, NJ: Prentice Hall
22. Olson, J. (1984) What makes teachers tick? Considering the routines of teaching. In R. Halkes & J. Olson (Eds.), *Teacher Thinking* (pp35-42) Lisse: Swets & Zeitlinger
23. Petrie-Brown, A. (1990). Freedom of speech: The concepts of guidance and dialogue in intelligent tutoring. In M. Elsom-Cook (Ed.), *Guided discovery tutoring*, (pp132-140) London: Paul Chapman
24. Reigeluth, C., & Schwartz, E. (1989) An instructional theory for the design of computer-based simulations. *Journal of Computer Based Instruction, 16*, 1-10
25. Riesbeck, C., & Schank, R. (1989). *Inside Case-Based Reasoning*. Hillsdale, NJ: Erlbaum
26. Rivers, R, & Vockell, E. (1987) Computer simulations to stimulate scientific problem solving. *Journal of Research in Science Teaching, 24*, 403-415
27. Self, J. (1974) Student models in computer-aided instruction. *International Journal of Man-Machine Studies, 6*, 261-176
28. Shrager, J., & Klahr, D. (1986) Instructionless learning about a complex device. *International Journal of Man-Machine Studies, 25*, 153-189
29. Shuell, T. (1988) The role of the student in learning from instruction. *Contemporary Educational Psychology, 13*, 276-295
30. Richmond, B. (1987) *A user's guide to STELLA*. Lyme, NH: High Performance Systems

31. Voss, J. (1988) Problem-solving and reasoning in ill-structured domains. In C. Antaki (Ed.), *Analysing Everyday Explanation*. (pp74-93) London: Sage
32. Wenger, E. (1988). *Artificial Intelligence and Tutoring Systems*. Los Altos, CA: Morgan Kaufmann
33. Wilson, S., Shulman, L., & Richert, A. (1987) 150 ways of knowing: Representations of knowledge in teaching. In J. Calderhead (Ed.), *Exploring Teachers' Thinking* London: Cassell

Learning and Instruction with Computer Simulations: Learning Processes Involved*

Ton de Jong[1] and Melanie Njoo[2]

[1] Department of Philosophy and Social Sciences, Eindhoven University of Technology, PO Box 513, 5600 MB Eindhoven, The Netherlands &
Department of Social Science Informatics, University of Amsterdam, Roetersstraat 15, 1018 WB Amsterdam, The Netherlands

[2] Department of Philosophy and Social Sciences, Eindhoven University of Technology, PO Box 513, 5600 MB Eindhoven, The Netherlands

Abstract: Nowadays prevalent learning theories state that in the study process the learner is actively involved in constructing and reconstructing his/her knowledge base. This conclusion is reflected in modern approaches to teaching that have abandoned viewing the learner as an 'empty box' into which knowledge could be poured, and stress the active role of the learner and the importance of his/her foreknowledge. Some forms of Computer Assisted Instruction are well suited for this teaching approach. The use of hypertext-like systems, in which learners are encouraged to explore a domain, is such an example. A second example of CAI that elicits exploratory behaviour is simulation-based learning.

It is, however, also evident that exploratory learning puts a high cognitive demand on the learner. Instructional support is needed if learning from simulations is to be effective. In practice this support is often provided by human tutors. The topic of the SIMULATE project is to investigate how this support can be given by a computer learning environment. We have termed

---

*Part of the research reported was conducted in the project SIMULATE. SIMULATE is part of SAFE, a R&D project partially funded by the CEC under contract D1014 within the Exploratory Action of the DELTA programme. In total 17 partners are involved in the SAFE consortium of which Philips TDS is the prime contractor. The partners directly involved in SIMULATE are: Philips TDS (Germany), University of Leeds, University of Lancaster (UK), TIFSA (Spain), University of Amsterdam, Eindhoven University of Technology, Courseware Europe (The Netherlands).

Wouter van Joolingen (EUT) contributed to a number of the ideas expressed in this paper. We would also like to thank all our colleagues from the SIMULATE project for their comments on our work, and Wim Vaags (EUT) for commenting on a previous version of this paper. Lynda Hardman has put some polish to the English. Finally, we have greatly appreciated the extensive comments of an anonymous reviewer.

environments that combine a simulation with (intelligent) support: Intelligent Simulation Learning Environments (ISLEs).

In our analysis we identified four characteristics of instructional use of simulations: presence of (simulation) models, presence of instructional goals, elicitation of exploratory learning processes and possibility of learner activity. The significance of these characteristics for designing an Intelligent Simulation Learning Environment is assessed by combining these characteristics with the four 'classical' design components of Intelligent Tutoring Systems: the domain, learner, instruction, and learner interface component. Combining components and characteristics leads to a descriptive framework in which ingredients necessary for ISLEs can be placed. The present chapter summarises these findings and puts an emphasis on 'exploratory learning processes'.

Keywords: Exploratory learning, computer simulations, learning processes

Introduction

There is a clear tendency in contemporary instructional design to create learning environments in which learners are not offered 'ready made', directly consumable knowledge, but in which they have to create their own knowledge. Self [33] has labelled these environments 'cognitive gymnasia'. One example of such a type of learning environment can be found in *computer simulations*.

Computer simulations are indeed a popular type of computer assisted instruction. A recent survey [3] shows that about 50% of the 300 CAI programs listed in a database of CAI used in Higher Education in the Netherlands is indicated as being a simulation or a combination of a simulation and another type of CAI (e.g., drill or tutorial).

The reasons that simulations are so popular may vary with the goals for which they are used. First, simulations used for teaching *skills* or *procedures* offer very practical advantages, such as being able to introduce catastrophes (e.g., in nuclear power plants), the reduction of stress (e.g., in treating patients in medical simulations) and cost effectiveness (e.g., flight simulators or laboratory experiments).

Second, when the simulation is about *dynamic models*, it allows natural *time scale* to be altered, so that processes may be speeded up or slowed down in order to make them more visible to the learner.

Third, simulations may be used to *simplify models* from the real world in order to have them match the prior knowledge and level of cognitive development of the learner.

These reasons for using simulations are all subordinate to the fact that simulations offer the opportunity for *exploratory* (or *scientific discovery*) learning, a way of learning regarded as having advantages by many authors [amongst others 1, 23, 24, 34] though some studies [31, 36, 38] don't find discovery learning more effective, merely more efficient than non-discovery learning.

The present chapter concentrates on exploratory learning. Learning processes involved will be described and compared with learning processes that we identified in a non-exploratory environment: studying texts. It is generally recognized that exploratory or discovery learning is a complex and demanding task and it therefore seems necessary to offer the learner support in order to ensure an effective and efficient study process. After describing some of the problems exploratory learning might pose to learners we will give some examples of support for learners involved in exploratory learning. Research into providing this support by a computer environment is the topic of a DELTA project called SIMULATE.

The SIMULATE Project: Providing Support to Exploratory Learning

Our work is part of a project called SIMULATE. SIMULATE (SIMulation Authoring Tools Environment) is an EC sponsored DELTA project that started in April 1989. The project has as its ultimate goal developing an authoring tool that will enable an author to create what we have called an Intelligent Simulation Learning Environment (ISLE). An ISLE is a simulation embedded in an environment that includes a diversity of types of support.

The present phase of the project (with a duration of two years) is envisaged as a preparatory phase that aims at gathering information and defining requirements and (global) specifications of both SIMULATE (the authoring tool) and ISLEs (the systems that will be built). A description of the project set-up is given in [17].

One of the starting activities in the project has been to make an inventory of potential elements of ISLEs. Since we see ISLEs as Intelligent Tutoring Systems we classified elements of ISLEs into one of the four 'classical' components of ITSs: *domain*, *learner*, *instructional strategy* and *learner interface*. As a second organiser we identified four characteristics of instructional use of simulations: *simulation models*, *learning goals*, *exploratory learning processes* and *learner activity*. By combining characteristics and components we could assess the consequences of having simulations as the core of an ITS. The results of the inventory are organised in a matrix of which the four characteristics of instructional simulations make one axis, and the ITS design components are at the other axis [for the complete overview see 6].

A second starting activity in the project was to create an inventory of authoring tools and means that could be used for creating different components of future ISLEs (for example, an inventory of all kinds of simulation construction tools has been made) and is reported in [39].

The present chapter will offer a selection and summary of the overview on elements of ISLEs and will concentrate on one of the characteristics of instructional use of simulations: the elicitation of exploratory learning processes. First, however, we will outline the four characteristics of instructional use of computer simulations.

Computer Simulations in an Instructional Context

The term 'simulation' is used for describing a wide variety of different situations, also within an instructional context. It seems necessary therefore to start with a more precise view on the instructional use of computer simulations. As described elsewhere [5], instructional use of computer simulations can be characterised by the following four features:

a. *Presence of formalised, manipulable underlying models*
Computer-based simulation means that a phenomenon, a process, a system or an apparatus (or whatever it is that is being simulated) is *formalized* into a model and implemented as a computer program. This model may have a qualitative character, or a quantitative one, or both. It is essential that the output of the program is *calculated* or *inferred* from the implemented model in response to input from the learner.

b. *Presence of learning goals*
Second, the simulation has to be used in the context of reaching a certain *learning goal*. These goals can be of different types: *conceptual knowledge*, *procedural knowledge*, which might be cognitive skills (e.g., problem solving in a specific domain) or skills with a psychomotor aspect (e.g., learning to fly), and *knowledge acquisition skills* related to the exploratory study process that takes place while learning with the simulation.

c. *Elicitation of specific learning processes*
Third, the simulation must be used to invoke specific *learning processes* characteristic of exploratory learning. The path to the learning goal thus leads through these learning processes, such as hypotheses generation, predicting, and model exploration.

d. *Presence of learner activity*
Fourth, there must be some level of *learner activity*. This means that the learner must actually *manipulate* something within the simulation, for example setting input variables and parameters, collecting data, making choices in a procedure, setting data presentations, or controlling simulation time.

Together, these four characteristics describe the instructional context of interest to us. A closely related type of instruction/learning is *modelling*. In modelling the learners are not only allowed to change values in the underlying model, but they may also interfere with the properties of the underlying model.

Exploratory Learning

Despite the fact that simulations are popular there is a general feeling that still much needs to be known about the exploratory (or scientific discovery) study process, or as Lesgold [23,

p. 325] says: '*We lack a good theoretical account of the process of scientific discovery....*' Fortunately, there is a number of recent studies that help chart this process [11, 19, 21, 23, 32, 36].

Learning processes

What became clear from Langley's et al. [21] detailed analysis of the process of discovery learning is its complicated and demanding nature. Klahr and Dunbar [19] and others stress that discovery encompasses two basically different processes: hypothesis generation and experimentation. Schauble et al. [32] emphasise that in scientific discovery learning, knowledge structures and discovery processes interact in a complicated way. This is important, given the fact that learners often have an incomplete and faulty knowledge base [see also 30]. If we are going to try to help learners in exploratory learning environments, we first need to know what is going on in the exploratory study process.

To start we distinguish two levels: the complete process of acquiring knowledge in a specific situation and the more detailed processes from which this knowledge acquisition process is composed. In this contribution we will denote the complete process as the *study process* and the detailed processes as *learning processes*. In describing the learning processes a similar approach to the one followed by [10] is taken, and detailed processes are categorised into more global and comprehensive processes.

The inventory of learning processes involved in an exploratory study process that we adopted in the SIMULATE project was developed on the basis of existing studies [e.g., 19, 22, 30], and two empirical studies [26, 27]. In the empirical studies we observed learners working with a computer simulation on control theory in mechanical engineering. Subjects (second and third year University students) had to think aloud while learning with the simulation. Along with the simulation the students received an assignment that prescribed a number of steps to be taken.

We do not see the resulting inventory as *the* conclusive set of exploratory learning processes, but it is regarded as a sensible basis for further work within SIMULATE. The list of processes will receive elaboration, more detail and possibly restructuring as research is continuing. Table 1 presents a summary of our inventory of exploratory learning processes.

Table 1
Exploratory learning processes

TRANSFORMATIVE PROCESSES **Analysis:** Analysis concerns charting the domain information. This learning process is subdivided into: • *Looking for or finding information* The learner tries to find domain information by searching in text books, additional material, asking a tutor etc. • *Model exploration* Identifying and relating variables and parameters in the model and indicating general properties of the model. This can be done on the basis of prior knowledge, additional material, but of course also from running the simulation. **Hypotheses generation:** Hypotheses generation is the formulation of a relation between one or more variables (input and output) and parameters in the simulation model. A hypothesis is stated with the intention to test it. **Testing:** Testing involves those activities that are necessary for furnishing data on which the learner expects to be able to accept or refute a new hypothesis, or to create an hypothesis. It consists of the following subprocesses: • *Designing an experiment* Indicating what will be changed in a simulation model and in which order. • *Making predictions* A prediction states the expectation of a simulation run outcome as the result of designated value attributions to variables. • *Learner activities*, such as changing variable values, etc. • *Data interpretation* Interpreting the data without a direct reference to model relations. The learner can do this in a local manner (noticing specific characteristics of the output, for example: this is an a-symptotic relation) or at a conceptual level by comparing output to other output that is known or to information from other sources. **Evaluation:** In evaluation results are put into a more general context. This process consists of: • *Evaluating/Judging* Here the actions of the learner and the results thereof are evaluated (e.g., "I shouldn't have done this experiment."). • *Generalising* In generalising the learner puts his/her actions and the results thereof in a broader context both as learning processes or as domain information (e.g., "This is an approach I think I can use more often."). REGULATIVE PROCESSES **Planning:**

In Table 1 a main distinction is found between *transformative* and *regulative* processes. The three *regulative* processes that we distinguish are *planning*, *verifying*, and *monitoring*. Transformative processes relate to processes in which domain information is transformed into knowledge. In the transformative class we distinguished: *analysis*, *hypotheses generation*, *testing*, and *evaluation*. These classes of learning processes are similar to those identified by others [for example 11, 12].

The learning processes in Table 1 are arranged in a certain more or less logical order. This is not necessarily the order in which they occur in exploratory learning. The inventory of learning processes should be regarded as a list of exploratory learning processes that can be applied in different sequences. So, for example, a learner who is an 'experimenter' [see 19] would first apply the process 'learner activity' and subsequently the process 'model exploration'.

A comparison to non-exploratory learning processes

We can get an impression of crucial aspects of the exploratory study process by comparing the learning processes, as identified and applied in our studies on exploratory learning (see Table 1 and [26, 27]) with the results that were obtained by Ferguson-Hessler and de Jong [10] in their investigation of learning processes when studying text. In the latter study subjects were offered a ten page text on a physics topic (the Aston mass spectrometer). The text was divided into small parts and after each part learners were asked to tell what they had done in the period before it. Analyzing these statements led to the identification of 32 different learning processes classified into main categories of which *superficial processing* (such as 'taking for granted'), *integrating* (bringing structure into new knowledge) and *connecting* (relating new knowledge to previous knowledge) were the important ones.

The first difference that becomes apparent from comparing the two learning processes inventories is that in the text processing list no explicit reference to *regulative* processes is made, whereas this is a main category for exploratory learning. Njoo and de Jong [27] report that up to 38% (with a mean of about 33%) of student's learning processes fall into this category. One of the subprocesses of regulation, *verifying*, as identified for exploratory learning has its counterpart in text processing where learners may check or verify derivations as they are given in the text. The subprocess *monitoring* from the exploratory learning processes list can be retraced to, for example, a study text learning process such as 'deciding that one can follow a deduction or derivation', but no general monitoring categories are indicated in the list of text study learning processes. *Planning* is a process that is absent in the text processing inventory. The only process that somewhat suggests planning is 'Quickly glancing through the text at the beginning of the studying to gain a first impression of major points and general structure of the text.' Although our text processing study was not directly aimed at finding regulative processes, we may conclude that, apparently, study texts can be rather imperative and prevent learners from making their own strategy of learning. In our studies on learning with a simulation on control theory we found that (even with learners who were generally reluctant to show exploratory behaviour) planning processes are quite important. The percentage of planning processes compared with the total number of processes applied ranged from 7 to 20% [27].

Another important aspect of exploratory learning is *model exploration* and *hypotheses generation*. Here learners *actively* construct their view of the domain. In our text processing inventory we find a number of processes that are characteristic of knowledge structuring. These are, for example, 'Confronting the text with other ideas or arguments, doubting the correctness of the text, generating alternatives for information from the text', 'Drawing conclusions', 'Finding relations oneself', and 'Imposing structure not given in the text'. A text processing category such as 'finding contradictions between one's own conclusions or between one's own conclusions and information from the text' comes very close to the exploratory process of testing an hypothesis.

There is some overlap between our inventories of exploratory learning processes and learning processes involved in studying text. In the latter study we compared the behaviour of good and poor performing students. It is interesting to see that the processes from studying text that seem to be present in exploratory learning (and that we labelled 'deep' processes in our study text research) are mostly found with good performers, whereas poor performers perform these processes less frequently [10].

Summarising, we can say that exploratory learning compared with studying text seems to call upon a broader range of learning processes. Also, the processes involved in studying text that have an analogous character to exploratory learning processes were found to be performed more frequently by good performers. This implies that a learning environment that asks this kind of behaviour explicitly from learners may produce major obstacles for the less proficient learners.

Potential problems in exploratory learning

A careful analysis of the exploratory study process will reveal the necessary ingredients of exploratory learning. In addition to this, in order to design effective and efficient support, we require an analysis of unproductive exploratory behaviour. Many studies report only that exploratory learning is difficult, but fail to indicate why and how. The comparison of text studying and exploratory learning from the preceding section suggested that weak students in particular will experience difficulties in exploratory learning, and this assumption is supported in literature [e.g., 23, 38].

Potential problems can be categorised with the help of the inventory of exploratory processes as listed in Table 1. Some problems that students encounter are related to a single *transformative learning process*. It is, for example, generally recognised that processes such as *model exploration* may benefit from the learner's prior knowledge [see for example 15]. This is also the reason that we found that a large percentage (means of 12 and 20 %) of students' learning processes may be concerned with looking up information [26, 27]. Schauble et al. [32] found indications that learners' knowledge also influences the strategy that they use in exploratory learning. Lavoie and Good [22] report unsuccessful prediction behaviour correlates with low prior knowledge.

Hypothesis generation is another process that might pose problems to learners. Shute and Glaser [36] state that learners do not recognise that apparent regularities in data need to be tested with hypotheses. Van Joolingen and de Jong [42] found that learners are prone to state either too global or too complicated hypotheses.

According to Reimann [30], who observed learners interacting with a microworld on optical refractions, learners do not *design experiments* systematically. Shute and Glaser [36] found that learners (especially the poor ones) do not vary one variable at a time and thus fail to detect important domain relations, a finding also reported by Lavoie and Good [22], and Schauble et al. [32]. Schauble et al. [32] also found that unsuccessful and successful exploratory learners are equally active in their behaviour, a finding that was also reported for studying text [10].

Data interpretation may also provide major obstacles to students. Many computer simulations present their data as graphs, and graph interpretation appears to be problematic, especially for children [24, 25]. We also found that data interpretation is still not easy for university students [26].

Another group of problems relates to the *connection between (the results of) different learning processes*. One of the most remarkable findings is that learners seek for confirmative evidence. They tend to keep hypotheses despite disconfirming information, and they even reject hypotheses for which they have confirming evidence [19]. In a study of students working with a computer simulation in chemistry (chemical titration), van Joolingen and de Jong [42] found that students quite often do not design experiments suited to test the hypothesis that they stated. Shute and Glaser [36] found that good performers tend to base their conclusion on adequate and sufficient evidence, whereas poor performers tend to rely on inadequate data. They also conclude that poor performers quite readily accept a at first sight correct statement about their data, even if this statement is not completely adequate. This is in line with the finding of Ferguson-Hessler and de Jong [10] that poor learners more easily accept a finding without taking the trouble of questioning it.

Finally, an important source of difficulties that learners experience is in the area of *regulation*. Some authors emphasise that exploratory learning with computer simulations is a complex process and students seem to underestimate the complexity [32]. This is a phenomenon we also found for student learning with a computer simulation on decision support theory. where students made no notes despite the fact that they were offered dedicated forms to do so [7]. Shute and Glaser [36] comparing poor and good performers in learning with a microworld on economics, found that successful students use more thinking and planning skills. These learners use more global (across a number of experiments) planning, whereas the planning of poor performers has a more local (within one experiment) character. Similar results are reported by Schauble et al. [32]. We found that learners even tend to restrict planning to stating what they will perform as an immediate next action, and fail to perform any form of global planning at all [26, 27].

Supporting Learning Processes

The preceding section indicated some difficulties that students may encounter when working in an exploratory, computer simulation, environment. A next step is to search for ways to overcome these difficulties.

First, one may explicitly train learners in exploratory or scientific inquiry skills. Examples of such training programmes are given in [11] and [12].

A second approach is to support the learner while s/he is working with the simulation. Generally, this support is given by a human tutor, but it can be provided by a computer environment as well. In this way, the plain simulation is combined with other forms of instruction as is advocated by [9]. There is now a number of systems that present the learner a simulation along with some kind of support. Some of these systems are: QUEST (troubleshooting of simple electronic circuits; [44]), STEAMER (operating a steam propulsion plant; [18]), MACH-III (maintenance and trouble shooting of a complex radar device; [20]), IMTS (troubleshooting in complex devices; [40]), and Smithtown (micro-economics; [36, 37]).

In providing support we can make a distinction between *directive* and *non-directive* support. *Directive support* (or guidance) steers the learner in a certain direction. In this way it guides the learner, for example when the learner gets direct feedback and/or hints on directions to follow. Especially when the learning goal involves some kind of procedure, the learner's action sequence can be traced. When it diverges from some normative sequence, the learner can be corrected [see for example 20]. Hints such as 'it is better to change only one variable at a time' [31, 36] and hints on where in a device to search for a fault [40] are also examples of directive support. A survey of directive support can be found in [41].

Non-directive support, does not steer the learner in a certain direction, but helps with accomplishing what s/he would have done in a completely free exploratory environment. We have called these kinds of tools *learner instruments*, of which examples are *dedicated scratchpads*: scratchpads that are meant to support a specific learning process. Hypotheses scratchpads are such dedicated scratch pads. Different forms exist. The more structure the 'scratchpad' has (e.g., a list of predefined hypotheses) the more directive it is [for a discussion of different types of hypotheses scratchpads see 4]. Other examples of learner instruments are 'notebooks' for recording intermediate results [36]. Finally, we have genuine 'scratchpads', empty spaces the learner can use to note down anything s/he likes. These scratchpads do not support specific learning processes but have the general aim of reducing the working memory load.

The distinction between directive and non-directive support is not clear. Somewhere in between directive and nondirective support is *putting restrictions on the simulation environment*, for example when certain parts of the underlying model are not accessible to the learner, when the range of possible values to give to variables is restricted [2], or when learners are presented with a sequence of qualitatively different models to explore [13, 29, 44]. A recent study by Veenman and Elshout [43], however, doubts the advantage of structuring exploratory environments in this way.

The space of possible actions of learners can also be restricted by offering them an *assignment* that suggests a particular problem (e.g., optimisation or fault diagnosis) to solve with a computer simulation. We have, however, the experience that this sometimes inhibits exploratory learning [26] which might be difficult to overcome even by introducing hints that stimulate exploratory behaviour [27].

Finally, learners can be supported in planning and monitoring their study process. In a present study [28] we offered learners, who were using a computer simulation in control theory, forms to work on that had separate areas for the different transformative processes as given in Table 1. In this way, learners are encouraged to follow a complete exploratory study process, and they can easily monitor their own study process. In another ongoing study we are developing ways to provide learners with overviews of their interaction with a simulation on decision support theory [7]. These overviews present the learners with a summary of their interaction strategy with the simulation, and thus they offer a means of monitoring the study process. Providing overviews or navigation tools is currently a major research topic in hypertext based learning [14].

Conclusion

Exploratory learning with computer simulations is in the centre of contemporary interest. Despite this and despite the fact that numerous simulations are used in instruction, there is still not enough research that describes exploratory learning processes and even less research on how to support learners in exploratory learning. The present chapter attempted to give a summary of our present work on these topics, much of which is being performed within the context of the SIMULATE project.

There are two directions that we will follow in our future (empirical) research. First, we will continue to elaborate and restructure our inventory of exploratory learning processes. This will be done mainly by observing learners involved in exploratory learning in a number of domains. This also provides the opportunity for discovering empirically problems that learners encounter. Second, in accordance with the first objective, we will evaluate the effectiveness of a number of support mechanisms on exploratory learning. Examples of these are different forms of hypotheses scratchpads [42], study process guidance forms and ready-made hypotheses [28], and overviews [7].

In the SIMULATE project an effort is currently being made to find a more formal representation language for describing the information as gathered in our inventory work (of which a part was reported in the present chapter). In addition, we are attempting to identify basic conceptual units that have to be described in this language (or languages). This means that we will identify *generic* elements, relations and structures for each of the components of an Intelligent Simulation Learning Environment. The next step will then be to create building blocks for each of these components that authors may actually use to create ISLEs by means of *selecting*, *specialising* and *instantiating* [for a more detailed description see 8].

The final SIMULATE system is envisaged to consist of a permanent *shell*, filled with *default information*, containing the functionality of an ISLE that is always present, and *author generated information* [for a description see 16]. For providing this author generated information the author is supported by a library of building blocks of ISLE components as mentioned above, and further by a set of rules that constitute *recommendations* of good instructional design and a *methodology*. Information as collected in our inventory will play a crucial role in designing SIMULATE. As should be clear, however, creating a true SIMULATE authoring work bench lies beyond the reach of the present phase of the project.

References

1. Ausubel, D.P., Novak, J.D., & Hanesian, H. (1978). *Educational psychology: A cognitive view*. New York: Holt, Rinehart & Winston.

2. Böcker, H.D., Herczeg, J., & Herczeg, M. (1989). ELAB - An electronics laboratory. In D. Bierman, J. Breuker, & J. Sandberg (Eds.), *Proceedings of the Fourth International Conference on AI & Education* (pp. 15-25). Amsterdam: IOS.
3. de Jong, T. (1990a). *Ontwikkelingen in Computer Ondersteund Onderwijs in het Nederlands Hoger Onderwijs in de jaren 1988/1989.* [Developments in computer-assisted instruction in the Netherlands for the years 1988/1989] (OCTO report 90/01). Eindhoven University of Technology.
4. de Jong, T. (1990b, November). *Learning and instruction with computer simulations: interface aspects.* Paper presented at the NATO AETW 'Cognitive modelling and interactive environments', Eindhoven, The Netherlands.
5. de Jong, T. (1991a). Learning and instruction with computer simulations. *Education & Computing, 6*, 217-229.
6. de Jong, T. (Ed.) (in press). *Computer simulations in an instructional context.* Amsterdam: Elsevier.
7. de Jong, T., de Hoog, R., & de Vries, F. (1991). SUPER-MIDAS. A computer simulation for learning and instruction decision support systems. In H. Hijne & J. van Berkum (Eds.), *Prototype/mock-up of an Integrated Simulation-based Learning Environment* (DELTA project SAFE P7061; SAFE/SIM/CE-rep.). Courseware Europe BV.
8. de Jong, T., Tait, K., & van Joolingen, W.R. (in press). Authoring for intelligent simulation based instruction: A model-based approach. *Proceedings of the DELTA and Beyond conference*, The Hague, The Netherlands.
9. Farr, M.J., & Psotka, J. (1989). Introduction. *Machine-Mediated Learning, 3*, 1-6.
10. Ferguson-Hessler, M.G.M., & de Jong, T. (1990). Studying physics text. Differences in study processes between good and poor performers. *Cognition and Instruction, 7*, 41-54.
11. Friedler, Y., Nachmias, R., & Linn, M.C. (1990). Learning scientific reasoning skills in microcomputer-based laboratories. *Journal of Research in Science Teaching, 27*, 173-191.
12. Germann, P.J. (1989). The processes of biological investigations test. *Journal of Research in Science Teaching, 26*, 609-625.
13. Hamburger, H., & Lodger, A. (1989). Semantically constrained exploration and heuristic guidance. *Machine-Mediated Learning, 3*, 81-107.
14. Hardman, L. (1990). User interface tools to support learning (DELTA project SAFE, working paper SAFE/HYP/OWL-pap/user_i/f_tools). Edinburgh: Office Workstations Ltd.
15. Hartley, J.R. (1988). Learning from computer based learning in science. *Studies in Science Education, 15*, 55-76.
16. Hijne, H., & van Berkum, J.A. (1990, October). *Authoring for intelligent simulation learning environments.* Paper presented at the DELTA & Beyond conference, The Hague, The Netherlands.
17. Hijne, H., & de Jong, T. (1989, September). *SIMULATE: Simulation authoring tools environment* (OCTO report 1989/2). Paper presented at the EARLI Conference, Madrid, Spain.
18. Hollan, J.D., Hutchins, E. L., & Weitzman, L. (1984). STEAMER: An interactive inspectable simulation-based training system. *AI Magazine, 5*, 15-27.
19. Klahr, D., & Dunbar, K. (1988). Dual space search during scientific reasoning. *Cognitive Science, 12*, 1-48.

20. Kurland, L.C., & Tenney, Y.J. (1988). Issues in developing an intelligent tutor for a real-world domain: Training in radar mechanics. In J. Psotka, L.D. Massey, & S.A. Mutter (Eds.), *Intelligent tutoring systems: Lessons learned* (pp. 119-181). Hillsdale, NJ: Erlbaum.
21. Langley, P., Simon, H.A., Bradshaw, G.L., & Zytkow, J.M. (1987). *Scientific discovery, computational explorations of the creative process*. Cambridge: MIT Press.
22. Lavoie, D.R., & Good, R. (1988). The nature and use of prediction skills in a biological computer simulation. *Journal of Research in Science Teaching, 25*, 335-360.
23. Lesgold, A. (1990). Tying development of intelligent tutors to research on theories of learning. In H. Mandl, E. De Corte, S.N. Bennett, & H.F. Friedrich (Eds.), *Learning and Instruction. European research in an international context* (vol. 2.1, pp. 321-337). Oxford: Pergamon Press.
24. Linn, M.C. (1990). Perspectives of research in science teaching: Using the computer as a laboratory partner. In H. Mandl, E. De Corte, S.N. Bennett, & H.F. Friedrich (Eds.), *Learning and Instruction. European research in an international context* (vol. 2.1, pp. 443-460). Oxford: Pergamon Press.
25. Mokros, J.R., & Tinker, R.F. (1987). The impact of microcomputer based labs on children's ability to interpret graphs. *Journal of Research in Science Teaching, 24*, 369-383.
26. Njoo, M., & de Jong, T. (in press). Learning processes of students working with a computer simulation in mechanical engineering. *Proceedings of the EARLI conference*, Madrid, Spain.
27. Njoo, M., & de Jong, T. (in press). *Stimulating exploratory learning with a computer simulation for control theory: The effect of hints* (OCTO Report). Eindhoven University of Technology.
28. Njoo, M., & de Jong, T. (1991, April). *The effect of offering study process planning support, learning process information, and ready-made hypotheses, on learning with a computer simulation on control theory.* Paper presented at the AERA conference, Chicago.
29. Plötzner, R., Spada, H., Stumpf, M., & Opwis, K. (1990). *Learning qualitative reasoning in a microworld for elastic impacts* (Forschungsbericht nr. 59). Research group on cognitive systems, University of Freiburg.
30. Reimann, P. (1989). Modelling scientific discovery learning processes with adaptive production systems. In D. Bierman, J. Breuker, & J. Sandberg (Eds.), Artificial intelligence and education; synthesis and reflec-tion. *Proceedings of the Fourth International Conference on AI and Education* (pp. 218-227). Amsterdam: IOS.
31. Rivers, R.H., & Vockell, E. (1987). Computer simulations to stimulate scientific problem solving. *Journal of Research in Science Teaching, 24*, 403-415.
32. Schauble, L., Glaser, R., Raghavan, K., & Reiner, M. (1990, April). *Causal models and experimentation strategies in scientific reasoning*. Paper presented at the AERA conference, Boston.
33. Self, J. (1989, May). *The case for formalising student models (and Intelligent Tutoring Systems generally).* Paper presented at the AI & Education conference, Amsterdam, The Netherlands.
34. Shulman, L.S., & Keislar. E.R. (Eds.) (1966). *Learning by discovery: A critical appraisal.* Chicago: Rand McNally.
35. Shute, V.J. (1990, April). *A comparison of inductive and deductive learning environments: Which is better for whom and why?* Paper presented at the AERA conference, Boston.

36. Shute, V.J., & Glaser, R. (1990). A large-scale evaluation of an intelligent discovery world: Smithtown. *Interactive Learning Environments, 1*, 51-77.

37. Shute, V. J., Glaser, R., & Raghavan, K. (1989). Inference and discovery in an exploratory laboratory. In P.L. Ackermann, R.J. Sternberg, & R. Glaser (Eds.), *Learning and individual differences* (pp. 279-326). New York: W.H. Freeman.

38. Swanson, J.H. (1990, April). *The effectiveness of tutorial strategies: An experimental evaluation.* Paper presented at the AERA conference, Boston.

39. Tait, K. (Ed.) (1990). *Towards the specification of support tools for authors constructing simulation-based intelligent learning environments* (DELTA project SAFE (P7061), deliverable SIM/22). University of Leeds, Computer-Based Learning Unit.

40. Towne, D.M., Munro, A., Pizzini, Q.A., Surmon, D.S., Coller, L.D., & Wogulis, J.L. (1990). Model-building tools for simulation-based training. *Interactive Learning Environments, 1*, 33-50.

41. van Berkum, J.J.A., & de Jong, T. (1991). Instructional environments for simulations. *Education & Computing, 6*, 305-358.

42. van Joolingen, W., & de Jong, T. (1991). A prototype scratchpad for hypothesis formation and experimental design. In H. Hijne & J. van Berkum (Ed.), *Prototype/mock-up of an Integrated Simulation-based Learning Environment* (DELTA project SAFE P7061; SAFE/SIM/CE-rep.). Courseware Europe BV.

43. Veenman, M.V.J., & Elshout, J.J. (1990). De meerwaarde van een goede probleemaanpak. [The surplus value of a proper problem approach]. *Tijdschrift voor Onderwijsresearch, 15*, 337-347.

44. White, B.Y., & Frederiksen, J.R. (1990). Causal model progressions as a foundation for intelligent learning environments. *Artificial Intelligence, 42*, 99-157.

Two Uses of Computers in Science Teaching: Horizontal Motion Simulation and Simulation Building

Magnus Moar, Fiona Spensley, Tim O'Shea, Ronnie Singer, Sara Hennessey and Eileen Scanlon

The Institute of Educational Technology, The Open University, Milton Keynes, MK7 6AA, UK

Abstract: Computers possess a unique range of properties to assist the learning of Mathematics and Science. Of particular importance is their interactivity, exemplified by supporting Direct Manipulation, Memory Augmentation, Qualitative Reasoning, Conflict Resolution and the presentation of Counter-Factual examples. This paper describes two ways in which these properties have been explored. Firstly, in the design of an 'alternate realities' simulation for exploring horizontal motion. Secondly, by focusing on the importance of internal and self-generated external visual representations in problem solving, the design of a combined animation / modelling system is suggested.

Keywords: Simulation, horizontal motion, visual representation

Introduction

The use of computers in mathematics and science teaching has tended to focus on two capabilities. Firstly, computers can provide an adaptive educational tool, giving guidance, instruction or remedial help as a result of constructing and analysing a model of the student's state of understanding. This is exemplified by the Intelligent Tutoring System approach. Secondly, computers can provide the student with a means of formally expressing and inspecting their conceptualisation of some domain, for example by constructing a LOGO or PROLOG program, or by using a modelling system.

Expressing reservations about the usefulness of these approaches to Physics and Maths instruction, O'Shea, O'Malley and Scanlon [21], have focused on 5 properties computers may possess which can facilitate, in particular, exploratory learning: direct manipulation, memory augmentation, the support of qualitative reasoning, counter factual reasoning and conflict resolution. This paper will describe two ways in which some of these properties may be incorporated into the design of computer based educational systems.

Five Functions Supported by Computers

Direct manipulation (DM)

Given the benefits of physically manipulating concrete objects which can be interpreted mathematically (e.g., Dienes' blocks), it is possible that similar benefits may be realised using computer based systems which allow on-screen manipulation of simulated objects. The means by which a user gains information about the functionality of a computer based system is through the interface. By minimising the intrusiveness of the interface to a simulation, in other words increasing its transparency, the represented worlds may become more directly present, cognitively. This makes it possible for the student to experimentally directly manipulate objects, constraints and variables. The degree of interface transparency is a function of the size of the gaps between a student's goals and knowledge on the one hand, and the level of description provided by the system on the other. These gaps are the 'Gulfs of execution' and 'Gulfs of evaluation' referred to by Norman [20], the former being bridged by attempting to make the commands and mechanisms of the system match the thoughts and goals of the user, and the latter being bridged by making the output of the system represent easily the actions and state of the system. It is proposed that in designing computer based systems for use in education, direct engagement should be encouraged by reducing the gulfs of execution and evaluation.

Memory augmentation

By recording an interaction and supporting note-taking and hypertext systems, the computer can act as 'externalised memory', or memory prosthesis. This may help pupils to reference and organise their learning and also to reflect upon their problem-solving processes, so encouraging metacognitive skills. This point is echoed in [29] where such considerations are used as design principles for environments which can encourage such processes as abstracting relevant information from notes and creating overviews. There is also reason to suppose that the support of self-generated external graphic imagery in problem solving may also be desirable (e.g., The construction of appropriate diagrams, sketches, illustrations etc.). In Waller's classification of graphical communication [33], a category is reserved for such external visual representations as a tool for self-enquiry and thought. He reasons:-

> *"... the construction of a ... diagram, is itself part of the process of organising and structuring ideas. That is , it generates a particular kind of critical self analysis that it is impossible for most people to carry out in their heads"* (p. 220).

Additionally, the computer can permit the construction of *dynamic* graphical displays, which may be useful in making explicit and reflecting upon such things as state transitions, temporal sequencing and causal relationships.

Qualitative reasoning

There is some evidence to support the notion that expert problem solving begins with a qualitative causal analysis and understanding of a problem, with quantitative solutions being developed later [e.g., 2,4,14]. It follows that computer based learning environments may be improved by providing means of expressing and evaluating a domain in qualitative terms.

Counter factual reasoning

The computer can provide an environment for exploring a simulated world where objects and forces are different from our perceived reality. For example, exploring a violated law and seeing the consequences of the violation, may help to establish the existence and effect of that law in reality.

Conflict-resolution

Students may develop multiple, often conflicting models of a domain. The computer can be a means of displaying and running these models, inviting comparison and insight into inconsistencies between them [e.g., 1].

This paper will describe two ways in which these five properties can be used as guiding principles in the design of computer based learning environments. Firstly, an interactive simulation which helps students to explore the horizontal motion of objects will be described. Secondly, by focusing on the issues of memory augmentation and externalisation of visual imagery, the design of a combined animation / modelling system is suggested.

Horizontal Motion Simulation

Children hold many informal theories and beliefs about force and motion [e.g., 7]. For example, many children account for the slowing down of a moving object in terms of that object running out of energy, or, where explanations use the notion of opposing forces, in general only air resistance is referred to. It was felt that providing a simulation of movement over surfaces of differing co-efficients of friction would help students to explore the role of friction in horizontal motion.

In considering the design of such a simulation, The Alternative Reality Kit [31] would seem like a suitable starting point in that it utilises many of the educationally useful properties of computers outlined above. It is a Direct Manipulation system for creating interactive simulations which is based on a physical metaphor, that is, objects have a physical representation, velocity and mass. The student's actions are achieved using a mouse operated 'hand' icon that can grab and manipulate objects. The physical laws that act on the system, such as gravity, can be increased or decreased using a 'slider control'. Also, the student can create several, simultaneously viewable 'worlds' or 'alternate realities' which have different settings for physical laws (Gravity can be stronger or weaker for example). The effects of these settings can be compared with each other and the real world. This can provide a context for the student to explore conflicting models and to consider counter factual examples. In becoming familiar with and thinking about the behaviour of the simulated objects, taking measurements and performing involved calculations may not be immediately necessary and students are afforded the possibility of reasoning qualitatively about the alternate realities they have created. Should the student require it, quantitative information about the simulation parameters can be displayed and measurements can be provided automatically without the student being distracted from the underlying problem by becoming involved in the process of measurement itself.

O'Shea and Smith [22] have explored student's conceptions of the laws of kinematics using the Alternative Reality Kit (ARK). They found that pairs of subjects were able to identify several of their own misconceptions about the laws relating to conservation of energy and conservation of momentum. It was felt that a similar simulation could be constructed without requiring the relatively expensive, complex hardware and software used to build ARK. Also it was proposed that for simplicity, subjects should be presented with a set of simulated worlds rather than constructing them, and that these alternate realities should be based on students' prior misconceptions. This would invite comparison between any conflicting models the students had about the motion of the simulated objects. There being no directly relevant literature, a simulation was firstly constructed which represented the 'real world' in order that data about student misconceptions could be gathered.

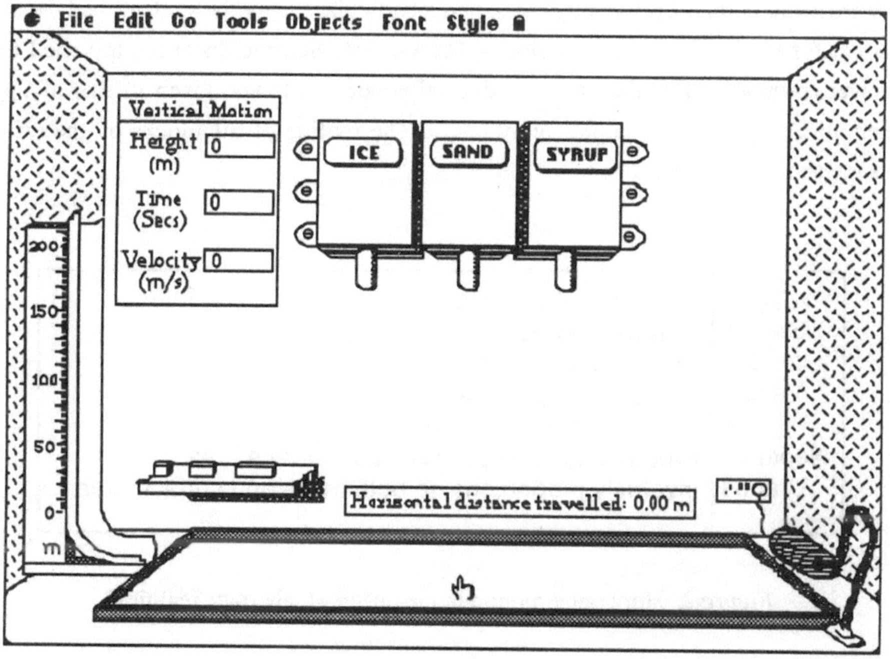

Figure 1. Initial screen for horizontal motion simulation.

The interface design is presented in figure 1. The interface was designed according to direct manipulation principles with continuously viewable objects that can be manipulated by click/dragging with a mouse pointer, which in this case is represented by a grabbing hand icon. Using this hand, the student is required to grab one of the blocks which are of differing shapes and weights. The block is then released, drops down the chute, and slides horizontally on one of the three available surfaces. These surfaces have different coefficients of friction:- ice, sand or syrup. As a block falls, the height from which it was dropped, it's mass, the time it took to fall, it's initial velocity at the beginning of its horizontal slide and distance travelled are all recorded and displayed. The user selects which surface s/he wants to experiment with by clicking on the appropriate hopper which then dumps it's contents onto the test area. The test area is cleared by clicking on the plug of the vacuum cleaner which sucks the test material away. All the animations are accompanied by realistic 'sampled' sounds.

After initial testing with Open University students (aged 25 - 60 yrs), it was found that the simulation was reported to be intuitive and easy to use. In addition, as a result of some student's intuitions about surface area, vertically-oriented blocks were added. Furthermore, a momentum gauge was added to the simulation

From this, and further evaluations of the interface, a set of alternate realities were designed in which the blocks behaved as if the misconceptions the students had shown in the original study were actually true. Each reality embodied a particular theory the students had developed to account for the behaviour of the blocks, for example that the distances travelled by the blocks were proportional to their masses. Each alternate reality was given a name of a planet between which the students could freely travel. The motion of the blocks on the different planets are shown in Figure 2:

Mars	Proportional to mass
Saturn	Inversely proportional to mass
Jupiter	Proportional to surface area
Venus	Inversely proportional to surface area
Pluto	Real world (minus air resistance)
Neptune	Proportional to both mass and surface area
Uranus	Inversely proportional to both mass and surface area

Figure 2. Horizontal motion in the different 'alternate realities'.

It should be noted that some alternate realities were not based on the students' misconceptions but were included for the sake of completeness.

The simulation was then used by 51 students who were asked to identify in which reality the blocks behaved as they would in the real world. They were also required to give a qualitative account of their how they arrived at this conclusion. The results demonstrated an enthusiastic response to using the environment, with the students generating widely different theories and approaches to the problem. In fact, 11 students managed to correctly identify the 'real' world. It is interesting to note that most students thought that the problem was initially trivial until actually going to another alternate reality. It was also observed that many groups of students constructed ad-hoc experiments (for example using rulers and pieces of paper), to test out theories about the real world, and this shows a possible future use in promoting off computer experimentation in a classroom setting.

The success of the simulation is perhaps an indication of the benefits of using the five properties of computers mentioned previously as guiding design principles. The simulation was reported as being easy to learn and use, which may demonstrate the power of the direct manipulation approach to educational software design. The students didn't need to continually disrupt their 'train of problem solving thoughts' to attend to the task of operating the system. Similarly, the fact that the students were continually updating their theories as they moved from one alternate reality to another may be indicative of the the potential benefits of supporting counter factual reasoning and conflict resolution in learning environments. Although the students made notes about the experiments they performed, this activity was not explicitly supported by the system. A useful extension to the simulation environment would be to incorporate some note-taking and annotation tools as memory augmentation devices.

Memory augmentation and the use of qualitative descriptions are themes picked up again in the next section of the paper which looks at the issues concerning the design of a simulation building system based on the importance of generating external visual representations in problem solving.

Visual Representation in the Design of a Simulation Building System

Making and using mental representations is widely regarded as the fundamental property of human cognition. Much of the attention devoted to understanding cognition has tended to focus on linguistic or propositional representations. However, there has been a growing awareness of the need to investigate different kinds of representation, particularly visual imagery, and to integrate these into theories of cognition [e.g., 12,11]. The traditional importance given to non-visual cognition has apparently been reflected in the design of computer based modelling tools which attempt to facilitate students' expression and self-examination of their conceptualisation of some domain. In view of the introductory remarks on the importance of bridging the gulfs of execution and evaluation, and the possible beneficial use of graphical expression as a memory augmenting device, this lack of support for visual expressivity and interpretation would appear to be a significant omission from such systems in their role as educational tools. It will be argued here then, that such modelling environments may be enhanced by encouraging visual expression in the form of pictures, illustrations and diagrams, and by integrating such representations with linguistic and formal descriptions of that domain.

Modelling and simulation building tools

Using a modelling or simulation building tool involves a wide range of problem solving skills, such as discovering the critical objects, identifying transitional states, inferring and expressing their causal relationships and evaluating the competence and internal consistency of the constructed model. As de Kleer and Brown [3] suggest, this may involve not just the construction of a formal model of the domain, but also the running 'in the mind's eye' of a mental simulation. It would seem that any modelling tool which did not permit the expression of such imagery could possibly be making the construction task unnecessarily difficult. Many existing educational general purpose modelling and simulation toolkits tend to be highly formal or linguistically oriented (e.g., Spreadsheets or knowledge-based 'shells') or allow only the manipulation and not design of the icons and graphical objects [e.g., 27]. This may lead to problems with younger students, for example, when trying to map between objects in a mentally modelled system and the abstracted textual label or icon. Riley [28] notes an instance of this difficulty in icon-object mapping

> *"I'm thinking of a picture in a textbook to remember what these boxes represent" (p. 255).*

This sort of problem may be obviated and a different quality of understanding of the domain may be achieved by encouraging children to create their own images of the state changes of the objects as the simulation progresses.

The use of visual imagery

Using different representations can bring new insights into the problem domain being modelled. Visual imagery has several unique properties which can support a different kind of representation of information from more formal or verbal representations. For example visual imagery allows the simultaneous apperception of large amounts of information, is significantly isomorphic, is easily retrieved [13] and can be memorable and emotive [11,30]. Such empowering properties may account for the empirical [e.g., 6,9,11,17] and anecdotal [e.g., 18, 30] evidence of the facilitative effects of visual imagery in learning and problem solving.

However, building a simulation in which the domain is to be described visually requires the use of self-generated *external* imagery (producing pictures, diagrams etc..), and there is less research which specifically addresses this issue. Many of the empirical studies cited above note the apparent symbiotic relationship between generating and manipulating internal imagery and constructing a picture or more abstracted diagram of the problem domain. Also, general guidelines on mathematical problem solving tend to advocate the use of drawings and diagrams which again points to their possible facilitative effect [e.g., 24,35]. Similarly, the reported anecdotal use of imagery by scientists is accompanied by the observed extensive use of technical drawing and diagrams in scientific and technical literature [e.g., 5]. The intimate relationship between the creation of external visual images and the generation of hypotheses is further explored by MacDonald-Ross [15] who also notes the paucity of literature on such a significant subject. The properties of external visual images which makes them useful in problem solving would appear to be similar to those of internal imagery. In their role as memory augmentation devices, self-generated external images can reify mental images, and increase the amount of material that can be maintained and transformed when thinking about a domain.

However, as Kosslyn [13] notes, one of the most noticeable aspects of mental imaged objects is that they often mimic the properties of real objects. When we imagine a ball thrown in the air, we can see that it acts as if it had weight, that it's bounce depends on it's material and so on. In a novel context, our prediction of the ball's behaviour cannot be due to replaying some previous experience. This presumably is due to our having deeper, more abstract semantic knowledge about the world with which we can reason and form causal relationships. Clearly, if we want to express this information, the use of images would not be optimal, and textual or formal representation of this information would be more appropriate. Of course, visual representations can be supplemented by explanatory text in the form of an annotated diagram or picture with explanatory text and this is clearly useful [34]. In this respect, it might be predicted that providing a friendly computer environment for the creation of animated, annotated displays may facilitate exploration of the represented domain, a point to be developed later. However, in specifying a computer based simulation the causal relationships and parameters need to be expressed explicitly in some formal, machine processable form.

Towards the design of a simulation building environment

Given the possible benefits of combining visual and linguistic representations, providing a computer environment for the creation of animated, annotated displays may assist exploration and understanding of the represented domain. Systems which support this, such as Apple's HyperCard™ or MacroMind's Director™, are already available. They provide the facility to create a picture or animated sequence of pictures using on-screen paint tools, or by importing a scanned or video-grabbed image. All other features being the same, a system which allows the production of animations would be expected to produce a different quality of understanding from that of a static only system. Constructing an animated display can bring a student into an enactive intimacy with the transitions, removing, replacing and transforming the key objects. This may facilitate the annotation of such transitions and possibly promote reflection on how the behaviour of the transformed object affects the other objects being depicted. Also, replaying two successive frames of an animation can make the changes of state more apparent than a side-by-side comparison, again possibly assisting reflection on the significance or appropriateness of the transformation. However, static views are also important, giving time for contemplating the whole system. In practice though, the construction of an animated display necessarily involves the generation of a set of static views, and the use of a storyboard assists overviewing of the whole sequence.

However, there would seem to be benefits not just for combining text and animation, but for combining a formal model and animation, where a model of a domain is taken to be one in which elements of that domain are represented in a machine processable way. Producing such models can demand a greater precision than a description using text alone. Also, because the resulting model is runnable, the inconsistencies in a poor model may become apparent. However, this means that the student must firstly learn the formal modelling language, which in the case of a general purpose programming language, may not be trivial. The alternative, is to integrate a modelling system, such as STELLA™ or an Expert system shell with an animation system, and STELLAStack [27] (a STELLA/HyperCard system) represents such an endeavour. The child quoted by [28] cited earlier, would clearly benefit from designing his/her own icons to represent the STELLA boxes, perhaps even going on to design a 'quantity space' or range of permissible state changes for that object. However, it is not yet clear how educationally useful this system is, and like its parent, STELLA, may be too complex for widespread use for children under the age of 12 for example. There is a case then for providing a computer based learning environment which combines a simple, possibly qualitative, modelling language with an animation construction system which could be used to produce models of sufficient detail as to be educationally useful.

An object-oriented basis

In view of the importance given here to cognitive considerations in designing a simulation building system, there are several reasons for supposing that it may be worthwhile to base the design on an Object- Oriented Programming approach (OOP):

- As argued by Williams, Hollans and Stevens [36], our everyday, intuitive reasoning about objects may involve the use of 'Autonomous Objects' - a kind of conceptual primitive or entity, which is characterised as having definite boundaries, an explicit representation of state, a set of internal parameters, a set of internal rules, and a representation of connectedness. So, for reasons of reducing the gulfs of execution and evaluation, and maximising congruency between domain, mental and expressive representation, an OOP approach may be appropriate.

- A computer based object whose behaviour is defined by the modeller, and which has a visual representation of it's state, can be an instance of a *Transitional* object as described by

Winnicott [e.g., 37]. Such objects are claimed to be powerful tools for understanding by having a metaphoric quality (one can see other objects in terms of them) and by lying at the boundary between what is known and familiar, and what is unknown and explorable. Hodgkin [8] has elaborated this notion, and sees transitional objects as toys which are intermediaries between external tools and internal symbols. Similarly, Papert [23] has used the concept of transitional object to explain the power of computer based objects (e.g., the 'Turtle' in LOGO).

- If the computer based objects are defined and seen as autonomous actors, they can take on the empowering quality of anthropomorphic metaphors. Seeing things in terms of a body image has been argued to be an important aid to understanding an unfamiliar domain, indeed Papert [23] has partly explained the power of the LOGO turtle as a tool for understanding geometry in these terms, a point reiterated by Pufall [25]. The importance (indeed the centrality) of the body image in cognition is further expounded by Johnson [10].

- As well as the above cognitive considerations, it should be noted that a large number of professional animation systems use the OOP approach, which suggests it's suitability for the task. [e.g., 26,16,32]. Intriguingly, this shift towards OOP is matched by a move towards a knowledge-based and qualitative animation specification.

Appropriate level of qualitative modelling

As indicated previously, developing a competent qualitative causal model may be important to the process of understanding a domain, so it may be useful to provide a means of expressing and exploring such models. Qualitative causal models are often informally expressed in small linguistic 'chunks' of how one part of the domain, in our case an 'object', affects another part [e.g., 36]. One way in which such descriptions may be accommodated by a computer based modelling system and made machine processable, is by using a Knowledge based approach whereby the 'chunks' describing the causation are embodied in user-defined production rules (e.g., If *condition a* Then *action b*). Given that an Object-Oriented approach is to be desired, a hybrid Knowledge Based / OOP system may be a good basis for an educational animation / modelling system, with the production rules being seen as an object's methods.

A prototype kernal has been developed [19], and ways in which children may create such knowledge-based animations are being investigated. Currently, the structural model of interacting objects is being explored, with the objects modelled being seen as Actors whose behaviour is specified by methods or rules depicted in a 'Thinks' bubble. These actors talk to each other, sending messages via 'Talk' bubbles. If a message matches the precondition of a rule, then that actor sends as a message, the action part of the rule, and so the next link in the causal model is generated.

Initial tests with a small group of 10-12 year olds, using a 'cardboard computer' to simulate the display, have been encouraging. After a 10 minute explanation and example, the forward chaining control and propagation of causality was rapidly understood, but not all children were able to construct new methods and states to achieve a desired system goal state. More promising perhaps, is the ease to which the children took to constructing animations using MacroMind's Director™ animation package. Future research will investigate the effects of constructing an annotated / non-annotated animation on the understanding of a domain. Also, the ability of 10 - 12 year olds to construct qualitative actor based models of a variety of domains will continue to be investigated.

Summary

Computers provide rich environments for learning and problem solving and we have found that focusing on their enabling properties can be useful in considering the design of learning environments. For example by exploiting direct manipulation interfaces to engender intimacy with the domain, allowing the examination of counter-factual examples, encouraging conflict resolution and facilitating reasoning at the qualitative level have proved useful in providing a simulation for exploring horizontal motion. Similarly, by looking at the issues surrounding memory augmentation in the form of self-generated external visual images has proved useful in considering the design of a simulation building environment.

References

1. Anzai, Y., & Yokoyama, T. (1984). Internal models in physics problem solving. *Cognition and Instruction,* 1, 397-450.
2. Chi, M., Feltovitch, P., & Glaser, R. (1981). Categorisation and representation of physics problems by experts and novices. *Cognitive Science,* 5, 121-152.
3. de Kleer, J., & Brown, J.S. (1983). Assumptions and ambiguities in mechanistic mental models. In D. Gentner & A.L. Stevens (Eds.), *Mental Models* (pp. 155-190). London: Lawrence Erlbaum.
4. de Kleer, J., & Brown, J.S., (1984). A qualitative physics based upon confluences. *Artificial Intelligence,* 24, 7-83.
5. Ferguson, E.S. (1977). The mind's eye. Non-verbal thought in technology. *Science,* 197, 827-836.
6. Gambrell, L.B., & Bales, R.J. (1986). Mental imagery and the comprehension monitoring performance of fourth and fifth grade poor readers. *Reading Research Quarterly,* 21, 454-464.
7. Gunstone, R., & Watts, M. (1985). Force and motion. In R. Driver, E. Guesne & A. Tiberghien (Eds.), *Children's ideas in science* (pp. 85-104). Milton Keynes: Open University Press.
8. Hodgkin, R.A. (1985). *Playing and exploring: Education through the discovery of order.* London: Methuen.
9. Hortin, J.A., Ohlsen, R.L., & Newhouse, B.S. (1985). Research for teachers on visual thinking to solve verbal problems. *Journal of Educational Technology Systems,* 13, 299-303.
10. Johnson, M. (1987). *The body in the mind: The bodily basis of meaning, imagination and reason.* Chicago: University of Chicago Press.
11. Kauffman, G. (1980). *Imagery, language and cognition. Toward a theory of symbolic activity in human problem solving.* Bergen: Universitetsforlaget.
12. Kosslyn, S.M. (1980). *Image and mind.* Cambridge, MA: Harvard University Press.
13. Kosslyn, S.M. (1983). *Ghosts in the mind's machine. Creating and using images in the brain.* New York: Norton.
14. Larkin, J.H., McDermott, J., Simon D.P., & Simon, H.A. (1980). Expert and novice performance in solving physics problems. *Science,* 208, 1335-1342.
15. Macdonald-Ross, M. (1979). Scientific diagrams and the generation of plausible hypotheses: An essay in the history of ideas. *Instructional Science,* 8, 223-234.
16. Magnenat-Thalmann, N. (1990). The problematics of facial animation. In N. Magnenat-Thalmann & D. Thalmann (Eds.), *State of the Art in Computer Animation: Proceedings of Computer Animation '89.* (pp. 47-58). London: Springer-Verlag.
17. McIntosh, W.J. (1986). The effect of imagery generation on science rule learning. *Journal of Research in Science Teaching,* 23, 1-9.
18. Miller, A.I. (1984). *Imagery in scientific thought.* Boston: Birkauser.
19. Moar, M. (1989). *A simple simulation builder for children.* Unpublished Master of Science Thesis. Kingston-upon-Thames Polytechnic.
20. Norman, D. (1986). Cognitive engineering. In D.A. Norman & S.W. Draper (Eds.), *User centred system design.* (pp. 31-69). London: Lawrence Erlbaum.
21. O'Shea, T., O'Malley, C., & Scanlon, E. (1989). *Magnets, martian and microworlds: Learning with and learning by OOPS* (CITE report No 53O). Institute of Educational Technology, Open University, Milton Keynes, Great Britain.
22. O'Shea, T., & Smith, R. (April 1987). *Understanding physics by violating the laws of nature: Experiments with the Alternate Reality Kit.* Paper presented at the Computer Assisted Learning Symposium (CAL'87), University of Strathclyde.
23. Papert, S. (1980). *Mindstorms: Children, computers, and powerful ideas* . London: Harvester Press.
24. Polya, G. (1971). *How to Solve it* . London: Penguin.
25. Pufall, P.B. (1988). Function in Piaget's system: Some notes for constructors of microworlds. In G. Forman & P.B. Pufall, (1988). *Constructionism in the computer age* (pp.15-35). Hillsdale, NJ: Lawrence Erlbaum.
26. Reynolds, C.W. (1982). Computer animation with scripts and actors. *Computer Graphics,* 16, 289-296.
27. Richmond, B., & Peterson, S. (1988). *A user's guide to stellastack,* Lyme, NH: High Performance Systems.
28. Riley, D. (1990). Learning about systems by making models. *Computers in Education,* 15, 255-263.

29. Scardamalia, M., & Bereiter, C. (September 1990). *A knowledge-building architecture for computer-supported learning*. Paper presented at the NATO Advanced Research Workshop on "Computer-Based Learning Environments and Problem Solving", Leuven, Belgium.
30. Sheppard, R.N. (1978). Externalization of mental images and the act of creation. In B.S. Randhawa & W.E. Coffman. (Eds.),*Visual learning, thinking and communication*. (pp. 133-189). New York: Academic Press.
31. Smith, R. (1986). The Alternate Reality Kit: An animated environment for creating interacive simulations. *Proceedings of the IEEE Workshop on Visual Languages* (pp. 99-106). Dallas, Texas.
32. Steele, B. (1989). OOP: Gets the message across. In *Computer Graphics '89. Proceedings of the Conference on Computer Graphics* (pp. 207-216). London: Blenheim Online.
33. Waller, R. (1979). Four aspects of graphic communication: An introduction to this issue. *Instructional Science*, 8, 213-222.
34. Waller, R. (1989). *Typograpghy and discourse* (CITE report No. 97), Institute of Educational Technology, Open University, Milton Keynes, Great Britain.
35. Wickelgren, W. (1974). *How to solve problems*. San Francisco: Freeman.
36. Williams, M.D., Hollans, J.D., & Stevens, A.L. (1983). Human reasoning about a simple physical system. In D. Gentner & A.L. Stevens (Eds.), *Mental Models* (pp. 131-153). London: Lawrence Erlbaum.
37. Winnicott, D.W. (1971). *Playing and Reality*. London: Tavistock.

Direct Manipulation of Physical Concepts in a Computerized Exploratory Laboratory

Vitor Duarte Teodoro

Faculdade de Ciências e Tecnologia, Universidade Nova de Lisboa
2825 Monte de Caparica, Portugal

Abstract: Computer microworlds are becoming more and more powerful for learning and teaching science. However, even a powerful computer microworld is not enough, by itself, to enable students to explore and learn about a formal domain. It is argued that it should be integrated with other media, especially books. In a computer microworld, direct manipulation techniques allow the implementation of direct manipulation of physical concepts, as it is shown with an example from physics – NEWTON, a computerized exploratory laboratory. In this computerized exploratory laboratory — a conceptual laboratory — the user can explore and experiment with concrete-abstract objects, confront multiple representations and pose and devise strategies to learn about the most fundamental phenomenon of Nature: motion.

Keywords: computer simulation, physics, software development.

An Approach to Computer Use in Science and Mathematics Education

In this section I introduce the concept of *metabook* as a possible answer to the felt need to change the role of software and books. It is argued that a metabook is a learning instrument, which intimately relates a book with one or more pieces of software of the genre computer conceptual laboratory. These are pieces of software based on a direct manipulation metaphor. Such software has special interest for exploring formal mathematics and scientific domains. The modalities of use of conceptual laboratories combine exploratory learning, direct teaching, a qualitative/semi-quantitative view of science and mathematics education and an inductive view of learning, supported by a constructivist approach to the process of learning.

A new role for computers and software in education? The concept of *metabook*

There is a general agreement that the promises of enthusiasts of computer use in education have not been fulfilled [3]. Nevertheless, there is also a general agreement that this it is not due to any lack of potentialities of the computer itself but rather to the lack of "organizational facilities" in schools and, sometimes, to the lack of quality software or to insufficient teacher education regarding the role of the computer in the curriculum.

I believe that the remarkable capacities of computers cannot be completely explored in schools, specially in the classroom. If we think of the computer as a personal tool — "a tool to think with" [19] — we can easily see that a student needs plenty of time to use it *personally*, outside the classroom, most of the time alone (as he uses his books or his pencil) and sometimes working together with colleagues. The rigidness of the organization in schools, specially high schools, constitutes a strong obstacle to a frequent and intimate computer use. We can, then, understand why the use of computers is most of the times limited to the classes of the computer enthusiast teacher, who brings his class to a special classroom, filled with computers, and who spends lots of time preparing materials, disks, etc. But we can imagine another relevant place for the computer: the home (or a school resource center, where a student can work alone), as a companion to books, paper and pencils.

This "view" on when and how students use computers for learning purposes stresses the fact that students must have a stronger *personal* link with the computer. But this view demands a new type of book: books that are accompanied by software, in a very *intimate* way. We can call this new genre of books *metabooks*. A metabook is an integrated package (written material and computer software), in which the software is considered an integral part of the book, *allowing the student to do everything he cannot do with the book*. For example, it is easy to imagine a metabook for high school Basic Geometry composed of the book itself and a piece of software, like Geometric Supposer [18], or a metabook on Newtonian mechanics composed of book plus software such as NEWTON [23] (also presented in the second part of this paper). I can imagine that metabooks will be very familiar to students by the end of this decade, when the personal computer will be so widespread as videocassette recorders are now. This is a challenge to book and software authors.

The changing role of the computer in education

There is now another general agreement about the role of computers in education: the capacity of the computer as a tutor (either as an "intelligent tutor" a "classic tutor") is very limited. It is possible to consider at least two different classes of arguments to support this idea. First, "it is too difficult to get a model of the student" [2, 13], an essential pre-requisite to intelligent tutoring. Secondly, what can software do with what it "knows" about the user, if that knowledge is almost always very circumstantial and partial [19]? So, I believe that the use of the computer as a tutor will always be very limited, as limited as the use of books as "self-teaching tutors" — one of the objectives of the programmed education movement of the early sixties.

What remains, then? Apart from all the now classic software tools that can really change the way students relate themselves to knowledge, knowledge retrieving, knowledge communication, knowledge search, etc., such as word-processors, database management systems, spreadsheets, telecommunication programs (and networks and communication services), etc., we are now witnessing the development of new kinds of software tools, specially in science and mathematics education. These new tools allow the user to explore a formal piece of knowledge in a personal way. They are, in a certain way, the development of early ideas presented in the beginning of the eighties by authors such as Papert. I call these tools computer conceptual laboratories or just conceptual laboratories. Similar terms are *microworlds* — computer representations of domains [11] — and *intellectual mirrors* — a "genre of software environments in which users, be they students or teachers, can explore an intellectual domain" [18, p. 58]. A conceptual laboratory, together with an intimately related book, can constitute a metabook.

The general idea is that the software must give the student full control to *express* and *test* his ideas about formal objects and representations. This was, in a certain way, one of the assumptions under Logo [14]. But the early work of Papert and others [e.g., 5] was based on a computational metaphor — *programming* — that is too elementary to allow the exploration of many fields. The new software tools for learning need to be based on a more powerful metaphor — a metaphor that allows the direct manipulation of *concrete-abstract objects* [10]. This metaphor is the *direct manipulation metaphor* [21].

A direct manipulation metaphor for software development in science and mathematics education: the notion of concrete-abstract objects

We are now all so familiar with friendly computer interfaces that we have almost forgotten "old" interface styles, such as the command-based ones, with strict syntax. The WIMP (Windows, Icons, Mouse and Pull-Down Menus) interface style combined with "What You See Is What You Get" (WYSIWYG) and "Multiple Windows" (MW) started a revolution only six or seven years ago. It seems now clear that this revolution has had more impact on general computer use, and specially on education, than work on artificial intelligence. But WIMP-WYSIWYG-MW interfaces are, sometimes, not enough. There are situations in which we need objects that can be directly manipulated on the screen, usually with a mouse and sometimes with the keyboard.

The *direct manipulation interface* concept includes all the concepts of WIMP-WYSIWYG-MW interfaces and the concept of *screen objects that represents either physical objects or psychological objects*. A "psychological object" is a psychological construct that can have physical meaning, such as *velocity*, *momentum* or *force*.

A direct manipulation interface is the only interface that allows the creation and manipulation of *concrete-abstract objects*: these objects are "concrete" because they react to the user's actions, using the mouse or the keyboard [10] and are "abstract" because they do not exist as real objects — only as screen objects — or because they frequently represent physical or mathematical constructs or representations, such as *vectors*, *graphs*, etc.

Whereas with a command-based interface actions are mediated by an *artificial process* — a language with strict syntax — with a direct manipulation interface actions — on objects on the screen — *are not mediated by a language*. This enables the user to establish a much closer relationship with actions, which is fundamental for experiential learning.

One of the problems related to the development of direct manipulation software for specific domains in science and mathematics is to define the primitive operations the user can do. As Schwartz [18, p. 60] pointed out, "primitive operations are not the most parsimonious constructs that can be concatenated to formulate the discipline it deals with, but rather are a more richly elaborated set of primitives". The definition of the primitive operations can only be based on research on learning in traditional environments and in computer environments. These operations can be very complex, as the example presented in the second part of this paper show.

Developing science and mathematics software: a multi-dimensional space

Levin and Waugh [11, p. 73] pointed out six fundamental *dimensions* in computer-based environments (simulations, tools, games and microworlds) for learning: *perceptual fidelity* ("the extent to which the computer is seen in a way similar to the situation being modeled"), *manipulative fidelity* ("the extent to which the learner´s actions correspond to the actions to be taken in the domain being modeled"), *functional fidelity* ("measures the correspondences between the internal structure of the model — the parts of the simulation that mediate between the learner´s actions and the perceptual scene presented to the learner — and the internal structure of the domain being modeled"), *dynamic support* ("sequence of systematically decreasing the amount of assistance provided to learners as they progress from novices to experts"), *breath of scope* and *learner control*.

These six dimensions account for all the relevant aspects of science and mathematics software — and, in particular, conceptual laboratories — but it is necessary to consider that in a conceptual laboratory some dimensions can change their "values". For example, in certain programs, it is possible to consider *different levels of perceptual fidelity*: the perceptual fidelity can decrease as the student knows more and more abstract concepts related to the theme of the program. As shown in the second part of this paper, this is one of the most important features of a piece of software when its domain is highly complex and when plain "common sense" is far away from it.

Table I shows a possible "value" in each dimension of a computer conceptual laboratory.

The Psychology of learning and its contribution to software development

The now dominant constructivist theoretical perspective of learning assumes that learning is an active, constructive (i.e., it changes over time) and interactive process. I prefer to use the term "theoretical perspective" because, as Fischbein [8, p. 11] pointed out, the constructivist

Table I

"Value" in each dimension of a conceptual laboratory

Dimension	"Value"
perceptual fidelity	from low to high
manipulative fidelity	from low to high
functional fidelity	from low to high
dynamic support	none or low
breath of scope	variable
learner control	high

approach to mathematics education (and to science education) "still lacks 'law' and 'instrumentation'". This perspective emerged in the last two decades from the work of psychologists and other social scientists, computer scientists and educationalists. The following is a brief summary of the constructivist view of learning, quoted from Forman and Pufal [9, p. 236].

For these authors, the "properties of constructivism" are

> (...) epistemic conflict, self-reflection, and self-regulation. Epistemic conflict involves two knowing systems. These systems may originate in different individuals, and it may be that in early development we are more dependent on externally induced conflict than we are subsequently. Whatever the source of conflicting epistemic stances, if there is a resolution it is within the individual experiencing the conflict, that is, it is an individual construction. If the resolution is developmental, in the strict sense, it means constructing a new way of thinking about reality and is marked by logical necessity.
>
> Self-reflection can be construed as a response to conflict. Perhaps paradoxically, it means objectifying our knowledge by specifying our assumptions about reality, about procedures for solving a problem, and so forth. Reflectivity is an opportunity to explicitly construct or transform our way of representing reality. At the same time, and embodied in the same intellectual act, there is the potential for developmental restructuring of thought. Developmental restructuring is self-regulation. Through our thinking about our practical and conceptual actions, these actions can be restructured to a "higher" level of knowing. These developmental transformations, though materially linked to empirical transactions, reflect self-organizing properties of the knower that reflectively, but not consciously, abstract structure from action.
>
> In sum, conflict and self-reflection are, of necessity, conscious, whereas the last step, self-regulation, that is to say, developmental restructuring, is in all likelihood unconscious in process and outcome. That is, we do not explicitly construct new logical structures of mind and do not need to be aware of them when they function in our day-to-day problem solving.

Educational software — and conceptual laboratories, in particular — can be a powerful source for cognitive conflict. Using a conceptual lab, students can explore a formal domain, usually a domain which they are not familiar with (e.g., Euclidean geometry, trigonometry, Newtonian dynamics). They can explore it by confronting their intuitive ideas with the formalism of the domain.

As Schanck [16, p. 5] pointed out, understanding "is not an all-or-none affair. People achieve degrees of understanding in different situations depending upon their level of familiarity with those situations". My view is that understanding is just this: *to be familiar with*. If you are *familiar with*, you *understand*. If you are not familiar with something, you don't understand it (e.g., now most young people understand perfectly that the Earth moves around the Sun: that is, they are familiar with the solar system's picture on TV, in comics and books...). Becoming familiar with something is an on-going process in which the environment plays a fundamental role. If you have rich environments, where complex ideas can be explored in a very informal way at early stages, you can become increasingly familiar with those ideas, i.e., you can understand them.

Conceptual labs create rich environments where students can increase their familiarity with most of the "strange" scientific ideas. For example, the concept of *inertia*, that took thousands of years to develop, can now become completely familiar to a young student who came across a software tool like Newton or with some videogames.

Schanck [16] considers the "spectrum of understanding" as composed of a continuum where the relevant points are "making sense — cognitive understanding — complete empathy". Usually, learning science in school is limited to making sense or, in the best hypothesis, to cognitive understanding. By giving students the opportunity to explore, i.e., to "play", with ideas and relations — conceptual labs can induce the necessity of cognitive understanding. For example, with NEWTON, a student can easily explore how the height of a projectile relates to its initial velocity — it is easy to explore this problem and "make sense" of the relation — but only with formal manipulation of dynamic and kinematic laws can he have formal proof — that is, a cognitive understanding — of the relation. These cognitive understandings can easily be found by some students, after having explored and having become familiar with the relation without formal knowledge of dynamics and kinematics.

Qualitative versus quantitative science and mathematics

In a recent article, Joanne Striley [22], following DiSessa and other authors, argued that the almost exclusive use of mathematical explanations seems to give students an opaque vision of physical laws and drive them to maintain their previous misconceptions or intuitive ideas, even when they can solve traditional physical problems using equations.

DiSessa [4, p. 70] formulated a notion of science teaching that:

> (1) is contiguous with the previous knowledge that children have (...) (2) develops largely using the same mechanisms as their intuitive knowledge so that we can expect it will feel easily and natural

(...) (3) incorporates one of the substantial structures changes that distinguish science from intuition.

This notion stated the importance of a shift in science and mathematics education: *from learning symbols and algorithms without meaning* to *learning meaning before symbols and algorithms*. That is, before students learn formal science and mathematics, they should be completely familiar with the fundamental ideas in a verbal way, without knowing them in an explicit and formal way.

Qualitative science and mathematics is, in a certain way, a "game" of words, verbal descriptions and relations, sketches of graphical representations and sketches of graphs. It is postulated that discussing and using the right words to describe phenomena and relations, sketching a phenomenon in different instants, sketching a graph of property against time or another property compel the student to become more and more familiar with scientific thinking. Only after one is completely familiar with *qualitative* (in fact, *semi-quantitative*) ways of describing and relating physical phenomena and mathematical ideas, is one able to become familiar with formal science and formal mathematics.

A conceptual lab is a tool for stimulating verbal and holistic discussions about a formal domain, using qualitative reasoning. But it is not sufficient, as I shall comment below. Books, teachers and peer students are fundamental.

An Example of a Conceptual Laboratory for Exploring Newtonian Dynamics

NEWTON, a conceptual laboratory in the domain of particle dynamics

In this section, I will present a computer conceptual laboratory (NEWTON[1]) for learning and teaching elementary dynamics. NEWTON is a prototype developed for the Apple Macintosh; it tries to address the difficulties in learning dynamics, at different levels of conceptual development. These difficulties are well documented in literature (see, for example, [24]). Some of the characteristics of NEWTON are the following:

(i) NEWTON has levels of complexity (accessed through passwords) related to the cognitive development phases of learning in dynamics; what we can see in a simulation in the first level is very close to what we perceive from real movements (Figure 1) — in Levin and Waugh's [11] words, there are different levels of *perceptual fidelity* — as the level increases, there are more and more possibilities beyond what one can perceive from observation, such as vectors, graphs of a physical quantity against another physical quantity, values, vector components, etc. (Figure 2).

1. NEWTON was designed by the author with the collaboration of João Tavares da Silva.

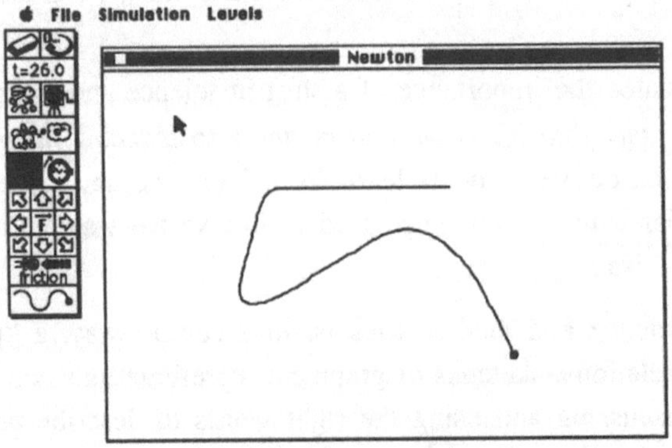

Figure 1. Observing the trajectory of a particle acted by forces in different directions in different instants (done in level 1).

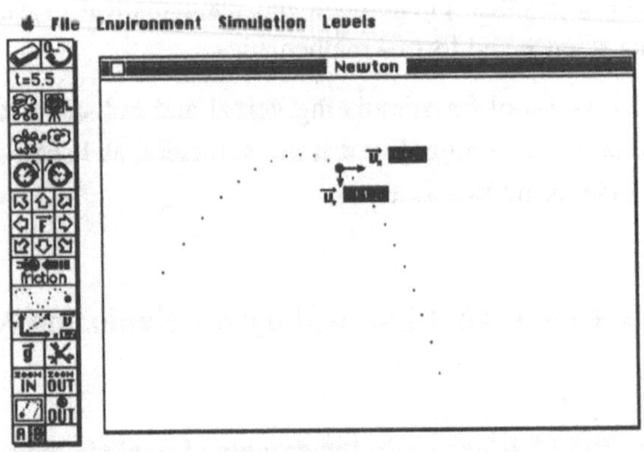

Figure 2. Investigating the change in the velocity components in a projectile motion with friction (done in level 3).

(ii) NEWTON allows the direct manipulation of physical constructs — physical variables — such as velocity (Figure 3), force, momentum, etc.;

(iii) when observing a movement, the user can watch the "real object" and/or what science "adds to the real thing" describing the movement. For example, it is possible to watch an object moving to the left and to the right or watch the same movement with the radius vector connected to the object and moving in real time — this allows the control of perceptual fidelity;

(iv) students can study the effect of friction and gravity on a movement, by "connecting" or "disconnecting" friction and/or gravity after the movement has started — see Figure 4 and Figure 5. These features allow them to explore "real worlds" and "artificial worlds";

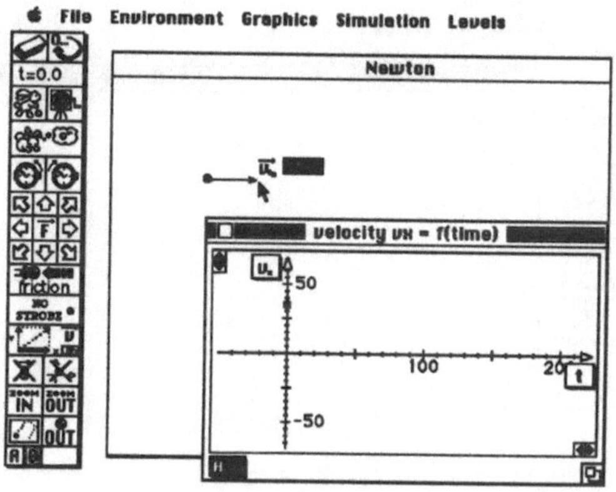

Figure 3. It is possible to manipulate directly (using the mouse) physical constructs, such as velocity.

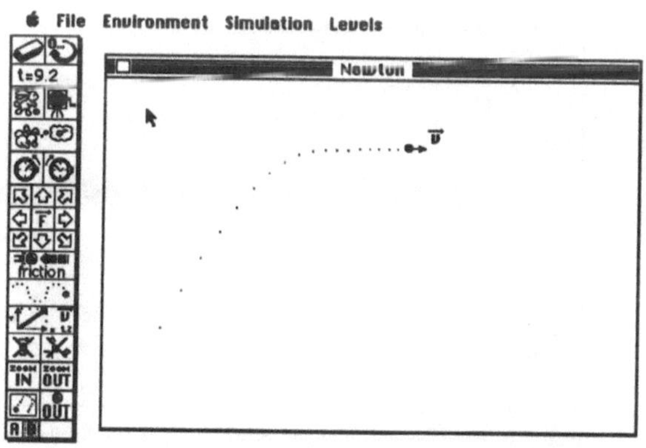

Figure 4. Investigating the effect of "disconnecting" gravity after starting a projectile motion (done in level 3).

(v) with NEWTON it is possible to make a detailed study of how different physical quantities change with time: the program allows the user to repeat an "experiment" as many times as he wishes, automatically or having full control of time, either backwards or forwards;

(vi) the user can control most of the simulation parameters, such as the time step of the stroboscopic observation, the intensity and the direction of gravity, the origin of a reference frame, etc.;

(vii) it is also possible to zoom in or zoom out, change scales on graphs, see multiple graphs simultaneously, change the physical quantities shown in a graph, etc.;

(viii) it is possible to confront in real time multiple representations, such as the trajectory of an object that moves right, then left, then stops, then... (see Figure 6);

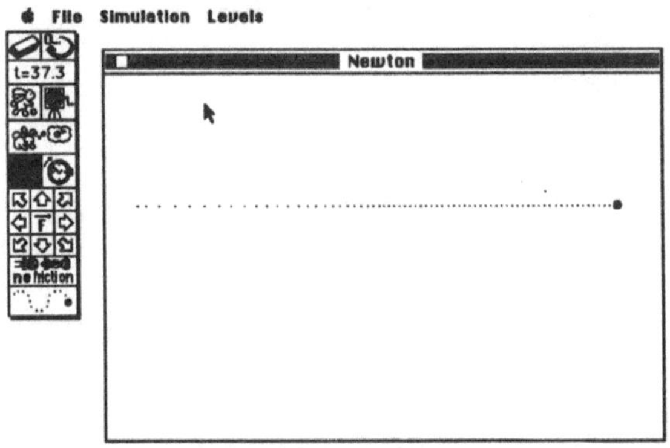

Figure 5. After an initial push, a particle is abandoned with friction and then friction is "disconnected" a few seconds later (done in level 1).

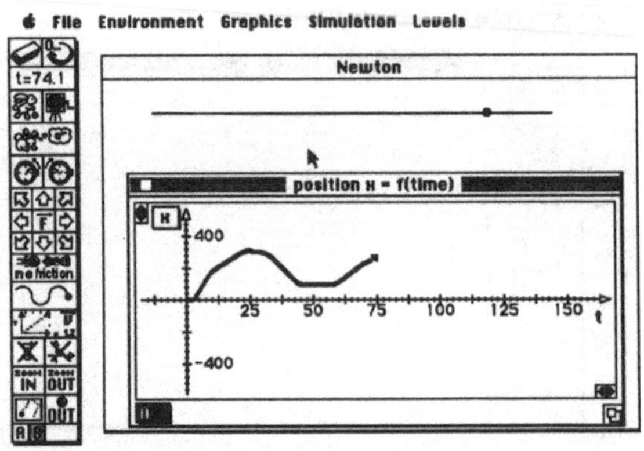

Figure 6. Confronting the representation of a trajectory with a graph of position as a function of time (done in level 4).

(ix) one of the most interesting features of NEWTON is that the user can hide the window where the movement occurs and see only graphs that show what is going on (see Figure 7);

Some notes about the modalities of use of NEWTON

The modalities of use of NEWTON should be bounded by the following statements:

(i) Newton was conceived to be used in conjunction with experiments with real objects, such as small metallic balls;

(ii) the use of the program by individual students or small groups of students should be combined with the use of the program by the teacher in front of the whole class with an LCD screen and an overhead projector;

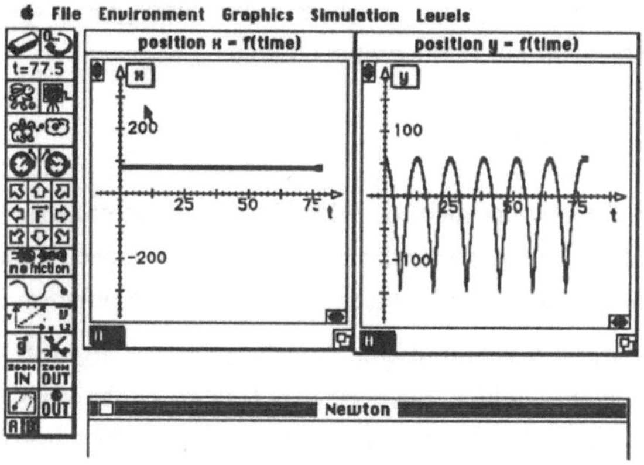

Figure 7. What is happening with the particle? (done in level 4).

(iii) the use of the program should be always based on questions leading to investigations or problems to be solved. The program should not be used just for browsing about its features.

The program stresses the role of stroboscopic views as the most important representation of movement. Such representations allow the student to get an immediate perception of the trajectory and the velocity and, for more advanced students, an indirect perception of acceleration and force. Discussions about movements should always include stroboscopic views in order to get that immediate perception. With NEWTON, students can see movement, movement and stroboscopic views, or movement and trajectory.

I hypothesize that most of the difficulties and misconceptions in the learning of dynamics are related to the impossibility to see vectors in association with moving objects in real time. In order to give students the opportunity to surpass these difficulties, it is possible for them to "connect" different vectorial quantities to the particle and see how they change with time under different conditions.

After an experiment with Newton, time can be reverted to zero and the experiment can be repeated by advancing time (or rewinding time) with the student in full control. It is also possible to replay the experiment observing different physical variables. For example, the user can make an experiment and see what is happening to the velocity and then replay the experiment seeing what happened with force.

I also assume that most of the difficulties and misconceptions around graphs are a consequence of the fact that students can't watch graphs being constructed while objects are moving. With NEWTON, students can observe one or more graphs being constructed at the same time as a particle is moving, or when it is at rest. They can also replay the experiment with different graphs and control the advance of time during the replay to analyze the correspondences on specific instants or positions.

Most of these facilities allow the student to make conceptual experiments that can significantly contribute to bring about a conceptual change.

A direct manipulation interface

The key problem in the development of NEWTON was the construction of an interface that would allow the student to manipulate the concepts and the representations in real time. I think that this was successfully achieved by means of a direct manipulation interface.

The NEWTON's direct manipulation interface uses a control panel, which is always present on the screen and which possesses increasing capabilities and complexity as the "level" increases. Most actions can be carried out only from the control panel (see Figure 8) or with "buttons" placed in other windows, as in a graphical window (Figure 9).

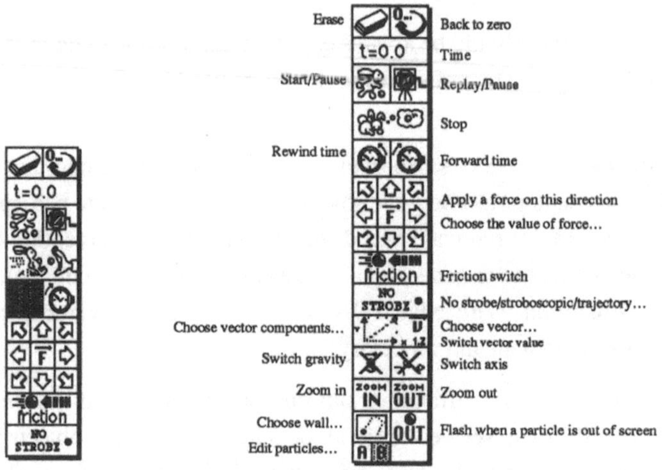

Figure 8. The control panel (level 1 and level 5).

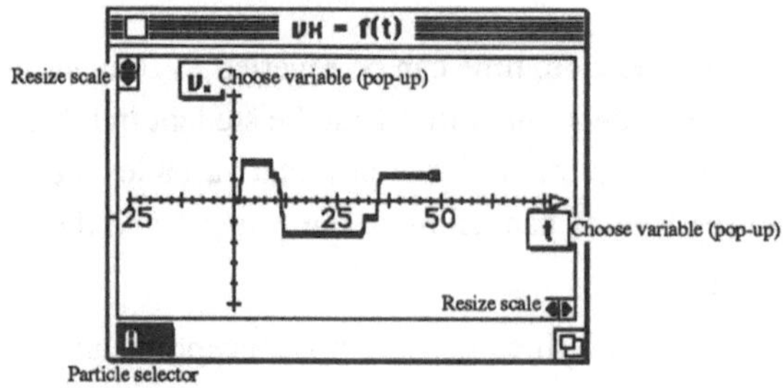

Figure 9. Buttons in a graphical window.

In the control panel there are icons that correspond to buttons with two conditions, like the gravity button — gravity can be "connected" or "disconnected" — or buttons that present a graphical pop-up where the student can choose an option — like the vectors button or the components button — Figure 10. There are also buttons for actions — start, stop, replay, return to zero, erase, forces, etc. — when the user clicks them on, the action is done.

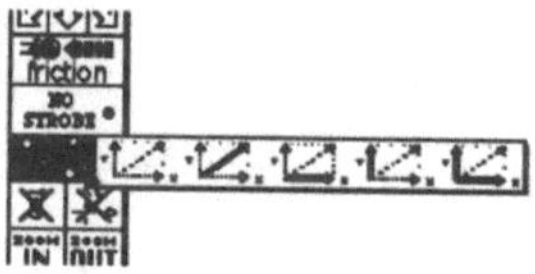

Figure 10. A button that presents a graphical pop-up.

When the development of NEWTON was in an advanced phase, I noticed that it would be very difficult for a child with no knowledge about physics to manipulate the control panel. There were options which were meaningless to the child or which could be associated with strong conceptual difficulties. Because of this, the program was redesigned with "levels", accessed by means of passwords. In the first level, the students have no access to vectors, gravity, graphs, frame of reference or other complex and highly conceptual features. In this level, it is only possible to see almost exactly what happens in reality.

The actions that are allowed at the different levels are based on research on conceptual evolution of learning dynamics [12,24]. For example, in level two, students can represent only velocity and force, two physical concepts that should come first in concept development in dynamics. Force is often associated with velocity. As Viennot [25] pointed out, students tend to use an intuitive "law" that can be expressed as "force proportional to velocity". In this level, students can make meaningful experiences about the relation between force and velocity, between force and change in velocity, between force, velocity and trajectory, and become more and more familiar with these physical constructs and the relations among them.

In level four, students can start using graphs of physical variables against time. For example, they can investigate the relations between trajectory and one or more graphs.

In level five, "everything" is allowed. For example, it is possible to investigate relations between variables using phase diagrams — one variable against the other. Figure 11 shows a graph of kinetic energy versus vertical velocity when an object is bouncing without friction in a vertical straight line (would that be a different graph if the object was also moving to the left? ...). This kind of graph can stimulate semi-quantitative / semi-qualitative discussions and investigations about relations between variables.

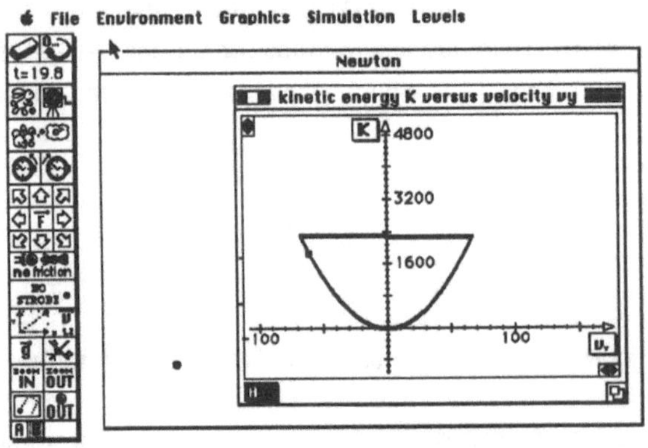

Figure 11. Kinetic energy of a bouncing ball versus vertical velocity, without friction.

Learning in an exploratory environment

Exploring a world and watching it evolve can be a remarkable learning experience. But real worlds are difficult to understand if one only uses common sense. As Driver [6, p. 9] pointed out:

> If we wish children to develop an understanding of the conventional concepts and principles of science, more is required than simply providing practical experiences. The theoretical models and scientific conventions will not be 'discovered' by children through their practical work. They need to be presented. Guidance is then needed to help children assimilate their practical experiences into what is possibly a new way of thinking about them.

Children need guidance and opportunities to explore scientific ideas, concepts and representations, and confront them with their prior ideas. Exploring and expressing scientific ideas — exploratory learning "permits pupils to investigate views of a given domain, which are different from theirs"; expressive learning "permits pupils to represent their own models of a domain and in this way reflect upon and explore their own models" [1, p. 37] — is something very different from just being told about an idea. When students explore and/or express ideas they can get a profound insight into the ideas they are exploring. Exploring and expressing ideas and representations, in a guided way, allows the student to build support for the not so obvious models that science uses.

Learning in a computer-based exploratory environment — what I call a conceptual laboratory — such as NEWTON, demands from the students both cognitive and metacognitive skills. Cognitive skills, in the sense that they must act to manipulate and to integrate information. Metacognitive skills, in the sense that they must regulate and control their actions and knowledge in order to build new knowledge.

NEWTON, like other exploratory environments, can only be of some use if the student devises and evaluates strategies. For example, in order to investigate the effect of gravity on velocity, students must design a series of experiments in which a particle is sometimes submitted and sometimes not submitted to gravity and see the effects on velocity vector components.

Posing problems, acting, verbalizing, controlling, predicting, devising strategies, evaluating, confronting, solving problems, are some of the cognitive and metacognitive processes that students must develop and use to construct their own knowledge with NEWTON.

It is not possible for a student to infer newtonian mechanics from observation. Newtonian mechanics is, in a sense, a consequence of the definition of force as the rate of change of linear momentum, a definition that is one of the first applications of differential calculus invented by Newton and Leibniz in the 18th century. But students can get a deep insight into the meaning and the consequences of this definition — without being aware of it — if they are allowed to do "direct manipulations" on all physical concepts related to the movement of a "particle" on a computer screen, e.g., velocity, force acting on the particle, position of the particle, etc., and watch the changes in real time.

One of the assumptions in the design of NEWTON is that knowing and understanding means *becoming familiar with different representations* of a phenomenon, as I assumed in the first section of this paper. With this program, students can have the opportunity to become familiar with the effects of forces, with vectors that represent physical constructs, with change in vectors during time, with relations between physical constructs and time (either in an analogical graphical representation or property vs. time graph), etc.

"Extending the range of manipulable objects" and "linking multiple representations" should be two "distinct contributions" that software must bring to science education with the aim of "revealing students' thinking" to provoke a conceptual change [7]. NEWTON is intended to extend the range of manipulable objects in the learning of dynamics, namely the abstract physical concepts of velocity, force, momentum, energy, etc. It also allows the student to confront multiple representations of movement in real time.

Research in science education, and specially in physics education, has produced an enormous body of knowledge about the ideas of students prior to teaching and about how students learn and develop concepts. I think that now it is time to use that knowledge and the knowledge about how new information technologies should be used in learning to develop and evaluate learning environments that are able to promote meaningful learning.

Future developments of Newton

NEWTON is an unfinished prototype. There are a few features which have not yet been implemented and which could contribute to significantly improve its potential use.

The following features are being implemented:

- two references frames, instead of one, with relative motion to each other; it would then be possible to explore physical quantities like velocity, force or momentum in different reference frames;

- elastic, inelastic or partial inelastic collisions between the particles;
- use of particles and rigid bodies instead of only using particles, to confront the dynamics of rigid bodies with the dynamics of particles;
- introducing the possibility of having massive bodies on the screen with gravitational interaction to explore the movement of celestial bodies and probes;
- introduce the possibility of having tensions applied to particles to study movements like the one of a pendulum.

The above features are relatively simple extensions of what is now implemented. But there are other possibilities that cannot be considered as simple extensions.

Linking simulation to data lodging

One of these new possibilities was suggested by Judah Schwartz (and used in one of his programs — Newtonian Sandbox): use of the mouse as an analogue input device to study movements. Schwartz [19] used the acronym MBL (mouse based laboratory) also used by other authors to mean Microcomputer Based Laboratory. With this feature implemented in the program it will be possible to move the mouse and lodge data. Then, the user can replay the movement on the screen, seeing all the physical quantities associated with the movement, displaying vectors, values or graphs.

Another feature that will be implemented is the possibility of using analog devices — two sonic sensors — to acquire data in two dimensions from the movement of a person, the movement of a ball or any other object and then reproduce it on the screen seeing the physical quantities that describe and explain the movement or seeing the graphs. The aim of this feature is to link data lodging to simulation. With this feature, students can explore not only movements created by the computer but also movements that they see in the real world.

The above feature represents a link between simulations and data lodging. There is another possible link that a program like NEWTON can establish: a link between simulation and modelling.

Linking simulation to modelling

The importance of modelling software in science teaching is growing [1,17] and research about its effects is starting to appear (e.g., [26]). Modelling software allows the user to fit or explore quantities related to a phenomenon in mathematical expressions or fit or explore mathematical relations to the known behavior of a phenomenon. There is also another kind of modelling, semi-quantitative or qualitative modelling [1]. In this type of modelling there are no mathematical expressions to explore: there are relations of the kind "if this variable increases, that variable increases a lot" or "if this decreases, that increases a little", etc. One of the features that will be implemented in NEWTON is a modelling module, with different characteristics in the different levels of the program. In the first and second level, the modelling capabilities will be qualitative: the student will have the possibility to choose a model for the movement of the particle (e.g., "force increases, velocity increases"; "no force, no velocity", etc.). With this possibility, the student will be able to explore his/her own conception about movement, expressed in a qualitative way, and confront his/her conception with what happens in reality. In the higher levels of NEWTON it will be possible to express models in a more quantitative way, either using verbal expressions (e.g., "force is proportional to velocity"; "force is proportional to distance", etc.) or mathematical expressions and confront the model with the consequences of it.

Questions for research

As Levin and Waught [11, p. 78] pointed out, research questions such as

> 'Will a computer-based science laboratory simulation promote student achievement as well or better than a traditional laboratory experience' have been far too common. Despite the fact that extensive data analysis may establish the superiority of the computer based simulation treatment group, little generalization of the experimental results is possible because the nature and the quality of the simulation experience which was examined may bear little resemblance to other computer-based laboratory simulations.

But, accordingly to the same authors, now "new and potentially important research questions can be asked". The range of questions suggested by the use of exploratory environments like NEWTON is enormous. There are questions related to modalities of use, to psychological effects, to teaching styles, to relations between learning in a simulated environment and learning in a real environment, to concept development supported by simulations, to use of exploratory environments as tools for problem solving, to effectiveness of the simultaneous use of multiple representations, etc. Some of these research questions — more specifically, the questions about problem solving and concept development supported by simulations — are being investigated by the author of this paper.

How Do Students Use a Conceptual Laboratory to Solve a Problem: Making a Movement that Describes a Trajectory with a Right Angle

In this section I will briefly describe part of a preliminary case study using NEWTON with junior high school students (aged 14) — group A, not yet introduced to Newtonian mechanics, and with students (aged 19) — group B — who had just finished a course on Physics in their first year of an university degree in Physics. Results presented were obtained with five students — three from group A and two from group B. None of the subjects was familiar with the Macintosh and the mouse and all the junior high school students were not familiar with computers at all.

The students were presented with simple problems that they should try to solve using NEWTON. One of the problems was: "how can we make an object describe a trajectory that makes a right angle, on a frictionless environment?". The aims of this study were: (1) to investigate how the students used the program: on an exploratory mode, by trial-and-error, or after planning actions; (2) to investigate how many tries they needed to discover how to solve the problem and what they did in order to solve it; (3) to investigate if there was any evidence of change in the ability to correctly answer physics questions about qualitative aspects of motion.

The students used the program alone — in level one (the closest level to what they can see with real objects) —, in the presence of a researcher, who explained to them in about 2 to 3 minutes how to use the program and what the problem was they would try to solve using the program. They showed no difficulty at all in using NEWTON. The session was videotaped — the camera recorded the dialogue between the researcher and the student and the screen of the computer. Each session took between 25 to 45 minutes. Before and after the session the students answered a paper-and-pencil test with 8 short essay qualitative questions about motion.

All the students used the program on an exploratory mode, even after the researcher suggested to them to "think first, do it later". They seemed not able to devise a plan of actions to be undertaken. Four of the five students (only one of group B was able to immediately do the correct actions to solve the problem), used trial-and-error strategies to solve the problem. These four students followed the same path to solve the problem: first, they tried to do the trajectory with a right angle applying a force that was perpendicular to the trajectory the object was following. They insisted three to eight times in using this process. After that, all four tried a force with an 90° plus 45° degrees angle with the trajectory (see control panel of the program Figure 8). Again, they insisted two to five times in using this process. The students, then, tried the beginning of a correct strategy: stopping the object. Two of them connected the "friction" bottom to stop the object and then apply another force. These two were asked by the researcher if there was another possibility to solve the problem. They immediately tried the three-stage strategy: first, stop the object, applying a force in an opposite direction to the movement, and then applying a perpendicular force.

Results on the pre- and on the post-test showed that none of the students changed their answers after the session. With the exception of one of the Group B students, all the other students had a weak and partial understanding of qualitative aspects of motion.

These results suggest that students used NEWTON on an exploratory mode, with trial-and-error strategies. They were not able to devise a plan to search for the solution, even after it was suggested to them to do that. The two-stage solution to the problem was discovered by four of the five students in an exploratory way, repeating most of the actions.

This very preliminary and partial work with NEWTON shows that students don't have any difficulty in using the facilities of the program — al least, in level one — and that they tend to use it only on an exploratory mode. This is an important point to focus on, because most of the investigations one can do with the program need a careful plan and more elaborated strategies. Students therefore need strong guidance and support from written materials and from the teacher, either on content, or on methodology of "research" using exploratory environments.

It is also important to stress the fact that none of the students changed his answers to the questions of the test after using the program. It seems that using the program during short periods of time doesn't have any cognitive effects.

Acknowledgments

The work reported in this paper is part of a research project conducted by the author under the supervision of Professor Maria Odete Valente. This research is partially funded by PROJECTO MINERVA and Portuguese Apple representative, Interlog s.a. The author acknowledges the assistance of Ana Laura Metelo in translating this paper into English.

References

1. Bliss, J., & Ogborn, J. (1989). Tools for exploratory learning. *Journal of Computer Assisted Learning*, *5*, 37-50, 1989.
2. Cabrol, D. (1990). Le rôle de l'intelligence artificielle dans l'enseignement: l'Example de la Chimie. *Proceedings of "Encontro Computadores no Ensino da Física e da Química"* (pp. 59-60). Coimbra, Portugal.
3. Cuban, L. (1989). Neoprogressive visions and organizational realities, *Harvard Educational Review*, *59* (2), 217-222.
4. DiSessa, A. (1987). Artificial worlds and real experience. In R. W. Lawler & M. Yazdani (Eds.), *Artificial Intelligence and Education*, Vol. 1 (pp. 55-77). Norwood, NJ: Ablex.
5. DiSessa, A. (1982). Unlearning Aristotelian Physics: A study of knowledge-based learning. *Cognitive Science*, *6*, 37-75.
6. Driver, R. (1983). *The pupil as scientist*. Milton Keynes: Open University Press.
7. Educational Technology Center (1988). *Making sense of the future*. Harvard: Harvard Graduate School of Education.
8. Fischbein, E. (1990). Introduction. In P. Nesher & J. Kilpatrick (Eds.), *Mathematics and cognition: A research synthesis by the International Group for the Psychology of Mathematics Education* (pp. 1-13) Cambridge: Cambridge University Press.
9. Forman, G., & Pufall, P. B. (1988). Constructivism in the computer age: A reconstructive epilogue. In G. Forman & P. B. Pufall (Eds.), *Constructivism in the Computer Age* (pp. 235-250). Hillsdale, NJ: Lawrence Erlbaum.
10. Hebenstreit, J. (1987). *Simulation et pédagogie, une rencontre du troisième type*. École Superieure d'Electricité, Gif-Sur-Yvette.
11. Levin, J. A. & Waugh, M. (1988). Educational simulations, tools, games, and microworlds: Computer-based environments for learning. In M. Rabinowitz (Ed.), Computer simulations as research tools, *International Journal of Educational Research*, *12*, 71-79.
12. McDermott, L. C. (1984). Critical review of research in the domain of mechanics. *Research on Physics Education: proceeding of the first international workshop* (pp. 137-182). Paris: CNRS.

13. O'Shea, T. (1990). Informal communication at the NATO Advanced Research Workshop Advanced Technologies in the Teaching of Mathematics and Science, Milton Keynes.
14. Papert, S. (1980) *Mindstorms, children, computers and powerful ideas*. NY: Basic Books.
15. Plomp, T., Pelgrum, W. J. & Sterneman, A. H. M. (1990). Influence of computer use on schools curriculum: limited integration. *Computer Education, 14* (2), 159-171.
16. Schanck, R. C. (1986). *Explanation patterns*. Hillsdale, NJ: Lawrence Erlbaum.
17. Schecker, H. (1990). *The didactic potential of model-building systems for Physics education*. Paper presented at the NATO Advanced Research Workshop Advanced Technologies in the Teaching of Mathematics and Science, Milton Keynes.
18. Schwartz, J. L. (1989). Intellectual mirrors: a step in the direction of making schools knowledge-making places. *Harvard Educational Review, 59* (1), 51-61.
19. Schwartz, J. L. (1990). Software to think with. *Proceedings of "Encontro Computadores no Ensino da Física e da Química"* (p. 51). Coimbra: Portugal.
20. Schwartz, J. L. (1990). Informal communication at the NATO Advanced Research Workshop Advanced Technologies in the Teaching of Mathematics and Science, Milton Keynes.
21. Shneiderman, B. (1983). Direct manipulation: A step beyhond programming languages. *IEEE Computer, 16*, 57-69.
22. Striley, J. (1988). Physics for the rest of us. *Educational Researcher*, August-September, 7-10.
23. Teodoro, V. D. (1990). *The computer as a conceptual lab: Learning dynamics with an exploratory environment*. Paper presented at the NATO Advanced Research Workshop Advanced Technologies in the Teaching of Mathematics and Science, Milton Keynes.
24. Thigs, G. (1988). How forceful is intuition? Students' difficulties in mechanics in relation to common sense. In Thijs, G. D., Boer, H. H., Mcfarlane, I. G. & Stoll, C. J. (Eds.), *Learning difficulties and teaching strategies in secondary school science and mathematics, Proceedings of a Regional Conference*, Botswana. Amsterdam: Free University Press.
25. Viennot, L. (1979). *Raisonnement spontané en dynamique élémentaire*. Paris: Hermann.
26. Wong, D. (1987). Teaching A level physics through microcomputer dynamic modelling: II. Evaluation of teaching. *Journal of Computer Assisted Learning*, 3, 164-175.

Multimedia Learning Environments Designed with Organizing Principles from Non-School Settings

Christina L. Allen

Institute for Research on Learning, 2550 Hanover Street, Palo Alto, CA 94304, USA[1]

Abstract: Learning organized by participants in activities outside of schooling is often uncommonly successful. In this chapter, research is described from the MultiMedia Works after-school club for underserved middle-school students from a multicultural community. The design of the club is guided by results of studies of non-school learning from anthropology and sociolinguistics, which are integrated for the purpose of identifying organizing principles important for engaging and sustaining successful learning. The club is modelled on properties of successful learning, and its activities include the use of state-of-the-art computer tools designed to exploit special properties of multimedia representations that foster conversation, representation and socially-focused learning. I contrast the organization of learning in the club to that found in traditional classrooms, and to conceptions of learning found in current developmental psychology theory. I conclude with characterizations of the successful dimensions of the activity for learners and their implications for the design of learning environments.

Keywords: Authentic activity, after-school club, multimedia

New learning technologies capture the imagination. While this may be a crucial aspect of why new technologies provide levers for educational reform and embodiments of new learning theories, it also brings its problems. For it is common to think that learning research involving new technologies is *about* the technologies. While computer-augmented

---

1This project is funded by IRL with additional support from Apple Computer, and the Carnegie Foundation. Collaborators include Christina Allen, Michael Chertok, Enrique Godreau, Roy Pea, Neils Veirum and the members of the after-school club. The theoretical perspective of this work have been greatly influenced by my colleagues at the Institute for Research on Learning.

learning environments may provide powerful tools for problem-solving, emerging studies of classroom computer ecology reveal that typically the social context motivating and sustaining their use is weak [13]. Uses of computers in most schools in underserved communities in the USA are doubly isolated - first because they are used only within the traditional classroom setting and with traditional text-based materials, and second, because their primary use is classroom drilling and testing.

I hold that the purpose of technology is to support different forms of learning and teaching that may be valuable to recognize and investigate. Thus, I am concerned with designing and advancing uses of computer tools that transcend current limitations through integration into learning which is connected with real world communities beyond the institution of schooling. Changes in conceptions of how learning is best achieved are needed to address current learning crises in schools, workplaces and society. These changes may include roles for technology, but fundamentally must re-interpret the relations between schools and communities, and knowledge and practice.

This paper describes the activities of the MultiMedia Works after-school club, presents the ways in which learning in this club is organized in ways different from traditional classroom settings, provides grounding for these intentional differences in anthropological research on successful learning in non-school settings, and characterizes the implications of this work for designing learning environments.

The MultiMedia Works Club

In this section, the activities and artifacts of the MultiMedia Works after-school club are described. The MultiMedia Works club is a learning environment and its supporting technologies which have been designed based on a perspective of *learning as participation* that will be developed later in this chapter.

The club met at the Institute for Research on Learning twice a week, for two-and-one-half hour sessions. The club ran for two years, and a total of 35 students joined in groups of ten for a semester-long activity. Seventh and eighth grade students from a local, public middle-school joined the club based on interest. The school was in an economically disadvantaged neighborhood, with a multi-cultural and multi-lingual population.

The MultiMedia Works club was held in a large room at IRL called the Studio, which had private workspaces for club members and staff, as well as public workspaces, technologies and meeting areas. There were presentation spaces, as well as resource libraries and white boards.

On a typical day, activity in the club concerned the design, research and production of MultiMedia Works research compositions. Unlike compositions in the traditional sense, these compositions included more than text. They also included video clips, animations, graphic images, and digitized or composed sounds. Using the MultiMedia Works System[2] (See Pea [12] for further details), these media elements were combined to form multimedia compositions using computers and other hardware, such as sound digitizers, videodisc, and videotape players. These compositions can be likened to multimedia slide shows. Each slide can display simultaneously the media types described above, and may be played automatically, or stepped through interactively.

In general, the process of creating a multimedia composition included activities such as: the selection of research topics; research to develop the topic; choice of working partners; group discussions of work in progress; research material collection and selection; design and building of research compositions; selection of audiences for critiques and reviews; and redesign of compositions based on feedback. In this section, a description of the research topics, resources and activities of the club in the last eighteen months will provide the basis for a later analysis of the potential of learning organized in ways quite different that in the traditional classroom.

Research topics

Research reports focused on the form of persuasive argument. Club members were encouraged to select topics that they wanted to learn more about, and for which they felt there were a number of points of view. At twelve and thirteen years of age, their topic choices included: a minor's right to an abortion without parental consent; the effects of drug gangs on their community; whether drugs should be legalized or further regulated; and the redevelopment of land in their community by corporate interests, particularly with regard to the racial and class tensions that would arise. Topic choices that eventually became final compositions included: "Freedom of Speech: Banning Rock Lyrics and Burning the Flag,"

---

[2] The Multi MediaWorks technology is built on a Macintosh-II platform and includes tools for digitizing videotape, scanning print and graphic images, laser printing, video archiving, graphics file management, animation, 3-D modelling, and sound digitizing and editing.

"Land Development in Whisky Gulch" (the commercial district in their neighborhood proposed for renewal), and "The Effects of Gang Violence on Neighborhoods."

Research strategies

Students of this age group are just learning to do research in school, and when asked to do research for a report, will usually go to texts such as books, newspapers and encyclopedias. With MultiMedia Works, strategies for learning about a selected topic included more than just reading. They involved seeking expertise and images from varied sources. For example, club members working on an argument about land development identified a local real estate agent and interviewed him a number of times about the effects of a planned office complex on their community. While doing this research, they were involved in detailed discussions about topics like eminent domain, the financial status of their community, the effects of redevelopment on their quality of life, as well as on the needed tax base for social services in their community. Students working on a composition about freedom of speech consulted their textbooks for information, and when they were disappointed, turned to the U.S. Constitution. They also recorded debates about banning rock lyrics from a nationally-syndicated network talk show (the Oprah Winfrey Show). When club members researched endangered species, field trips were planned beforehand with telephone calls to the zoo about what they would be seeing, and plans were made for filming the parts of the trip that would be of interest for the research.

Research resources

A wide variety of pre-packaged research materials were available in the Studio in such forms as videodisks, library resources, digitized sounds, on-line data bases and books, and magazines. The club members quickly exhausted these materials in favor of creating their own resources and seeking expertise from their schools and community. The opportunities provided by these other media types allowed them to incorporate many non-traditional forms of information into their compositions. This non-traditional information included interviews with community members, videotapes of local super-fund sites for an environmental impact research project, video and audiotaped materials from a field trip to the zoo, recordings from television, cable stations and commercially-available videotapes and videodisks, digitized sounds of traffic patterns, images scanned from books and magazines, and animations of processes such as the transformation of smoke from a smokestack into acid rain.

Club members and staff

There were a number of people involved in the day-to-day activities of the club: the research staff, which included system programmers, learning researchers, and the larger research community of IRL; club members from the middleschool; and for the last session of the club held during the summer, a group of five teachers from the middleschool. The work of researching and building compositions, learning about new technologies and data collection methods, organizing the daily activities, and designing the technologies themselves was distributed across these groups.

Collaboration

Most MultiMedia Works projects involved group work. Students chose to collaborate on a composition based on interest in the topic, and an understanding of the expertise needed to get a composition done. One of the more interesting examples of collaboration came between two girls in the summer session. One of the girls was researching the issue of flag burning. The other was working on a composition about banning the lyrics of rap groups. After an early critique in which both showed their work, they realized from the comments of others that they were working on the same, larger issue of freedom of speech. They decided to merge their compositions, and went on to do additional research about freedom of speech in order to integrate their work.

Club members working together to develop a composition adopted roles based on what they liked doing, or felt most comfortable doing. Some were enthralled with the computer and organized composition building in MultiMedia Works, some became experts in filming with the video cameras or in recording and editing sound, some were interested in developing argument in textual form as the basis for a composition, and others took lead roles in group discussions and in organizing the activities of the club with whiteboards and lists of plans. While all of the club members had some working knowledge of all of the roles, experts for particular skills became known, and club members would seek these experts when in need. This knowledge of expertise extended beyond the peer group of the club to include researchers at IRL, teachers from their school, parents and relatives, and members of the broader community. For example, one girl researching the effects of gang violence worked with an IRL social scientist to identify research on gangs, which led her to literature about the organization of Italian gangs in Chicago in the 1920s.

Compositions

The nature of the final compositions varied in terms of the research media used, the mode of expression which the club members chose, and the eventual purpose of the composition. One girl used only text and sound for her composition. She had won awards for poetry about the lives of black women in poor communities. When her research was complete, she composed an epic rap, detailing life with gangs. She digitized sound to record the rap, which also appeared on the screen as a poem. The composition on land renewal incorporated extensive club member-produced videotape of the area that was proposed for redevelopment, other vacant parcels of land available in the area that the club members thought would make better sites, and a videotape, carefully choreographed by the club members, in which one of the girls recited their ideas about why the area should not be developed. Students designed these compositions in order to stand by them, talk to an audience about them, step through them with intentional pauses, and to take questions at any point in the presentation. Throughout the research and design process, and during critiques, the compositions were the center of a great deal of talk and debate, but in almost no cases were they intended to stand alone.

Critique

Throughout a research and design cycle, club members were aware of one another's work through casual interactions, the search for expertise to support the necessary work, and through planned, public critiques. Critiques were organized by the club members and by the researchers throughout the course of the work. Reasons for critiques varied, and included times when there was a sense that people had a lot to show, were at an impasse and needed help and ideas, or when they had completed their work. The audience might be local for more informal critiques, or included anyone of relevance to the work. Critiques included teachers from the schools, researchers from IRL, members of the community such as real estate agents and corporate recruiters, and families and friends, and teachers. The critiques were held in a lounge area, and club members presented their compositions, as well as plans for future work, or frustrations they were having. The audience provided feedback on these needs, as well as ideas or comments that they had about any aspect of the work. At the end of each semester, there was a big party to which all of the teachers, researchers, friends, parents and siblings were invited. At this final critique, all of the club members presented their work printed to VHS format so that they could take it with them on inexpensive videotapes for future use at home or in school.

Summary

The multimedia nature of the research activities and the compositions accounted for some of how the MultiMedia Works club was organized -- the demands of the technology and the different mediatypes required varied expertise, the necessity of shared workspaces and shared resources encouraged collaboration, and the display-based nature of multimedia presentations, as well as the current lack of multimedia genres for research, encouraged public critique and review. But the shift in media alone did not account entirely for the nature of the club activities and interactions in the school sites. In the next two sections, I will describe how this club is different than conceptions of learning implicit in the organization of a traditional classroom, why this might be a good way for learning to be organized in schools, some of the research that grounds this learning environment design, and the indicators of success thus far.

The Club in Contrast to the Traditional Classroom

In contrast to traditional classroom situations in which there are clearly defined roles of expert and novice, members of the MultiMedia Works Studio, including club members, teachers and researchers, *played multiple roles in the research and design activities*. The identification of expertise was related to the needs of the situation, and in many cases, it was the club members, rather than the teachers or the researchers, who had the necessary expertise. Expertise appeared in unexpected ways and from unexpected sources as some club members thought of by teachers as "real problems" arose to particular challenges, or became facile in esoteric aspects of the work, such as sound digitizing. They were then viewed as resources and often called upon by others when help was needed. One salient example was when teachers from the school came to the Studio to work side-by-side with club members on their own agendas, projects and concerns. The club members, having worked within IRL and with MultiMedia Works before, took the task of teaching the teachers how to use the software and hardware, and provided examples of what could be done with it.

In traditional educational environments, classroom activities, content and organization are defined largely by institutions and people outside of the classroom. In the MultiMedia Works Studio, both club members and staff *helped to define the nature of the work through participation* in technology design, planning the activities of the group, developing the genres of multi-media composition, planning the use of their compositions, and selecting audiences for critiques. An excellent example of the extent to which both club members and adults contributed to the design of their environment was when the teachers won a grant to have a MultiMedia Works classroom in the school, and then successfully worked with club members and the school district to redefine the structure of the school day and relationships with students. They incorporated club-like activities into their curriculum and integrated different subjects traditionally taught separately (e.g., language, history, geography). Another example was in the process of technology development. The MultiMedia Works system was available to the club members as soon as it could be used, and it evolved based on what club members and staff wanted to do with it .

The role of evaluation in the MultiMedia Works club was different than is typical in the classroom. In the classroom, the product of the student's work, such as a paper or a test, is turned into the teacher for grading. It is rare that the results of the evaluation are used to produce a better paper. In the Studio, evaluation of the work was done by members of the local and wider community, and *evaluation was aimed at refining the activity* - improving the research, or the design of the composition. Norms for the research and genres of multimedia composition developed within the community through collaboration, critique, research and presentation. Given the topics chosen, there were direct lines to activity in their community, and an audience for their ideas that truly cared about the issues. The concern for communication to a relevant audience was evident in the types of compositions that resulted. They were designed to be interactive, and were intended to be the focus of conversation with the audience. Presenters, rather than letting the compositions play automatically, stepped through them, stopped to answer questions, and engaged in dialog with the audience.

Traditional classrooms rely on pre-packaged materials. In the MultiMedia Works Studio, club members continually integrated resources drawn from their larger communities and institutions into their research. Research materials were constructed and collected by the members, necessitating ventures into the "real world" to collect data and media for their compositions. Since the system allowed "printing" of the compositions onto inexpensive videotapes, club members found that distribution of the work to audiences outside of the Studio was simple. Moreover, the club members were able to harness what they already knew and cared about as a basis for further learning. This was evident in the selection of research topics that they knew from daily experiences to be important.

In summary, the MultiMedia Works club differed from a traditional classroom along a number of dimensions: There were many forms of participation, communication, and observation available from which members learned; students and the adults helped to define the nature of the studio through participation in the construction of activities and artifacts; evaluation of the work was done by members of the local and wider community, and evaluation was directed toward improving the activity; members continually integrated resources and audience drawn from their everyday experience, and from larger communities and institutions than the school; and there were multiple roles for members based on their expertises and needs, rather than predefined delimitations of expertise in terms of the common teacher-student asymmetry.

These differences, as well as the opportunities afforded by the additional media, represent some important new dimensions for the design of learning environments. For example, varying the expertise learners may contribute to a complex task provided multiple points of entry into the work, and many ways to contribute to successful, authentic activity. The diversity among students in backgrounds and interests became a resource for the group's learning rather than a problem. Kids had more options for expression of their ideas in the various media, and hence more opportunity to demonstrate their learning expertise. The multimedia composition tools allowed students to incorporate what they knew into the learning process because it allowed them to collect information, such as videotape, sound, graphics and text, and combine it into powerful presentations. An authentic audience became involved in the development of the research, which in turn, enriched the quality and depth of the on-going work. Finally, members of the club were able to observe one anothers' activities, including successes, failures, work in progress, sources of information and innovation, and forms of research and articulation of ideas, to come to understand the integrative process of using information in context and for an audience.

It seems that there are undiscovered opportunities for the advancement of learning in the organization of learning activities and artifacts such as exists in the MultiMedia Works club. These properties of the club activities were not accidental, but were designed in order to promote different forms of interaction and participation. These differences resulted from a learning environment design based on empirical work in the social sciences on learning organized in *non*-school settings. I turn to this research in the next section.

Social Science Contributions to Learning Environment Designs

The design of the MultiMedia Works learning environment was influenced primarily by research from anthropology and cross-cultural developmental psychology, some of which is reviewed in this section. I then compare this research to conceptualizations of learning and social relations in current developmental theories, in order to see whether these theories capture important aspects of the successful learning that are found to organize non-school systems.

Recall the general ways in which the activities and values of the MultiMedia Works club were different from those of a traditional classroom. In literatures describing learning that is organized outside of formal institutions of schooling, all of these differences are reflected. Let us explore how these differences are illustrated in practice, whether we look at the formation of musical clubs by Balinese children, at the emergence of various literacy skills in informal American play groups, at cognitive development of Brazilian street candy sellers, or at the development among the Hanunoo of Sanscrit script literacy in the absence of "formal" schooling.

McPhee [9] describes the formation of musical clubs *(gamelans)* by children in Bali. Throughout his account is evidence of a culture in which learning is integral with the practice of the larger community:

> Their early life is based on imitation of their elders; their play is partly reproduction in miniature of various adult activities, carried out with great regard for detail. They are devoted patrons of the arts and would rather die than miss a play or gamelan performance. (p. 74)

Within a population of people who love music, and have many traditions involving it, a group of small boys formed their own musical group. The gamelan was initiated by a few of the boys, and comprised members with different levels of expertise and enthusiasm. The group was motivated by the values and activities of the larger culture, and learning about both the music and the proper forms of performance occurred within the context of a desire for participation in the musical culture. Because learning was a means rather than an end, evaluation was focused on increasingly successful performance.

In *Sub-Rosa Literacy*, Gilmore [6] presents two types of play: 'Steps' by girls (a dancing and spelling activity), and boys' Dungeons and Dragons (a branching interactive fantasy world game). Within both groups, learning and assessment are based on continual evaluation of, and desire for excellence. What is notable in Gilmore's descriptions is the members' recognition of expertise within the group, and the role of 'captain' that is conferred as a result:

> Each girl does the performance with some embellishments and markers of individuality. The degree of oral composing varies, but performers who are creative are recognized for their virtuosity and often are designated as 'captains' who organize and instruct the others. (p. 161)

By identifying expertise through doing, and honoring it with a title and new roles, these girls are explicitly shaped the developmental course of their practice, and defined the ideal models in terms of which learning should occur. Moreover, the basis for assessment was multi-dimensional, including not only the oral performance, but also the dancing, attitude and understanding of the larger conventions that surrounded the activity.

The research on boys playing Dungeons and Dragons displays the extent to which they organized their community, play and work around a "counter-culture" activity serving to encourage learning and teaching of myriad skills also valued at school, but unrecognized because the boys do not demonstrate them in the school setting:

> Whether successful in the classrooms or not, these young males rush home to get their school work done so they can play D&D (often much more challenging reading than their school assignments). Students will get to school early, organize themselves at lunch (assigning roles for buying, etc.) so that they have enough free time to play. Many hours of individual reading and preparing written character and dungeon descriptions often precede the actual group play. (p. 165)

In both of the groups studied by Gilmore, normative concerns of 'what is good', 'who is good', and 'what should be taught', developed naturally, and broke through role definitions such as teacher/student, or master/apprentice.

Saxe [16] examines the cognitive development of Brazilian street candy sellers, and the relationship of school-based learning to mathematical reasoning in everyday situations. He provides examples of the internal workings of a group, particularly processes of expertise development by its members. Based on in-situ assessment, both candy sellers and clients modified their interactions to allow for the completion of authentic transactions with real clients by less expert sellers. Expert sellers and clients often designed sub-tasks in real time to lead less expert sellers successfully through skill-developing transactions. Thus, members of the community continually determined the shape of the teaching-learning practice for the candy sellers, and their ongoing assessment was in service of the development of new members of the community.

In *Bamboo Literacy on Mindora*, Conklin [4] describes how it is that 60% of adults, both men and women, read and write Sanscrit script with no "formal" schooling whatsoever. For the Hanunoo, reading and writing were not the end goals of learning, nor the motivation for seeking to learn. Literacy had the important purpose of facilitating successful courtship. Thus, when a child was of an age to be concerned with courtship, he or she was motivated to seek out those who could teach Sanscrit script, and hence provide access to the traditional love songs and stories that were preserved in writing. In order to court, reading and writing were but a small step towards success. For it was not only knowledge of the stories and songs, but appropriate timing, context and style of performance that were important to success. This integrated knowledge was not available only in the writing, but rather, long before the writing was salient to children, in their experience of the rituals of their elders year after year. Thus, the learner, desiring success in courtship rituals:

> ...spends considerable time with those of his older male relatives who are good singers, trying always to pick up additional verses which may be of use in serenading. On hearing his uncle sing a long and particularly pleasing song, he asks him to write down the words on a lime tube made from an internode of bamboo....While the boy's uncle is finishing the inscription he explains that each character stands for only one syllable. Then, while repeating the song aloud, they both go over the writing carefully, following with a finger the incised marks. (p.9-10)

Reading and writing are learned in the context of song performance, and learning succeeded to a greater or lesser extent in the performance of the rituals.

The groups described above - music-performing Balinese children, boys playing Dungeons and Dragons, Brazilian street candy sellers, girls performing 'Steps,' potential Hanunoo lovers - differ widely in the specific features of their culture and goals. However, they share general features in their organization of learning. These include learning through and for doing, assessing for increasingly successful practice by members, connections to other communities beyond their learning group, continually constructing and evolving the nature of their activities, and recognizing and utilizing the expertise of various members based on the performance needs of the group. Furthermore, normative standards exist within all of these groups, and are developed both internally, and in response to the practices and values of the larger community. For Balinese children learning music, cultural models are available from birth for the roles and practice of music. Imitation plays a role in forming their normative concerns, yet they also develop their own values and construct their own, unique interpretations of musical performance and plays.

In addition to specific cultural and community practices, these groups are learning many of the types of knowledge that are valued in formal education: reading, writing, spelling, musical composition and performance, mathematics and economics. Moreover, the lack of transfer of learning to life/work settings so common as a complaint about formal education is less of an issue since the knowledge is learned and motivated in the context of authentic practices.

Within major theoretical studies of learning, a number of perspectives share some of the emphases of the MultiMedia Works club, and emphases found in the literatures on non-school learning described above. But these theoretical perspectives are often importantly incomplete. For example, a focus on "active learning," and the role of the learner's involvement in the learning process is common to Dewey [5], Piaget [14], cognitive science theories of learning such as Anderson [2], and the "higher order thinking skills" movement in educational research [3]. But many of these perspectives restrict their concerns about active learning to the individual child, actively engaged in mental and problem-solving activity. The social grounding and motivations for such individual exploration and learning are left unexamined, and in the worst cases, ignored or relegated to social psychologists to worry about. The rich embedding of learning in cultural practice characteristic of the works described earlier in this section is generally lacking in these accounts.

The same foundational assumptions about the active learner characterize many theories of peer collaborative learning in the past decade, such as in the works of Slavin [17] and Webb [18]. The concerns that have motivated these studies arise from social psychology, and have to do with an attempt to optimize the design of groups of interacting individual agents, viewed as having separate intelligence quotients and cognitive styles, for the learning of the group. These groups are treated as top-down learning units designed by the researcher, which are, incidentally, not at all connected to broader social communities.

Similar limitations are evident in the neo-Piagetian studies by Murray [11], and by Genevans such as Mugny and colleagues [10]. Again, these learning groups are treated as social isolates - the result of a developmental psychological researcher's experimental designs to determine how social interactions with peers can be efficacious in bringing about cognitive change. The learning groups do not have any natural relation to activities of communities beyond the laboratory from which students could learn by participation.

There are many common concerns between our work and the experiential pedagogy and social philosophy linking school and society offered by John Dewey [5]. Dewey's approach to making connections for the learners to their communities included field trips, and "farm-like" activities in the classroom. However, these were toy-like, model activities, without any legitimate audience in the communities outside the classroom, and where the motivation for participation came from teachers and institutions, rather than being felt by the students.

Western interpretations [such as 19] of Vygotsky's approach to developmental psychology share many of the problems with other perspectives described above. Regarding the concept of the zone of proximal development, a widespread interpretation [15] is that what takes place between a child and a more mature individual eventually comes to take place inside the child's individual mind. In other words, what is social on the outside goes to mental on the inside. In this interpretation, the adult guides the child through social activities to higher stages of mental growth and solitary cognitive activity. However, this orientation neglects the authenticity of the learning that takes place, relegating the social interaction to the role of "training wheels" for individual development, rather than as the on-going context of the usefulness of learning, and a source of emerging values and norms as the society undergoes historical change.

Thus the researcher can construct classroom-based activities designed to move the child through the zone of proximal development for culturally and practically irrelevant problem solving work. Yet the dynamics of movement through the true zone of concern, that described by Lave and Wenger [8] as the trajectory from the periphery to the core of participation in the activities of a community-of-practice, are too easily omitted from this version of the concept of the zone of proximal development.

Lave and Wenger's theoretical work is largely influenced by the circumstances of apprenticeship in disciplines that have hardly changed in hundreds of years. Thus, their theories describe the increasing participation of learners in formal, stable practices to which the participants can aspire. What is not covered in the Lave and Wenger account, but is evident in the anthropological studies on learning in groups reviewed earlier, is the activity of the members of the group to define themselves, and to construct themselves and their activities in changing circumstances, and with emergent values over time. Yet it is exactly this process of natural evolution and change in the values and activities of learning groups that are of central concern to educational researchers and designers who would like to intervene, and introduce new environments, technologies, norms, and activities for learning. It is this *emergent* aspect of practice most particularly that can be seen in instances of learning organized by groups engaged in authentic activities.

Indicators of Success in the MultiMedia Works Club

With this background in mind, we can now consider the indicators of success of the MultiMedia Works club. In our work with the club to date, it has not been appropriate to collect quantitative data, in part because we (and the teachers who helped us in our liaison to the school) saw additional testing as betraying the boys' and girls' excitement about the club as a non-school activity. A project now beginning in the schools will focus on both qualitative and quantitative measures of learning with MultiMedia Works. Yet it is important to discuss the indicators of success along a number of important dimensions based on our analysis of the interactions and the projects produced by members of the MultiMedia Works studio.

The first indicator is *the general behavior, attitudes and motivations of the members of the club.* We had a consistently high attendance rate for a voluntary after-school club in every session of the club that was held at the MultiMedia Works studio. Club members also wanted to continue working with us after the clubs ended, and made special arrangements with us to come on the weekends and in the evenings. As a result of their interest, five club members were hired for the summer as multi-media producers with funding from a federal youth employment program. Two have been hired to provide support for the MultiMedia Works activities at the school.

The second indicator is the face validity of *learning outcomes of the processes of producing the research compositions*. What was immediately evident was that the club members did vastly more reading and writing than they were required to do for school. Their reading sources included the constitution, articles about freedom of speech in the newspaper, legal and media briefs about land development in the community, analysis of lyrics of popular songs and performances, etc. Moreover, these sources were read critically, because club members were looking for information relevant to their research projects. The writing elements that they produced in their compositions were often criticisms and counter-arguments to the reading materials they had encountered.

Beyond writing and reading, the possibilities of multiple forms of expression fostered critical and creative consideration of media choice and novel media creation and use. Club members were instantly able to see the value of the information available in their everyday experiences and knowledge of the their world. Their compositions demonstrated that they came to have a new view of the world as a place full of information and resources, which they could use and assess, and to which they could contribute valued ideas and points of view.

The third indicator of success was *the quality and the content of the research projects that they produced.* The MultiMedia Works compositions developed by club members ultimately taught things to adults, as well as to one another. For example, a group of girls and boys built a presentation about an IRL interdisciplinary project on physics learning in geometrical optics that has been used as a model by a national foundation for presenting rationale and results from computer-enhanced research projects.

The fourth indicator was that there was *incidental learning of a variety of valued skills*. In a traditional setting, these skills would be taught in a more isolated manner. Assessment of compositions and research was aimed at improving the work, and therefore always included relevant audiences for the topics. The critique structure focused on increased learning and success for the research projects, and the evaluations found their way into improvements in the next round of work on a project. The critiques by valid audiences rapidly improved attention to spelling, argument structure, sentence structure, proof, attribution, resources available for research, presentation formats and details. Future studies will seek to document the magnitude and properties of these learning gains.

Motivated by inquiry into topics of their choosing, the boys and girls were exposed to diverse disciplines. For example, they needed information about statistics to determine traffic flow into and out of their community with proposed construction that they were investigating. For the same composition, they did research on the tax base of their town to determine why their community would consider approving the proposed office tower. For another project, they interviewed research scientists in-depth about their work in optics, physics education, educational anthropology, and cognitive research. Incidentally, they learned to film with video cameras, edit videotape, build computer animations, record and digitize sound, digitize color graphics, use paint tools and word processors, and compose multimedia presentations.

The fifth indicator of success was *the extent to which the club members, staff, teachers and researchers participated in a community of peers* rather than in hierarchical relationships. Both the adults and the children came to the work with different sensitivities and strengths, and developed different roles and levels of expertise in various areas. In each of the clubs there developed recognized and appreciated experts in different areas who served as resources for the rest of the community. One high point came when a group of students developed the documentation and user help for the MultiMedia Works technology for the schools, teachers and other students. This pattern was also evident in the overlap and integration of the students' work with other communities and institutions. In their selection of research topics they aimed squarely at the issues that were relevant to their lives and communities.

Conclusions and Implications

In this paper, I have described the research and design of the MultiMedia Works club and its supporting technologies. I have articulated a background for its design in the research on the organization of successful learning in non-school groups. This approach has been compared with other theoretical perspectives on learning, and I have reviewed why we consider the learning augmented by the MultiMedia Works club to be successful. What are the broad implications of the research activities and theoretical perspective outlined here?

Perhaps the most important conclusion is the need to move from descriptions of successful learning outside of school and the failures of learning in school to designs for learning environments that partake of the organizing features of successful learning wherever it is found. Central dimensions of the organization of successful learning have been characterized in this chapter. These dimensions include:

- Many forms of participation, observation and communication available between members
- Authentic overlap with communities outside school and building on what members already know
- Member participation in the construction of activities and artifacts, goals and values
- The role of members' co-evaluation to point to improvement and new learning
- Multiple roles for members' participation, with distributed expertise

What has been the role of technologies in these newly-designed learning environments? Several summary observations may be made. Most fundamentally, the multimedia nature of the technologies provided a connection between the club members and diverse communities beyond the club site. These communities included local government, environmental groups, designers, researchers, and other potential contributors to and audiences for the childrens' work. Because the MultiMedia Works tools allowed club members to create *locally*-produced media as part of their multimedia compositions, they were able to bring into a communal space their own media for expressing ideas and expertise about the topics under development in their compositions.

Furthermore, the shareable nature of multimedia documents under design and review made for a different kind of community of learning - one in which peer commentary and critique was an ongoing influence contributing to the shape and quality of the products that emerged. Multimedia also allowed the capture and utilization of contemporary events and culture impossible in textbook-based learning, with its lengthy time cycle of production, state adoption, and prolonged school use. And this, in turn, provided a much less bland and outdated media ecology than could ever be accepted in the textbooks which must be approved by state curriculum committees. By providing multiple forms of expression, multimedia allowed students (particularly with minimal written English language skills) to demonstrate their learning successes for others.

Finally, I urge that practice and theory are too often isolated in their developments. Research on the successes and failures of non-school-based or school-based learning is essential, but sterile if it remains unconnected to the design of new learning environments. It will not pass without notice that many of our indicators of "successful" learning would not be represented by traditional assessments. In addition to learning history, mathematics, and even beyond the relevance of the subjects to practice which may come as a result of this new approach, these children (and teachers, and researchers) were learning constructive criticism, how to forge community links, collaboration and leadership skills, how to search for and ask for information and help, technological skills, and myriad other skills crucial to successful participation in communities, but under-valued in terms of many current conceptions of learning as retention of knowledge. Thus, it is important to recognize the need for the continuing evolution of learning environment design and values based on emergent consequences of cycles of research and intervention [3].

As a result of these concerns, we are now engaged in a study to understand the incorporation of these changes at a school site. We are focusing on the teachers' and students' work to negotiate the transformation of traditional curriculum into new genres of multimedia learning expressions allowed by the technologies, and to acknowledge and represent the additional forms of learning and expertise gained by the fundamental change in the activities and organization of social aspects of the learning environment.

References

1. Allen, C. L. (in press). Reciprocal evolution: A design and research strategy for the development and integration of basic research, design, and studies of work practices. In D. Schuler (Ed.), *Participatory design.* Hillsdale, NJ: Lawrence Erlbaum Associates.
2. Anderson, J. R. (1982). Acquisition of cognitive skill. *Psychological Review, 89*, 396-406.
3. Chipman, S. F., Segal, J. W., & Glaser, R. (1985). (Eds.). *Thinking and learning skills* (Vol. 1 & 2). Hillsdale, NJ: Erlbaum.
4. Conklin, H. (1949). Bamboo literacy on Mindora. *Pacific Discovery. 3*, 4-11.
5. Dewey, J. (1938). *Experience and education.* New York: Macmillan.
6. Gilmore, P. (1986). Sub-Rosa literacy: Peers, play and ownership in literacy acquisition. In P. Gilmore & B. Schieffelin (Eds.), *Acquisition of literacy: Ethnographic perspectives* (pp. 155-168). Norwood, NJ: Ablex.
7. Lave, J. (1988). *Cognition in practice: Mind, mathematics and culture in everyday life.* Cambridge: Cambridge University Press.
8. Lave, J., & Wenger, E. (in press). *Situated learning: Legitimate peripheral participation.* Cambridge: Cambridge University Press.
9. McPhee, C. (1955). Children and music in Bali. In M. Mead & M. Wolfenstein (Eds.), *Childhood in contemporary culture* (pp. 70-98) Chicago: University of Chicago Press.
10. Mugny, G., Perret-Clermont, A. N., & Doise, W. (1981). Interpersonal coordinations and social differences in the construction of the intellect. In G. M. Stephenson & J. M. Davis (Eds.), *Progress in Applied Psychology* (Vol. 1). New York: Wiley.
11. Murray, F. B. (1982). Teaching through social conflict. *Contemporary Educational Psychology, 7*, 257-271.
12. Pea, R. D. (in press). MultiMedia Works: Student learning through multimedia tools. *IEEE Computer Graphics and Applications.*

[3] Methods for on-going, in-situ social science research, and the integration of such research into the design of technologies and learning environments over time is further developed in Allen [1]

13. Pea, R. D., & Sheingold, K. (1987). (Eds.). *Mirrors of mind: Patterns of experience in educational computing*. Norwood, NJ: Ablex.

14. Piaget, J., & Inhelder, B. (1969). *The psychology of the child.* New York, NY: Basic Books, Inc.

15. Rogoff, B., & Wertsch, J. V. (1984). (Eds.). Children's learning in the "zone of proximal development". *New directions for child development, no. 23*. San Francisco, CA: Jossey-Bass.

16. Saxe, G. B. (1990). *Culture and cognitive development: Studies in mathematical understanding.* Hillsdale, NJ: Lawrence Erlbaum Associates.

17. Slavin, R. (1983). *Cooperative learning*. New York: Longman.

18. Webb, N. (1982). Student interaction and learning in small groups. *Review of Educational Research, 52*, 421-445.

19. Vygotsky, L.S. (1978). *Mind in society: The development of the higher psychological processes.* Cambridge, MA: Harvard University Press.

NATO ASI Series F

Including Special Programmes on Sensory Systems for Robotic Control (ROB) and on Advanced Educational Technology (AET)

Vol. 1: Issues in Acoustic Signal - Image Processing and Recognition. Edited by C. H. Chen. VIII, 333 pages. 1983.

Vol. 2: Image Sequence Processing and Dynamic Scene Analysis. Edited by T. S. Huang. IX, 749 pages. 1983.

Vol. 3: Electronic Systems Effectiveness and Life Cycle Costing. Edited by J. K. Skwirzynski. XVII, 732 pages. 1983.

Vol. 4: Pictorial Data Analysis. Edited by R. M. Haralick. VIII, 468 pages. 1983.

Vol. 5: International Calibration Study of Traffic Conflict Techniques. Edited by E. Asmussen. VII, 229 pages. 1984.

Vol. 6: Information Technology and the Computer Network. Edited by K. G. Beauchamp. VIII, 271 pages. 1984.

Vol. 7: High-Speed Computation. Edited by J. S. Kowalik. IX, 441 pages. 1984.

Vol. 8: Program Transformation and Programming Environments. Report on a Workshop directed by F. L. Bauer and H. Remus. Edited by P. Pepper. XIV, 378 pages. 1984.

Vol. 9: Computer Aided Analysis and Optimization of Mechanical System Dynamics. Edited by E. J. Haug. XXII, 700 pages. 1984.

Vol. 10: Simulation and Model-Based Methodologies: An Integrative View. Edited by T. I. Ören. B. P. Zeigler, M. S. Elzas. XIII, 651 pages. 1984.

Vol. 11: Robotics and Artificial Intelligence. Edited by M. Brady. L. A. Gerhardt, H. F. Davidson. XVII, 693 pages. 1984.

Vol. 12: Combinatorial Algorithms on Words. Edited by A. Apostolico, Z. Galil. VIII, 361 pages. 1985.

Vol. 13: Logics and Models of Concurrent Systems. Edited by K. R. Apt. VIII, 498 pages. 1985.

Vol. 14: Control Flow and Data Flow: Concepts of Distributed Programming. Edited by M. Broy. VIII, 525 pages. 1985.

Vol. 15: Computational Mathematical Programming. Edited by K. Schittkowski. VIII, 451 pages. 1985.

Vol. 16: New Systems and Architectures for Automatic Speech Recognition and Synthesis. Edited by R. De Mori, C.Y. Suen. XIII, 630 pages. 1985.

Vol. 17: Fundamental Algorithms for Computer Graphics. Edited by R. A. Earnshaw. XVI, 1042 pages. 1985.

Vol. 18: Computer Architectures for Spatially Distributed Data. Edited by H. Freeman and G. G. Pieroni. VIII, 391 pages. 1985.

Vol. 19: Pictorial Information Systems in Medicine. Edited by K. H. Höhne. XII, 525 pages. 1986.

Vol. 20: Disordered Systems and Biological Organization. Edited by E. Bienenstock, F. Fogelman Soulié, G. Weisbuch. XXI, 405 pages.1986

Vol. 21: Intelligent Decision Support in Process Environments. Edited by E. Hollnagel, G. Mancini, D. D. Woods. XV, 524 pages. 1986.

NATO ASI Series F

Vol. 22: Software System Design Methods. The Challenge of Advanced Computing Technology. Edited by J. K. Skwirzynski. XIII, 747 pages. 1986.

Vol. 23: Designing Computer-Based Learning Materials. Edited by H. Weinstock and A. Bork. IX, 285 pages. 1986.

Vol. 24: Database Machines. Modern Trends and Applications. Edited by A. K. Sood and A. H. Qureshi. VIII, 570 pages. 1986.

Vol. 25: Pyramidal Systems for Computer Vision. Edited by V. Cantoni and S. Levialdi. VIII, 392 pages. 1986. *(ROB)*

Vol. 26: Modelling and Analysis in Arms Control. Edited by R. Avenhaus, R. K. Huber and J. D. Kettelle. VIII, 488 pages. 1986.

Vol. 27: Computer Aided Optimal Design: Structural and Mechanical Systems. Edited by C. A. Mota Soares. XIII, 1029 pages. 1987.

Vol. 28: Distributed Operating Systems. Theory und Practice. Edited by Y. Paker, J.-P. Banatre and M. Bozyiğit. X, 379 pages. 1987.

Vol. 29: Languages for Sensor-Based Control in Robotics. Edited by U. Rembold and K. Hörmann. IX, 625 pages. 1987. *(ROB)*

Vol. 30: Pattern Recognition Theory and Applications. Edited by P. A. Devijver and J. Kittler. XI, 543 pages. 1987.

Vol. 31: Decision Support Systems: Theory and Application. Edited bv C. W. Holsapple and A. B. Whinston. X, 500 pages. 1987.

Vol. 32: Information Systems: Failure Analysis. Edited by J. A. Wise and A. Debons. XV, 338 pages. 1987.

Vol. 33: Machine Intelligence and Knowledge Engineering for Robotic Applications. Edited by A. K. C. Wong and A. Pugh. XIV, 486 pages. 1987. *(ROB)*

Vol. 34: Modelling, Robustness and Sensitivity Reduction in Control Systems. Edited by R.F. Curtain. IX, 492 pages. 1987.

Vol. 35: Expert Judgment and Expert Systems. Edited by J. L. Mumpower, L. D. Phillips, O. Renn and V. R. R. Uppuluri. VIII, 361 pages. 1987.

Vol. 36: Logic of Programming and Calculi of Discrete Design. Edited by M. Broy. VII, 415 pages. 1987.

Vol. 37: Dynamics of Infinite Dimensional Systems. Edited by S.-N. Chow and J. K. Hale. IX. 514 pages. 1987.

Vol. 38: Flow Control of Congested Networks. Edited by A. R. Odoni, L. Bianco and G. Szegö. XII, 355 pages. 1987.

Vol. 39: Mathematics and Computer Science in Medical Imaging. Edited by M. A. Viergever and A. Todd-Pokropek. VIII, 546 pages. 1988.

Vol. 40: Theoretical Foundations of Computer Graphics and CAD. Edited by R. A. Earnshaw. XX, 1246 pages. 1988.

Vol. 41: Neural Computers. Edited by R. Eckmiller and Ch. v. d. Malsburg. XIII, 566 pages. 1988.

Vol. 42: Real-Time Object Measurement and Classification. Edited by A. K. Jain. VIII, 407 pages. 1988. *(ROB)*

NATO ASI Series F

Vol. 43: Sensors and Sensory Systems for Advanced Robots. Edited by P. Dario. XI, 597 pages. 1988. *(ROB)*

Vol. 44: Signal Processing and Pattern Recognition in Nondestructive Evaluation of Materials. Edited by C. H. Chen. VIII, 344 pages. 1988. *(ROB)*

Vol. 45: Syntactic and Structural Pattern Recognition. Edited by G. Ferraté, T. Pavlidis, A. Sanfeliu and H. Bunke. XVI, 467 pages. 1988. *(ROB)*

Vol. 46: Recent Advances in Speech Understanding and Dialog Systems. Edited by H. Niemann, M. Lang and G. Sagerer. X, 521 pages. 1988.

Vol. 47: Advanced Computing Concepts and Techniques in Control Engineering. Edited by M. J. Denham and A. J. Laub. XI, 518 pages. 1988.

Vol. 48: Mathematical Models for Decision Support. Edited by G. Mitra. IX, 762 pages. 1988.

Vol. 49: Computer Integrated Manufacturing. Edited by I. B. Turksen. VIII, 568 pages. 1988.

Vol. 50: CAD Based Programming for Sensory Robots. Edited by B. Ravani. IX, 565 pages. 1988. *(ROB)*

Vol. 51: Algorithms and Model Formulations in Mathematical Programming. Edited by S. W. Wallace. IX, 190 pages. 1989.

Vol. 52: Sensor Devices and Systems for Robotics. Edited by A. Casals. IX, 362 pages. 1989. *(ROB)*

Vol. 53: Advanced Information Technologies for Industrial Material Flow Systems. Edited by S. Y. Nof and C. L. Moodie. IX, 710 pages. 1989.

Vol. 54: A Reappraisal of the Efficiency of Financial Markets. Edited by R. M. C. Guimarães, B. G. Kingsman and S. J. Taylor. X, 804 pages. 1989.

Vol. 55: Constructive Methods in Computing Science. Edited by M. Broy. VII, 478 pages. 1989.

Vol. 56: Multiple Criteria Decision Making and Risk Analysis Using Microcomputers. Edited by B. Karpak and S. Zionts. VII, 399 pages. 1989.

Vol. 57: Kinematics and Dynamic Issues in Sensor Based Control. Edited by G. E. Taylor. XI, 456 pages. 1990. *(ROB)*

Vol. 58: Highly Redundant Sensing in Robotic Systems. Edited by J. T. Tou and J. G. Balchen. X, 322 pages. 1990. *(ROB)*

Vol. 59: Superconducting Electronics. Edited by H. Weinstock and M. Nisenoff. X, 441 pages. 1989.

Vol. 60: 3D Imaging in Medicine. Algorithms, Systems, Applications. Edited by K. H. Höhne, H. Fuchs and S. M. Pizer. IX, 460 pages. 1990.

Vol. 61: Knowledge, Data and Computer-Assisted Decisions. Edited by M. Schader and W. Gaul. VIII, 421 pages. 1990.

Vol. 62: Supercomputing. Edited by J. S. Kowalik. X, 425 pages. 1990.

Vol. 63: Traditional and Non-Traditional Robotic Sensors. Edited by T. C. Henderson. VIII, 468 pages. 1990. *(ROB)*

Vol. 64: Sensory Robotics for the Handling of Limp Materials. Edited by P. M. Taylor. IX, 343 pages. 1990. *(ROB)*

Vol. 65: Mapping and Spatial Modelling for Navigation. Edited by L. F. Pau. VIII, 357 pages. 1990. *(ROB)*

NATO ASI Series F

Vol. 66: Sensor-Based Robots: Algorithms and Architectures. Edited by C. S. G. Lee. X, 285 pages. 1991. *(ROB)*

Vol. 67: Designing Hypermedia for Learning. Edited by D. H. Jonassen and H. Mandl. XXV, 457 pages. 1990. *(AET)*

Vol. 68: Neurocomputing. Algorithms, Architectures and Applications. Edited by F. Fogelman Soulié and J. Hérault. XI, 455 pages. 1990.

Vol. 69: Real-Time Integration Methods for Mechanical System Simulation. Edited by E. J. Haug and R. C. Deyo. VIII, 352 pages. 1991.

Vol. 70: Numerical Linear Algebra, Digital Signal Processing and Parallel Algorithms. Edited by G. H. Golub and P. Van Dooren. XIII, 729 pages. 1991.

Vol. 71: Expert Systems and Robotics. Edited by T. Jordanides and B.Torby. XII, 744 pages. 1991.

Vol. 72: High-Capacity Local and Metropolitan Area Networks. Architecture and Performance Issues. Edited by G. Pujolle. X, 536 pages. 1991.

Vol. 73: Automation and Systems Issues in Air Traffic Control. Edited by J. A. Wise, V. D. Hopkin and M. L. Smith. XIX, 594 pages. 1991.

Vol. 74: Picture Archiving and Communication Systems (PACS) in Medicine. Edited by H. K. Huang, O. Ratib, A. R. Bakker and G. Witte. XI, 438 pages. 1991.

Vol. 75: Speech Recognition and Understanding. Recent Advances, Trends and Applications. Edited by P. Laface and Renato De Mori. XI, 559 pages. 1991.

Vol. 76: Multimedia Interface Design in Education. Edited by A. D. N. Edwards and S. Holland. XIV, 216 pages. 1992. *(AET)*

Vol. 77: ComputerAlgorithms forSolving LinearAlgebraic Equations. The State of the Art. Edited by E. Spedicato. VIII, 352 pages. 1991.

Vol. 78: Integrating Advanced Technology into Technology Education. Edited by M. Hacker, A. Gordon and M. de Vries. VIII, 185 pages. 1991. *(AET)*

Vol. 79: Logic, Algebra, and Computation. Edited by F. L. Bauer. VII, 485 pages. 1991.

Vol. 80: Intelligent Tutoring Systems for Foreign Language Learning. Edited by M. L. Swartz and M. Yazdani. IX, 347 pages. 1992. *(AET)*

Vol. 81: Cognitive Tools for Learning . Edited by P. A. M. Kommers, D. H. Jonassen and J. T. Mayes. X, 278 pages. 1992. *(AET)*

Vol. 82: Combinatorial Optimization. New Frontiers in Theory and Practice. Edited by M. Akgül, H. W. Hamacher and S. Tüfekçi. XI, 334 pages. 1992.

Vol. 83: Active Perception and Robot Vision. Edited by A. K. Sood and H. Wechsler. IX, 756 pages. 1992.

Vol. 84: Computer-Based Learning Environments and Problem Solving. Edited by E. De Corte, M. C. Linn, H. Mandl and L. Verschaffel. XVI, 488 pages. 1992. *(AET)*

Vol. 85: Adaptive Learning Environments. Foundations and Frontiers. Edited by M. Jones and P. H. Winne. VIII, 408 pages. 1992. *(AET)*

Vol. 86: Intelligent Learning Environments and Knowledge Acquisition in Physics. Edited by A. Tiberghien and H. Mandl. VIII, 285 pages. 1992. *(AET)*

MIX
Papier aus verantwortungsvollen Quellen
Paper from responsible sources
FSC® C105338

If you have any concerns about our products,
you can contact us on
ProductSafety@springernature.com

In case Publisher is established outside the EU,
the EU authorized representative is:
Springer Nature Customer Service Center GmbH
Europaplatz 3, 69115 Heidelberg, Germany

Printed by Libri Plureos GmbH
in Hamburg, Germany